Forensic Neuropathology

Forensic Neuropathology

Edited by

Helen L Whitwell

Professor of Forensic Pathology, University of Sheffield,
Sheffield, UK

Hodder Arnold

A MEMBER OF THE HODDER HEADLINE GROUP

LONDON

First published in Great Britain in 2005 by
Arnold, a member of the Hodder Headline Group,
338 Euston Road, London NW1 3BH
http://www.hoddereducation.com

Distributed in the United States of America by
Oxford University Press Inc.,
198 Madison Avenue, New York, NY10016
Oxford is a registered trademark of Oxford University Press

Whilst the advice and information in this book are believed to be true
and accurate at the date of going to press, neither the author[s] nor the
publisher can accept any legal responsibility or liability for any errors or
omissions that may be made. In particular (but without limiting the
generality of the preceding disclaimer) every effort has been made to
check drug dosages; however it is still possible that errors have been
missed. Furthermore, dosage schedules are constantly being revised
and new side-effects recognized. For these reasons the reader is strongly
urged to consult the drug companies' printed instructions before
administering any of the drugs recommended in this book.

British Library Cataloguing in Publication Data
A catalogue record for this book is available from the British Library

Library of Congress Cataloging-in-Publication Data
A catalog record for this book is available from the Library of Congress

ISBN-10:   0 340 70004 1
ISBN-13:   978 0 340 70004 4

1 2 3 4 5 6 7 8 9 10

Commissioning Editor: Serena Bureau
Development Editor: Layla Vandenbergh
Project Editor: Zelah Pengilley
Production Controller: Lindsay Smith
Indexer: Laurence Errington
Illustrations: David Graham

Typeset in 10/13pt Sabon by Charon Tec Pvt. Ltd, Chennai, India
www.charontec.com
Printed and bound in Italy

What do you think about this book? Or any other Arnold title?
Please visit our website: www.hoddereducation.com

To my parents and Robin

Contents

Contributors

Peter H Dangerfield MD
Senior Lecturer in Clinical Anatomy; and
Director Year 1 MB ChB Course
School of Medical Education
University of Liverpool, Liverpool; and
Professor in Health Sciences
Staffordshire University
Stoke on Trent, UK

Alexander R W Forrest LLM FRCP FRCPath CChem FRSC RFP
Professor of Forensic Chemistry; and
Consultant, Clinical Chemistry and Forensic Toxicology
Sheffield Teaching Hospitals NHS Foundation Trust
Sheffield, UK

Jennian F Geddes FRCPath
Former Reader in Clinical Neuropathology
Barts and The London, Queen Mary's School of
Medicine and Dentistry
Queen Mary, University of London, London, UK

Christopher M Milroy MD ChB LLB (Hons) FRCPath
DMJ (Path)
Professor of Forensic Pathology,
University of Sheffield, Sheffield, UK

Philip D Lumb BMedSci (Hons) MBChB DipRCPath (Forensic)
DMJ (Path) MRCPath (Forensic)
Home Office Pathologist
Sheffield, UK

Damianos E Sakas
Professor of Neurosurgery
University of Athens Medical School; and
Chairman, University Department of Neurosurgery
Evangelismos General Hospital
Athens, Greece

Colin Smith MD MRCPath
Senior Lecturer in Pathology; and
Honary Consultant in Neuropathology
Western General Hospital, Edinburgh, UK

Paul Watson BA (Hons)
QC of Gray's Inn, Barrister
Paradise Chambers, Sheffield, UK

Helen L Whitwell MBChB FRCPath DMJ (Path)
Professor of Forensic Pathology
University of Sheffield
Sheffield, UK

Foreword

For many pathologists the complexities of the central and peripheral nervous systems are enigmatic, as an understanding of their pathophysiology requires knowledge of neuroanatomy, structure–function correlation and the appreciation that although there are many diseases/disorders that are restricted to the nervous system, there are many others in which the nervous system is secondarily affected because of disease or disorder elsewhere in the body. The situation becomes even more complex when observations require an explanation in a medico–legal setting when circumstances are uncertain and may reflect natural disease, suicide, be suspicious or overtly murderous. The identification and interpretation of involvement of the central nervous system in any of these settings is therefore central to the likely sequence of events and may provide critical information that allows successful conclusion of any court proceedings.

Although there are a number of texts of forensic pathology, there are only a few devoted to forensic neuropathology. Some of the more recent publications have the format of an atlas and text whereas this volume, edited by Helen L. Whitwell, is liberally illustrated and to that extent is a step change in approach: breaking out of the historical mould of opinion and ex cathedra statement to one that is evidence-based.

Forensic neuropathology has been practised for many years in those regions where there has been a close professional association between the forensic pathologist and neuropathologist. Some centres do at lot of this type of work, whereas others do less, but its importance is increasingly recognised by the procurement of such services through contractual arrangements. What is quite clear now is that the most efficient and effective provision of service is when there is a good working relationship between the two disciplines, with appreciation and recognition of each other's skills and the ability to work as a team sharing common goals through agreed standards and common protocols. The editor and contributors to this book clearly have such working relationships and in so doing have compiled a text of major importance by stressing the need for codes of good practice of the type required in modern practice. While not covering every contingency or scenario, there is so much information in the sixteen chapters of this book that the text should rapidly become an everyday bench book to which constant reference will be made.

The publication of this book marks a very important turning point in the practise of forensic neuropathology by providing guidelines that should raise the profile of both forensic pathology and neuropathology. By so doing, this book will provide strong encouragement and advocacy for a discipline that, although steeped in morbid anatomy, provides plenty of opportunity for the application of modern techniques of a cellular and molecular nature, particularly when used in conjunction with imaging (including spectroscopy).

This book will be invaluable to pathologists and neuropathologists who practise forensic medicine, science and neuropathology and, of course, to those in training, whose careers include both disciplines starting with the autopsy, the retention of tissues, sampling and the subsequent processing in the laboratory. However, the basic tenants remain a comprehensive autopsy and the retention of appropriate samples for histological examination, including immunohistochemistry, to achieve clinico–pathological correlation. This book is a must for those who aspire to good practise and therefore is strongly recommended.

David I. Graham
September 2005

Preface

Forensic neuropathology has become increasingly important within the broader speciality of forensic pathology, and over the past decade there have been considerable advances in the field. In the area of head injury in particular the introduction of immunocytochemical techniques for identifying traumatic injury has led to a greater understanding of head injury from the neuropathological aspect both in adults and in children.

In addition to traumatic brain injury, forensic neuropathologists must also consider cerebro–vascular disease and neurotoxicology as well as other conditions, such as epilepsy and other nontraumatic medical conditions, within the forensic setting.

This book is not intended to be a fully comprehensive forensic pathology or neuropathology text, of which there are currently many available, but instead has the unique aim of highlighting for the reader areas where difficulties may arise in the medico–legal setting. It includes descriptions of techniques that may be helpful either at the postmortem examination or on dissection of the brain, as well as blocking schedules for common disorders that necessitate detailed neuropathological examination. The book also introduces the biomechanical aspects of head injury: these are attaining increasing importance, and will no doubt require more detailed discussion in a future edition.

The book is intended for a range of readers including not only forensic pathologists but also neuropathologists with limited experience of forensic pathology, general pathologists, clinical forensic specialists and neuroscientists, neurologists and neurosurgeons. In addition, it is hoped that members of the legal profession will also find it to be a useful reference work.

Helen L. Whitwell
September 2005
Sheffield, UK

Acknowledgements

With thanks to all in the Department of Forensic Pathology Medico–Legal Centre at the University of Sheffield, in particular to Ian Newsome and Ann Cooper.

1

Anatomy of the head and neck

Peter H Dangerfield

The surface anatomy and features of the head and neck are derived from skeletal and soft tissue structures covered by skin and connective tissue that covers the underlying bony skull. In some areas, particularly the face, the skin is thin, allowing easy palpation of underlying skeletal features. It is also highly mobile as a result of the presence of a number of small but extensive subcutaneous muscles of facial expression, all supplied by a single cranial nerve (facial). The scalp is, in contrast, relatively tough and hidden from view in most individuals by hair. Within the skull, which acts as a protective shield, is the brain and brain stem, with its associated covering of tissue layers and blood vessels. The anatomy is complex and this chapter addresses an overview of the principal features. Cranio–facial anatomy (including the neuroanatomy) is included to show the anatomical link with the underlying structures. The neuroanatomy may be of particular significance in penetrating or blunt head trauma.

THE EXTERNAL FACE

The anatomical surface features of the face are never totally symmetrical. As facial expressions have evolved as a communication method, the underlying anatomy has evolved from the functional need. The skin is often marked by moles and freckles and may also present with scars consequential to cuts and other trauma.

The shape and size of the hairline varies with race and the eyebrows can also be highly variable.

The nose comprises underlying cartilaginous and bony structures. In coronal section, the nose is triangular in shape. The external nares (nostrils) are protected by coarse hairs (vibrissae) and serve to filter air entering the nose.

The anterior part of the nose is composed of flexible underlying fibrocartilage.

Inferior to the nose is the mouth, surrounded by the lips. Here, the size and shape of the mouth are very variable, both within races and between different racial groups. The lips have non-keratinized epithelium and thus appear pink as a result of the underlying blood vessels.

The eyes are set within the bony orbits of the skull but protected by the rim of bone. Their individual position relative to the nose is variable and can be close set or wide set in normal individuals. Attached within the orbits are the muscles that control eye movement while, superficially, the paired eyelids cover and protect the eye from potential damage. The lids normally permit only a portion of white sclera to appear laterally, with the transparent conjunctiva and cornea that cover the pigmented iris seen medially. The upper lid normally overlaps the iris, but the sclera may be seen between the iris and the lower lid. The shape of the lids can vary between individuals; in particular the elevator of the upper lid can be weak or damaged, leading to a drooping appearance.

Ethnic differences are often prominent. Mongoloid epicanthic folds and other minor folds of skin in the medial aspect of the orbit should be noted. Conditions such as exophthalmos associated with hyperthyroidism can result in prominent eyes. Facial fractures affecting the maxilla and inferior margin of the orbit can lead to a sinking of the eyeball. The shape of the lids themselves can lead to a wide range of different appearances of the eye within the orbit.

THE INTERNAL FACIAL STRUCTURES

Internally, the muscles of facial expression and their nerve and vascular supplies contribute to the facial structures.

In addition, the parotid gland is located within the lateral parts of the cheeks.

The facial expression muscles are supplied by the facial (cranial VII) nerve. Their function is to control and support the structures and openings in the face, such as the eyes and mouth. In humans, their functionality serves an important role in non-verbal communication as well as aiding actions such as screwing up the eyes and chewing. The mouth is surrounded by the sphincteric orbicularis oris muscle into which merge the fibres of the buccinator, the muscle of the cheek. The buccinator contracts during chewing and serves to prevent trapping of food within the space between the gums and teeth; it also acts to raise the pressure of air expelled by musicians playing wind instruments or by whistling. The orbicularis oculi surrounds the eye and serves to function in two ways. First, fibres that surround the eye serve to screw the eye up because they are attached to the bone on the medial aspect of the orbit. Second, the palpebral fibres attach to the lateral palpebral raphe and serve to close the eye when blinking. Additional fibres are attached to the lacrimal sac and serve to dilate the sac and keep the puncta in contact with the eyeball.

The facial nerve enters the face by passing through the tough fibrous capsule of the parotid gland and can be damaged during surgical procedures to that gland.

THE NOSE

The nose, as the upper part of the respiratory tract, is located superior to the hard palate and contains the organ of smell. It is divided into right and left nasal cavities by the nasal septum, with each nasal cavity having an olfactory and a respiratory area.

The external nose varies considerably in size and shape in individuals and races because of differences in the nasal cartilage structure. The inferior aspect is composed of two openings called the nares (nostrils), each separated from the other by the nasal septum. The nasal bones, the frontal processes of the maxillae, the nasal part of the frontal bone and the bony part of the nasal septum form the skeletal components of the nose whereas five main cartilages form the cartilaginous nose. These are two lateral cartilages, two alar cartilages and a septal cartilage that articulates with the bony septum.

The nasal cavities open through the choanae into the nasopharynx at the posterior. The nasal mucosa is bound closely to the periosteum and perichondrium of the nasal bones and cartilages, and lines the nasal cavities, except for the vestibule which is lined with skin. The olfactory area lies superior within the cavity and is the organ of smell,

with its nerve fibres passing through the cribriform plate to enter the olfactory bulbs, which lie against the inferior surface of the frontal lobe of the brain.

The narrow, curved roof of the nasal cavity is divided into frontonasal, ethmoidal, and sphenoidal parts, named by adjacent bones. The wide floor is formed by the horizontal plate of the palatine bone and the palatine process of the maxilla. Medially, the wall is the nasal septum, comprising the vomer, perpendicular plate of the ethmoid, septal cartilage, and the nasal crests of the maxillary and palatine bones. The lateral walls of the nasal cavity are made up of three nasal conchae or scroll bones, each forming a roof over a meatus connecting the nasal cavity to a sinus or the orbit. The superior meatus is between the superior and middle conchae, into which orifices from the posterior ethmoidal sinuses open. The middle meatus, inferior to the middle conchae, communicates with the frontal sinus via the frontonasal duct and the maxillary sinus at its posterior end. The inferior meatus is inferolateral to the inferior conchae and receives the nasolacrimal duct from the lacrimal sac into its anterior portion.

The nose receives arterial blood from many branches including the sphenopalatine artery, ethmoidal arteries and the facial artery. Kiesselbach's area, found on the anterior nasal septum, is rich in capillaries and is the site of profuse nose bleeding. The nerve supply of the nasal mucosa is by the maxillary nerve, nasal branches of the greater palatine nerve and the anterior ethmoidal nerves, and branches of the nasociliary nerve.

The paranasal sinuses are air-filled extensions of the nasal cavity within the frontal, maxillary, sphenoid and ethmoid bones, and are named according to each bone. The ethmoidal sinuses consist of ethmoidal cells located within the ethmoid bone between the orbit and nose. The sphenoid air sinuses are unevenly divided like the frontal air sinuses and separated by a bony septum. They occupy the body of the sphenoid bone and are separated by thin bone from the optic chiasma, the pituitary gland, the internal carotid arteries and the cavernous sinuses.

The maxillary sinuses are large pyramidal cavities within the maxillae. Their floor is formed by the alveolar part of the maxilla, with the roots of the maxillary teeth, particularly the first two molars, creating conical elevations (Figure 1.1).

THE ORAL CAVITY

The oral cavity consists of the oral vestibule and the oral cavity. The vestibule is the space between the lips and cheeks and the teeth and gums, and communicates with

Figure 1.1 The nasal cavity and associated structures.

the exterior through the orifice of mouth, the size of which is controlled by muscles including the orbicularis oris.

The oral cavity lies posterior and medial to the upper and lower dental arches and is limited posteriorly by the terminal groove of the tongue and palatoglossal arches, and anteriorly and laterally by the maxillary and mandibular arches containing the teeth. The roof is formed by the hard and soft palate, which also forms the floor of the nasal cavities. Posteriorly, the oral cavity communicates with the oropharynx. If the mouth is closed, the tongue fills the space of the oral cavity.

The hard palate forms the anterior component of the roof of the oral cavity, with its cavity filled by the resting tongue when it is at rest and formed by the palatine processes of the maxillae and the horizontal plates of the palatine bones. The incisive fossa and the greater and lesser palatine foramina open on the oral aspect of the hard palate.

The soft palate is the muscular posterior part, attached to the posterior border of the hard palate and extending as a posteroinferiorly curved free margin that terminates in the uvula. It is strengthened by a palatine aponeurosis formed by the expanded tendon of the tensor veli palatini and is attached to the posterior margin of the hard palate. Laterally, it is continuous with the wall of the pharynx and joined to the pharynx and tongue by the palatopharyngeal and palatoglossal arches. The masses of lymphoid tissue forming the palatine tonsil lie within the tonsillar fossa, bounded by the palatoglossal and palatopharyngeal arches and the tongue.

THE ORBIT

The orbit is a pyramidal, bony cavity in the face. It contains and protects the eye with its associated muscles, nerves and vessels, and the lacrimal apparatus. The roof is formed by the orbital part of the frontal bone, separating the orbit from the anterior cranial fossa and containing a small fossa for the lacrimal gland. The lesser wing of the sphenoid contributes to the roof at its apex. The medial wall is formed by the thin bone of the ethmoid, frontal, lacrimal and sphenoid bones. It is indented by the fossa of the lacrimal sac and nasolacrimal duct. The lateral wall comprises the frontal process of the zygomatic bone and the greater wing of the sphenoid, and is vulnerable to direct trauma. It serves to separate the orbit from the temporal and middle cranial fossae. The floor is made up from the maxilla, zygoma and palatine bones, with a thin inferior wall partly separated from the lateral wall by the inferior orbital fissure. At the apex of the orbit lies the optic canal, located medial to the superior orbital fissure, which carries the optic nerve and associated structures into the orbit.

BLOOD SUPPLY TO THE FACE

The arterial supply to the lower face is via the facial artery, a branch of the external carotid artery. This artery enters the face by looping across the mandible, almost to the midpoint of the ramus where it can be located by finding the small notch on the margin of the mandibular ramus in which it lies. It passes upwards and medially towards the margin of the mouth where it divides to give rise to superior and inferior labial arteries, supplying the lips. A further branch extends upwards towards the medial aspect of the eye and orbit alongside the nose.

The upper part of the face and scalp is supplied by the terminal branches of the external carotid artery. The deep facial structures are supplied by the maxillary artery, which passes deep to the mandible. The superficial temporal artery passes upwards to supply the temporal region. The transverse facial artery is a branch of this artery that runs medially across the face, supplying the cheek structures. Small supraorbital and supratrochlear arteries, branches of the ophthalmic branch of the internal carotid artery, supply the forehead and anterior scalp.

Posterior to the external ear can be found the occipital and posterior auricular arteries, which supply the posterior of the scalp and ear region. The extensive arterial supply to the face is highly anastomotic, and lacerations can and do bleed extensively from what might at first be assumed to be small vessels.

Venous return follows the same basic pattern as the arterial supply, except that the veins drain into either the internal or the external jugular veins.

Extensive lymph node distributions receive the lymph vessels, which follow the pattern of drainage of the views. Nodes can be located in the submental, mastoid, submandibular, parotid and occipital regions, and these in turn give rise to lymph vessels draining mainly into the deep cervical nodes.

THE NECK

The neck extends from the inferior boarder of the mandible to the superior borders of the clavicles and sternal manubrium in the chest anteriorly, and to a rather ill-defined border defined as a line between the acromia of the shoulder posteriorly.

Laterally, the neck is clearly marked by the sternocleidomastoid muscle, which is attached between the mastoid process and adjacent nuchal line on the occipital bone and the sternum.

The posterior neck region presents a muscular appearance as a result of the paraspinal cervical muscles, which lie deep to the trapezius muscle. Two folds are apparent within the hairline of the head formed by the nuchal ligaments.

The skin of the neck presents a series of horizontally arranged Langer's lines, important for scar formation and deep, to which is a thin superficial fascial layer. This layer also includes a thin muscle called the platysma, which forms part of the extensive muscles of facial expression of the face.

POSTERIOR STRUCTURES OF THE NECK

A layer of deep cervical fascia encloses the whole neck, attached superiorly to the superior nuchal lines, the mastoid processes and the lower border of the mandible, and inferiorly to the acromion process of the scapula, the clavicle and the upper border of the manubrium of the sternum. In the midline posteriorly, the investing layer attaches to the ligamentum nuchae. The investing facial layer splits to enclose the trapezius and sternocleidomastoid muscles.

Laterally, anatomical schemes create the anatomical posterior triangle, the posterior boundary of which is formed from part of the trapezius muscle, a very large, flat muscle located in the back of the neck, extending into the thorax, and having a long, linear attachment to structures in the midline. Its upper fibres are attached to the medial third of the superior nuchal line of the occipital bone, the external occipital protuberance of the occipital bone and the tips of all the spinous processes of the cervical vertebrae, except superiorly where it attaches to the fibrous ligamentum nuchae. The trapezius muscle receives its motor nerve supply from the spinal root of the accessory nerve. The inferior boundary of this triangle is formed by the middle third of the clavicle and its roof is formed by the deep cervical fascia of the neck. Within the floor is the prevertebral fascia which in turn covers the semispinalis capitis, splenius capitis, levator scapulae, scalenus posterior, scalenus medius and scalenus anterior muscles.

Semispinalis capitis and splenius capitis are part of a group of deep muscles of the back, functioning when movement of the vertebral column and the head on the cervical spine is initiated. It is important to appreciate that they are parts of the upper end of a long column of muscle that extends up the back of the abdomen, the thorax and the neck. This column occupies the vertical hollow, or groove, on either side of the spine, collectively referred to as the erector spinae muscle, a highly complex grouping in both its structure and function.

The levator scapulae muscles act to elevate the scapula. They, together with the three scalene muscles, are attached at their superior ends to the cervical vertebrae. The scalene muscles attach into the first and/or second rib. The scalenus anterior is a key to understanding the anatomy of the root of the neck as a result of its important relationships to other structures. These include the phrenic nerve, the subclavian artery and vein, the brachial plexus and the cervical part (i.e. dome) of the pleura.

THE SCALP

The scalp covers the cranium and comprises a thick hairy skin layer with a tough fibroadipose hypodermis and a deeper thick fibrous aponeurotic layer containing the neurovascular structures. The fibrous nature of this tissue prevents vessel constriction in lacerations, which can lead to extensive bleeding. A thin fascial areolar connective tissue layer deep to the aponeurosis covers the skull periosteum. This layer is only loosely held together and can allow the scalp to be stripped off from the bone of the skull with relative ease. Infection may also lead to scalp swelling.

THE SKULL BONES

The skull itself has two distinct parts. The vault is composed of relatively thin plates of bone that articulate with one another at a series of jagged-edged suture lines. In the neonates, these are relatively wide, with marked fontanelles delineating the junction of adjacent bones.

The anterior of the skull is made up of the frontal bone, which has left and right parts in the child, although these normally fuse by the end of growth. The frontal bones contribute to the roof of the orbit. Irregular frontal air sinuses are located in the medial aspect of the frontal bones.

Laterally, the vault is composed of the parietal and temporal bones whereas the posterior part is made up from the occipital bone. The occipital bone extends inferiorly into the base of the skull and is pierced by the large foramen magnum, the entry point of the spinal cord into the internal skull. The skull bones are united by the various sutures: coronal, sagittal and lamdoid.

Large surface palpable mastoid processes are located posterior to the pinna of the ear and are a prominent part of the occipital bone.

The base of the skull is a complex structure. The view typically seen at postmortem examination is illustrated in Figure 1.2a. The cranial vault is illustrated in Figure 1.2b.

THE BRAIN

The brain is protected from mechanical forces by the overlying cranial bones, the membrane layers called the meninges and the circulating cerebrospinal fluid (CSF). The nervous tissue is itself insulated from the general circulation by the so-called blood–brain barrier which serves to protect it from the majority of infectious agents and chemicals, including drugs. However, agents and drugs can and do breech this barrier from time to time.

THE MENINGES

The brain and spinal cord share the same layers, which are continuous with one another. Within the skull are found the dura mater, arachnoid and pia mater layers.

The dura mater has two distinct layers, namely an outer layer that is fused to the periosteum lining the inner surface of the skull and an inner fibrous layer. Consequently, no epidural space exists superficial to this layer, unlike the situation found within the spinal canal of the vertebral column. However, there is a potential space, termed the

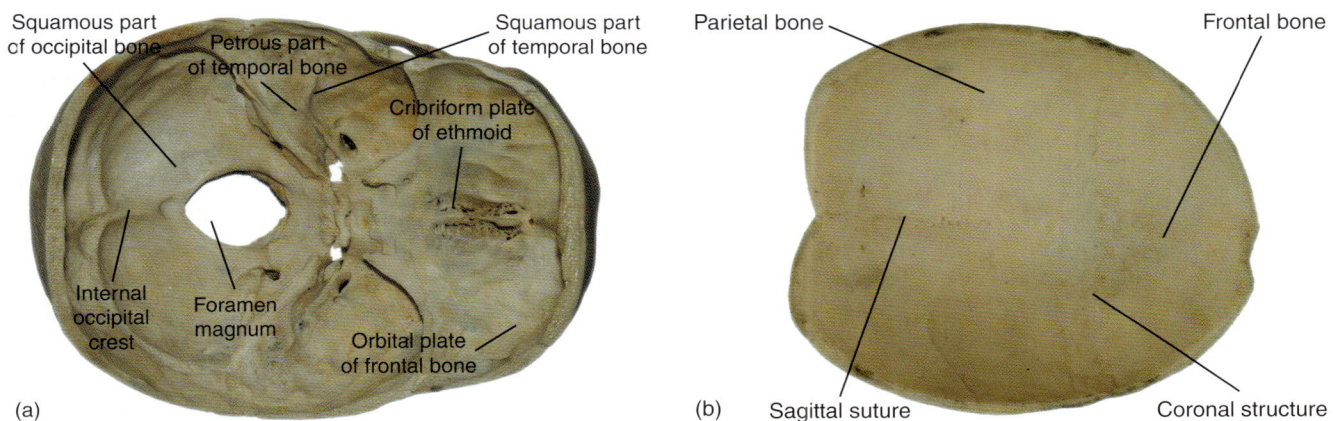

Squamous part of occipital bone   Petrous part of temporal bone   Squamous part of temporal bone   Cribriform plate of ethmoid   Internal occipital crest   Foramen magnum   Orbital plate of frontal bone

(a)

Parietal bone   Frontal bone   Sagittal suture   Coronal structure

(b)

Figure 1.2   (a) The base of the skull, (b) The cranial vault.

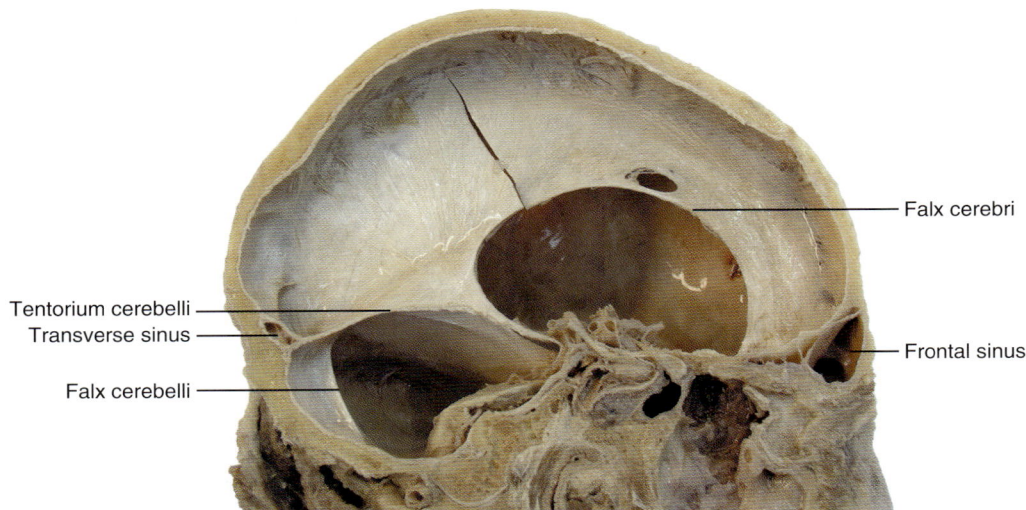

Tentorium cerebelli

Transverse sinus

Falx cerebelli

Falx cerebri

Frontal sinus

Figure 1.3   The tentorium cerebelli.

'extradural space', present that can serve as a reservoir for blood if the meningeal vessels become ruptured by trauma. The tightly adherent skull–dura layers serve to prevent spread of blood. Both layers of dura are separated by a thin gap layer, in which are found the major blood sinuses and other blood vessels. Arachnoid granulations project through the dura into the venous sinuses and serve to absorb CSF back into the venous system. All the venous sinuses drain eventually into the internal jugular veins of the neck.

The arachnoid layer is subdivided into a further three layers. The arachnoid membrane covers the brain and serves to smooth out the underlying gyri and sulci. This is underlaid by an epithelial layer, and the cells and fibres of the arachnoid trabeculae that cross the subarachnoid space, giving it the appearance of a spider web. These trabeculae join the arachnoid to the underlying pia mater, the deepest brain layer. All nerves and vessels passing into or out from the brain must traverse the subarachnoid space, which is filled with CSF.

The pia mater layer is firmly adherent to the surface of the brain and extends into all the folds of the brain surface; in addition it is adherent to cerebral blood vessels as they enter the substance of the brain. The pia mater is anchored by astrocyte processes.

Tough, fibrous extensions of the dura mater, the dural folds, hold the brain firmly in position and these, with the CSF in the subarachnoid space, serve to protect the brain from sudden shocks and other deceleration movements that often accompany cerebral trauma. Dural venous sinuses are located between layers of these folds. In trauma, traction on the small tributary superior cerebral veins as they enter the sagittal sinus may lead to their rupture, resulting in bleeding into the subdural space.

The dural folds also form a large fibrous sheet that extends deep within the brain and acts to support parts of the cortex and brain stem by providing additional stabilization. These are the falx cerebri, tentorium cerebelli and falx cerebelli.

The falx cerebri is located within the longitudinal fissure separating the left and right cerebral hemispheres. Inferiorly, it is attached to the crista galli and the internal occipital crest of the inner surface of the occipital bone. Both the superior and inferior sagittal venous sinuses lie within the falx cerebri. Its posterior margin is continuous with the tentorium cerebelli, which separates the cerebellar hemispheres from the cerebrum, but is at right angles to the falx cerebri. The transverse sinus is located within the tentorium cerebelli. Finally, the falx cerebelli separate the two cerebellar hemispheres in the midline but inferior to the tentorium cerebelli (Figure 1.3).

ARTERIAL BLOOD SUPPLY

Arterial blood reaches the internal cranial cavity via two major vascular supplies: the internal carotid artery and the vertebral artery.

The common carotid arteries are both branches of the brachiocephalic artery on the right and aortic arch on the left. They are enclosed within the tough protective fibrous carotid sheath that also encloses the internal jugular vein and vagus nerve to the posterior of the artery, and which lies lateral to the trachea and oesophagus. Both right and left common carotid arteries terminate at the level of the upper border of the thyroid cartilage, bifurcating to give rise to internal and external carotid arteries. A swelling at the division of the artery is the location of the carotid sinus and carotid body containing chemo- and baroreceptors.

The internal carotid artery has no branches within the neck and ascends deep to the styloid process of the skull and adjacent muscles, to enter the carotid canal of the petrous temporal bone. This canal passes forwards and medially to enter the internal cavity of the skull as the foramen lacerum. The internal carotid artery is now located to the lateral side of the sphenoid bone, enclosed within the cavernous sinus. Its path now proceeds anteriorly within the sinus and then it passes out of the sinus upwards to lie adjacent to the anterior clinoid process, where it enters the subarachnoid space by piercing the dura mater. This path is clearly seen on suitable radiographs and is know as the carotid siphon. At this point, the artery gives a branch called the ophthalmic artery, which passes through into the orbit, supplying the retina via the central artery of the retina and branches to the lacrimal gland and adjacent orbital structures.

The right and left internal carotid arteries are interconnected by an anterior communicating artery and also posteriorly by the posterior communicating artery to the basilar artery system, forming an important component of the circle of Willis on the inferior surface of the brain.

The carotid artery terminates within the subarachnoid space of the brain as the middle cerebral and anterior cerebral arteries. These supply the lateral aspect of the brain and the upper parts and medial aspect of the anterior of the brain.

The anterior cerebral artery passes above the optic chiasma and lies on the medial aspect of the cerebral hemisphere, where it forms an arch around the genu of the corpus callosum. It gives off a significant branch adjacent to the anterior communicating branch of the circle of Willis, which supplies the internal capsule.

The middle cerebral artery is the major terminal branch of the internal carotid artery, giving off deep branches to the internal structure of the brain, supplying the corpus striatum, internal capsule and thalamus. Occlusion of the lateral striate branch of this artery gives rise to classic stokes. The middle cerebral artery continues along the lateral fissure reaching the insula, where it divides into divisions that supply the frontal and parietal lobes, the temporal lobe and the midpart of the optic radiation.

The vertebral artery is a branch of the first part of the subclavian artery, itself a branch of the brachiocephalic artery on the right and of the arch of the aorta on the left. This artery passes superiorly and posteriorly to enter the foramen transversarium of the sixth cervical vertebra, lying at the apex of the triangle formed by longus cervicis medially and scalenus anterior laterally. The artery passes upwards within a canal in the transverse processes, giving rise to small twigs of vessels that supply the spinal cord in the neck.

The artery then continues across the superior aspect of the posterior arch of the atlas vertebra and enters the posterior occipito-atlanomembrane to enter the vertebral canal. It pierces the dura mater and continues upwards through the foramen magnum, joining its fellow from the opposite side and creating the basilar artery. This is located in the anterior aspect of the brain stem.

The basilar artery has a number of branches: the anterior spinal artery runs inferiorly in the anteromedian groove of the spinal cord; the posteroinferior cerebellar artery runs in the lateral aspect of the medulla and the cerebellum; and two posterior spinal arteries lie medial to the dorsal nerve roots and can arise variably from either the vertebral or the posterior cerebellar vessels. Lying on the ventral surface of the brain stem and the pons, the basilar artery supplies the brain stem and cerebellum through anteroinferior and superior cerebellar arteries. At the superior margin of the pons, the vessel divides into the posterior cerebral arteries, which supply the temporal and occipital lobes of the cortex. The posterior communicating artery (see above) forms the posterior part of the circle of Willis and links the basilar to the internal carotid artery system (Figures 1.4 and 1.5).

VENOUS DRAINAGE

The venous drainage of the brain is important because thromboses can give rise to many clinical syndromes. The cerebral hemispheres are drained by the superficial and deep cerebral veins. They lack valves.

The superficial veins lie within the subarachnoid space and empty into the venous sinuses. The upper part of the cortex drains into the superior sagittal sinus whereas the midpart normally drains into the cavernous sinus via the superficial middle cerebral vein. The lower parts drain into the transverse sinus.

The deep veins drain the choroid plexus, thalamus and corpus striatum. The various component branches drain into the great cerebral vein of Galen, which is located within the midpoint of the tentorium cerebelli, where it unites with the inferior sagittal sinus to form the straight sinus that in turn is a branch of the left transverse sinus.

The dural folds contain the major venous sinuses. The falx cerebri encloses the superior sagittal sinus within its upper part adjacent to the skull, whereas its lower free edge, lying within the longitudinal fissure of the brain, contains the inferior sagittal sinus. The straight sinus lies within the attachment of the falx to the tentorium and unites with the superior sagittal sinus at the confluence of the sinuses.

Anterior cerebral artery

Anterior communicating artery

Middle cerebral artery

Internal carotid artery

Posterior communicating artery

Posterior cerebral artery

Superior cerebellar artery

Basilar artery

Anterioinferior cerebellar artery

Anterior spinal artery

Posterioinferior cerebellar artery

Vertebral artery

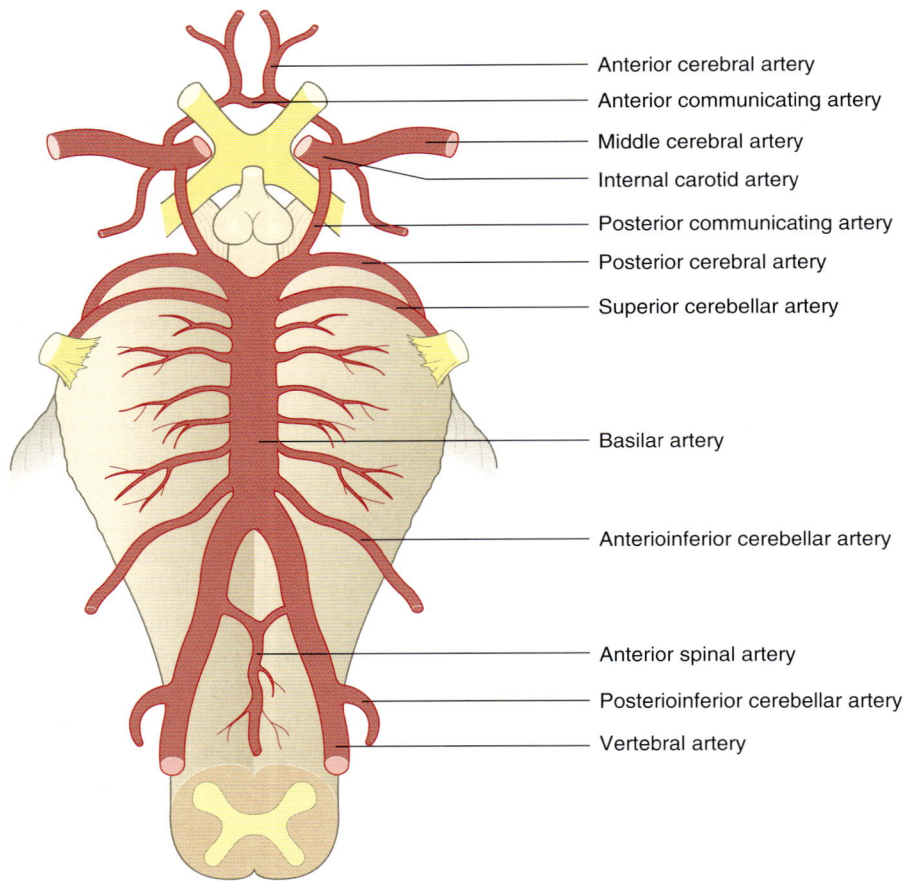

Figure 1.4　Circle of Willis.

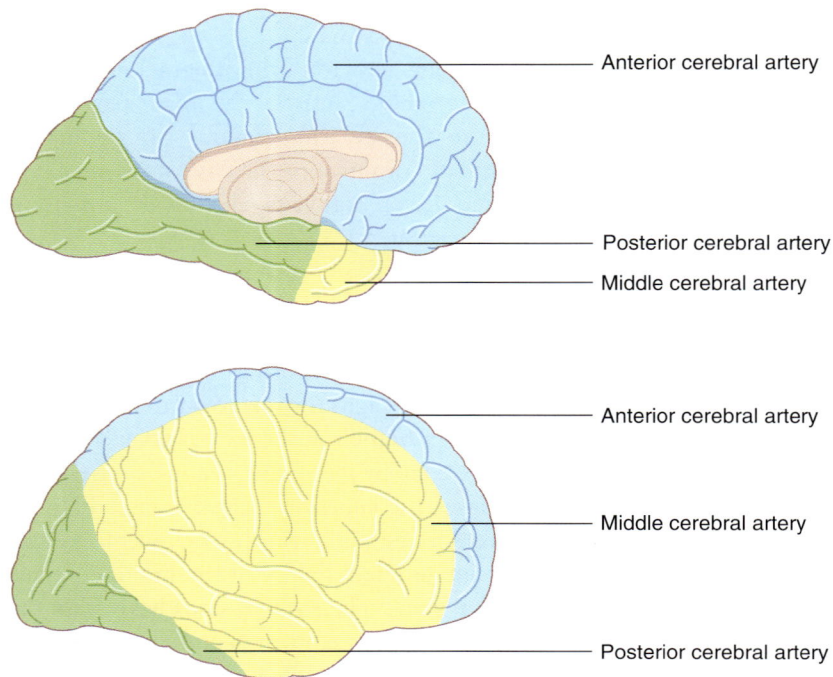

Anterior cerebral artery

Posterior cerebral artery

Middle cerebral artery

Anterior cerebral artery

Middle cerebral artery

Posterior cerebral artery

Figure 1.5　Arterial supply to the brain.

THE CEREBRUM

The cerebrum is the largest region of the brain. Paired cerebral hemispheres form the superior and lateral surfaces of the cerebrum and have cortical surfaces characterized by the presence of elevated ridges and grooves, serving to increase the surface area of the brain. These are the gyri, shallow grooves called sulci and deeper fissures. This folding is less in the neonate and develops into its characteristic pattern during childhood. Although the entire brain enlarged during human evolution, the cerebral hemispheres enlarged at a much faster rate than the other parts of the brain.

Each cerebral hemisphere is subdivided into regions, or lobes, named after the overlying cranial bones. On each hemisphere, the deep groove of the central sulcus divides the anterior frontal lobe from the posterior parietal lobe. The lateral surface of the frontal lobe comprises the precentral gyrus with the precentral sulcus in front. The inferior surface is marked by orbital gyri and is in direct contact with the forward extending olfactory tract and bulb.

Internally, the cerebral hemispheres contain the lateral ventricles (Figure 1.6) on the left and right and a central third ventricle. These are the site of CSF production.

On the lateral surface, a horizontally aligned lateral sulcus separates the frontal lobe from the temporal lobe. By retracting the lips (known as the opercula) of the lateral sulcus, the insula can be exposed. This effectively lies deep to the temporal lobe and is invisible from the surface.

A parieto-occipital sulcus posteriorly separates the parietal lobe from the occipital lobe.

Functionally, each lobe has less clearly defined regions dedicated to a range of motor and sensory roles, with very indistinct boundaries and considerable overlap.

The two cerebral hemispheres are almost completely separated by a deep longitudinal fissure containing the falx cerebri, remaining connected by the thick band of white matter called the corpus callosum, which forms a distinctive white C-shaped shape on sagittal magnetic resonance imaging (MRI) of the brain. By cutting the corpus callosum, the brain can be divided into two hemispheres, exposing the medial surface (Figure 1.7). The corpus callosum is a massive band of white matter comprising a trunk, an anterior genu, a posterior splenium and a narrow rostrum which extends from the genu to the anterior commissure.

Although the surface of the cerebrum is composed of grey matter, its interior consists primarily of nerve axon pathways or white matter. Association fibres interconnect areas of the cortex within each cerebral hemisphere. Shorter association or arcuate fibres pass from one gyrus to another whereas longer association fibres are arranged in bundles, or fasciculi. Longitudinal fasciculi link the lobes of each cerebral hemisphere. Commissural fibres link the two hemispheres and allow communication between them. These fibres are formed into densely packed bands of nerve axons, such as the corpus callosum and the anterior commissure.

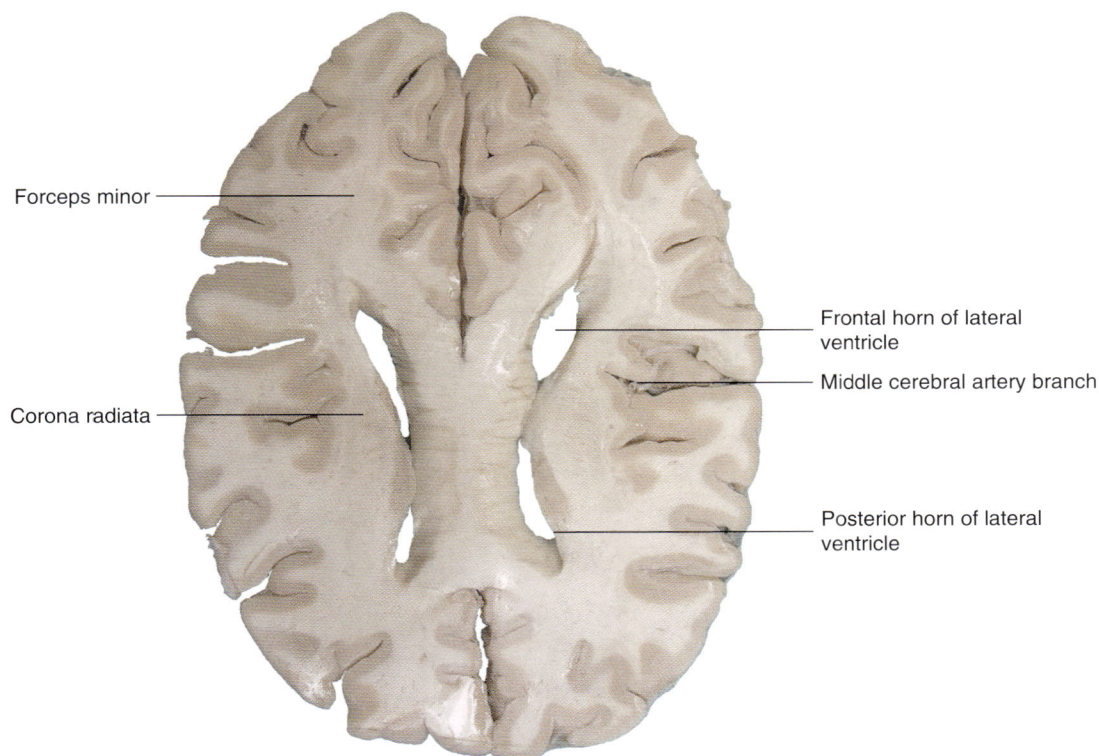

Forceps minor
Corona radiata
Frontal horn of lateral ventricle
Middle cerebral artery branch
Posterior horn of lateral ventricle

Figure 1.6 Sagittal section through the cerebral hemispheres.

The cerebral cortex is also linked to other parts of the brain and brain stem by projection fibres. These fibres traverse the diencephalon, where axons passing up to sensory areas of the cerebral cortex mix with axons descending from motor areas of the cortex within an area called the internal capsule. These ascending and descending nerve fibres look alike.

The central sulcus separates the motor and sensory areas of the cortex, with the precentral gyrus of the frontal lobe forming the anterior border of the sulcus. This region is the primary motor cortex. Specialist pyramidal cells within the primary motor cortex are responsible for voluntary movements by controlling somatic motor neurons in the brain stem and spinal cord.

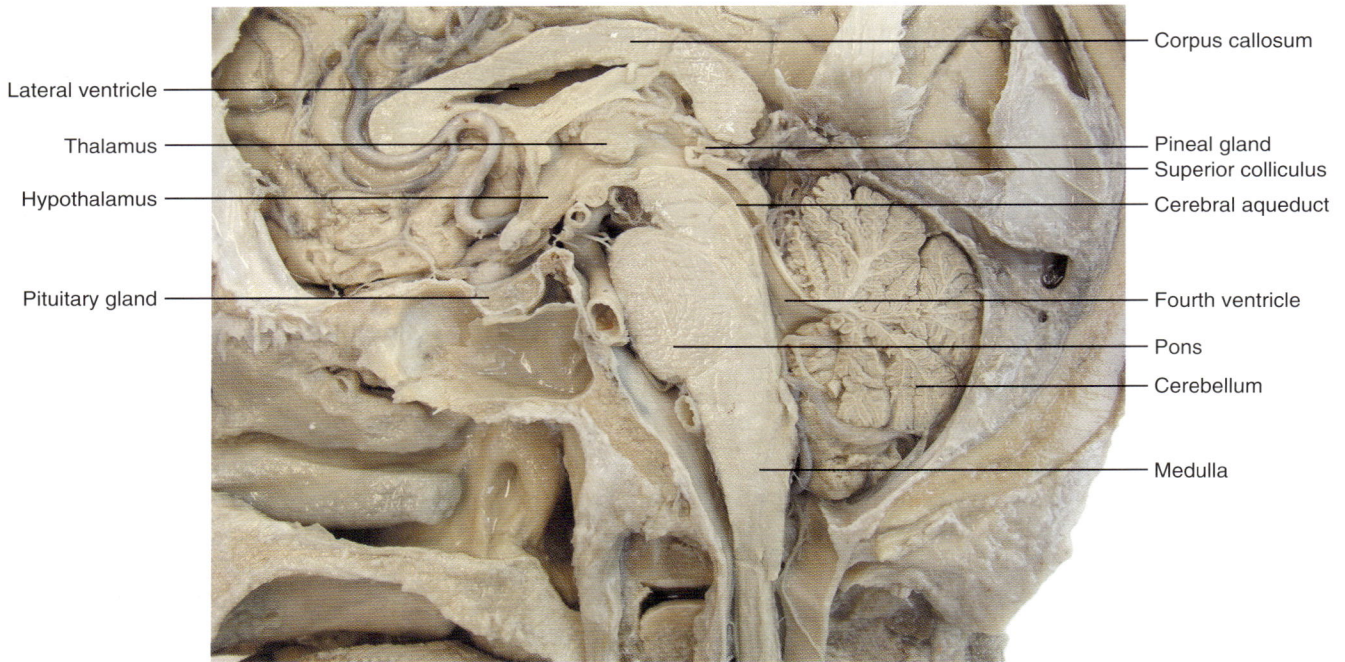

Figure 1.7   Sagittal section through the head.

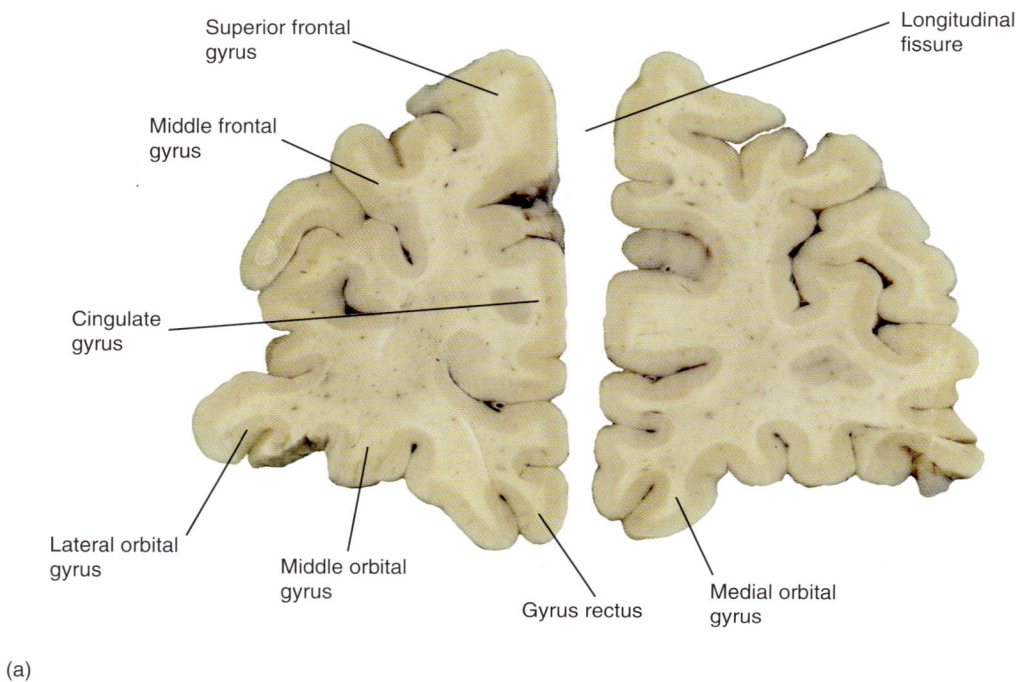

(a)

Figure 1.8   (a)–(g) Coronal sections through the cerebral hemispheres.

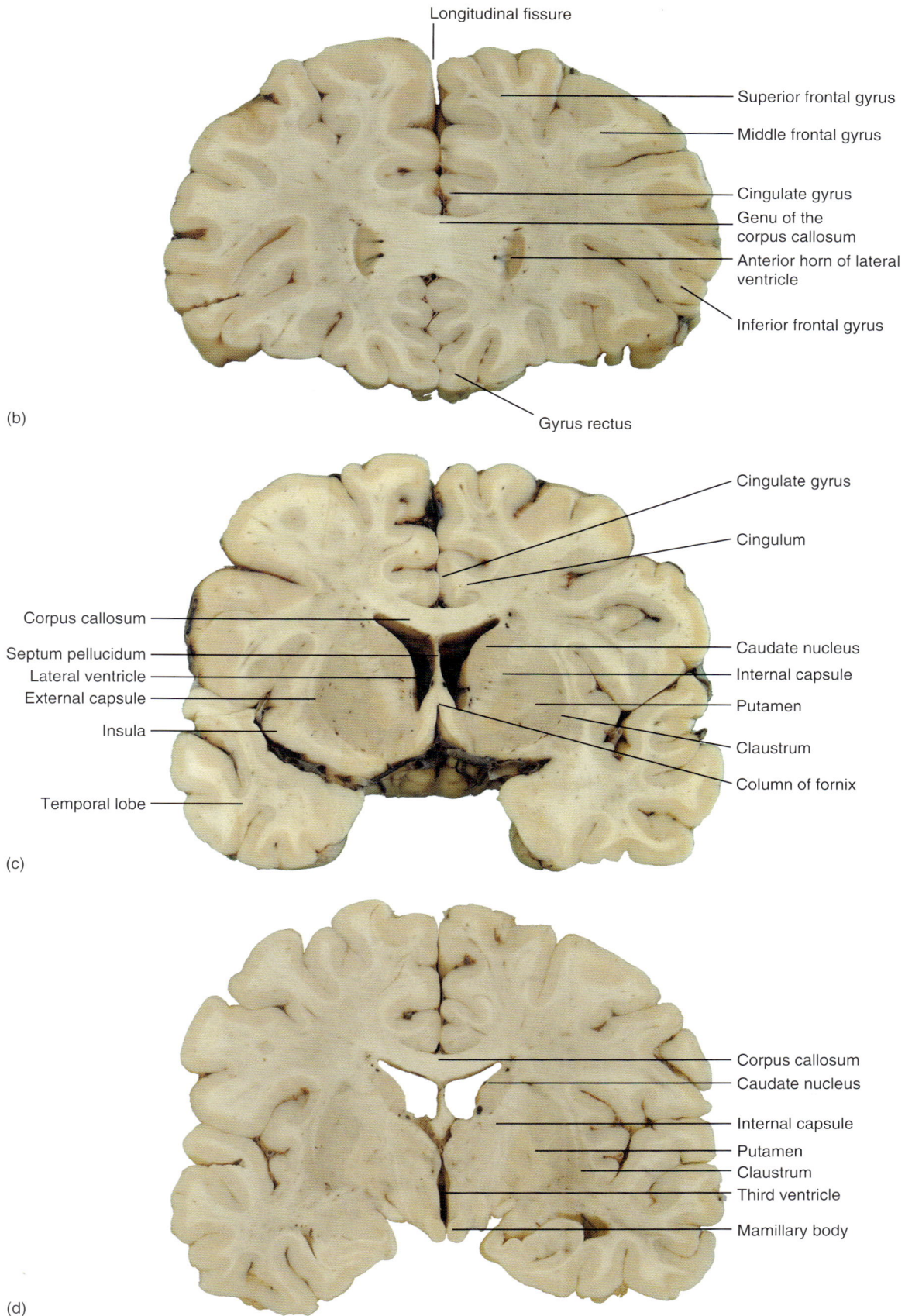

Longitudinal fissure

Superior frontal gyrus

Middle frontal gyrus

Cingulate gyrus

Genu of the corpus callosum

Anterior horn of lateral ventricle

Inferior frontal gyrus

Gyrus rectus

(b)

Cingulate gyrus

Cingulum

Corpus callosum

Septum pellucidum

Lateral ventricle

External capsule

Insula

Temporal lobe

Caudate nucleus

Internal capsule

Putamen

Claustrum

Column of fornix

(c)

Corpus callosum

Caudate nucleus

Internal capsule

Putamen

Claustrum

Third ventricle

Mamillary body

(d)

Figure 1.8 Coronal sections through the cerebral hemispheres (*continued*).

(e)

Longitudinal fissure

Cingulate gyrus

Lateral ventricle

Caudate nucleus

Body of the corpus callosum

Body of fornix

Thalamus

Third ventricle
Red nucleus

Substantia nigra

(f)

Cingulate gyrus

Lateral ventricle

Body of the corpus callosum

Fornix

Inferior frontal gyrus

Pulvinar

Retrolenticular limb of the internal capsule

Inferior horn of the lateral ventricle

Hippocampus

Temporal lobe

Superior colliculus

(g)

Longitudinal fissure

Precentral gyrus

Postcentral gyrus

Inferior parietal gyrus

Inferior horn of the lateral ventricle

Middle temporal gyrus

Optic radiation

Inferior temporal gyrus

Tapetum

Lingual gyrus

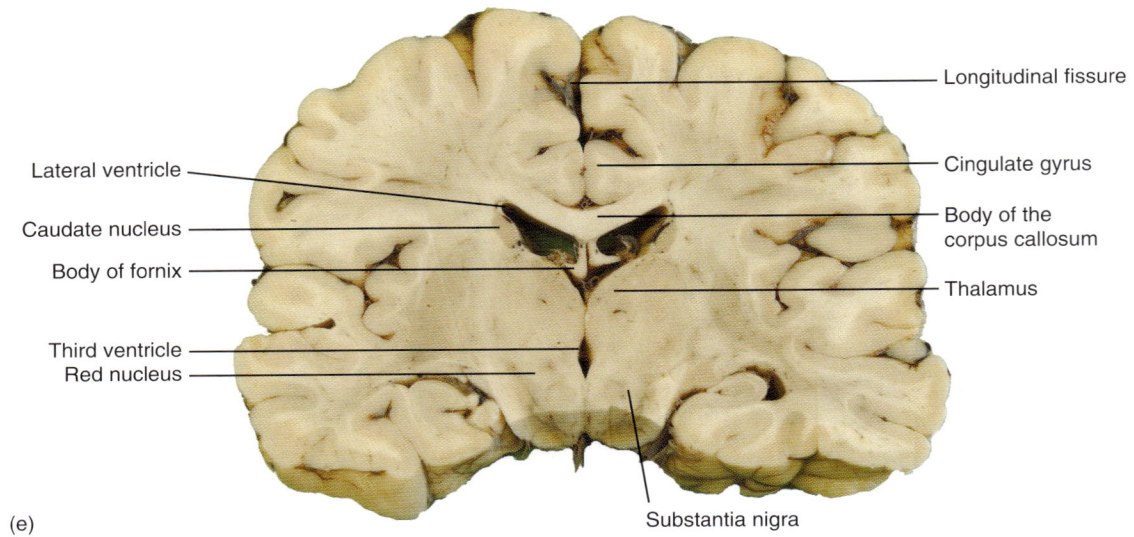

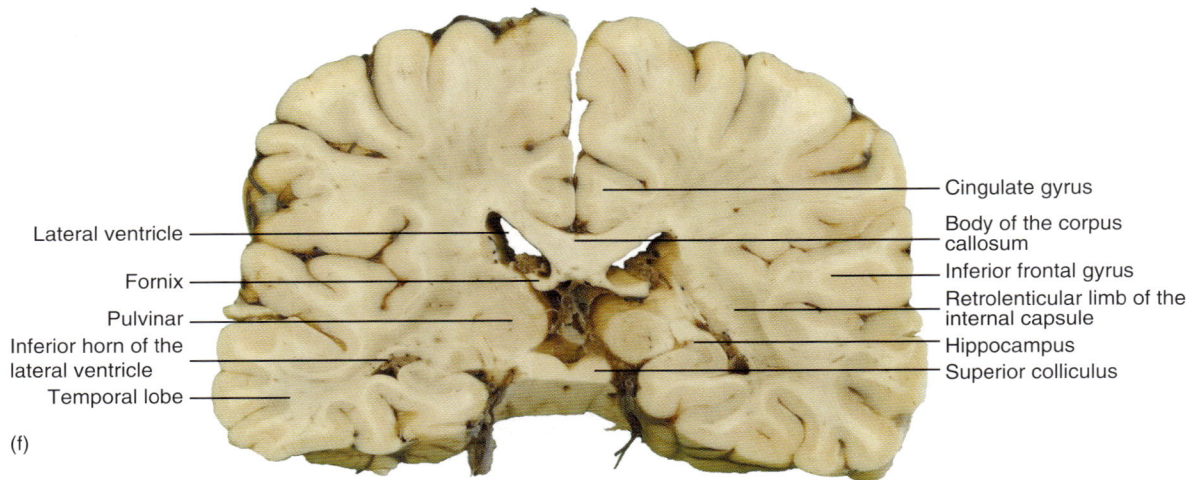

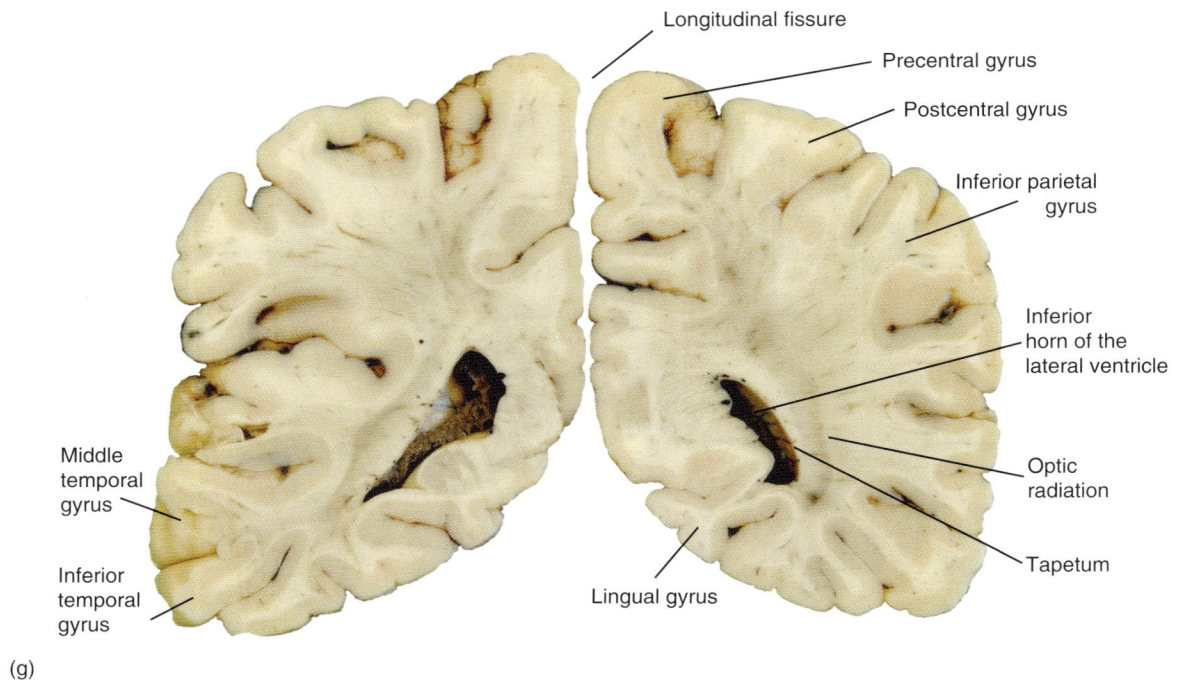

Figure 1.8   Coronal sections through the cerebral hemispheres (*continued*).

The primary sensory cortex is located in the postcentral gyrus of the parietal lobe, which forms the posterior border of the central sulcus. Neurons in this region receive somatic sensory information from sensory receptors responsible for pain, temperature, touch, pressure, vibration or taste.

The thalamus relays this sensory information to the primary sensory cortex. Other regions of the cortex have developed special sensory roles. The occipital lobe contains the visual cortex whereas the temporal lobe is responsible for olfactory and auditory sensations. A specialist area of the frontal lobe and the adjacent anterior part of the insula are the gustatory cortex. This processes information from taste receptors of the tongue and pharynx.

THE BRAIN STEM

The brain stem comprises the medulla oblongata, itself an upward continuation of the spinal cord, the pons and the midbrain; it is continuous with the diencephalon component of the forebrain. It is related to the basiocciput or clivus, and is connected to and overlaid by the posterior and lateral aspects by the cerebellum.

The stem itself can be viewed anatomically only if this overlying cerebellum is removed by cutting through the three pairs of peduncles, bundles of nerve fibres attached to either side of the stem.

The dorsal surface of the brain stem is marked by a continuation of the median sulcus extending upwards from the spinal cord into the medulla. Internally, within its caudal two-thirds, is a closed central canal. This canal moves to a more a posterior position in the more rostral medulla, eventually opening as the fourth ventricle, deep to the cerebellum. Its floor is the dorsal surface of the rostral medulla and, dorsolaterally, it is dominated by the inferior cerebellar peduncle. Nerve fibres pass through the inferior cerebellar peduncle from the medulla to the cerebellum.

The fourth ventricle is rhomboid in shape and is at its widest in the region of the junction between the medulla and pons. Nuclei of the vagus and other cranial nerves are located below the fourth ventricle, some of which may be identified using appropriate staining techniques. The postrema is an area where the blood–brain barrier is absent. It forms the most caudal aspect of the floor of the fourth ventricle. It is also the site of action of emetics. The fourth ventricle narrows superiorly and becomes the narrow cerebral aqueduct, which passes throughout the length of the midbrain. A small lateral aperture at the widest part of the fourth ventricle (also known as the foramen of Luschka) allows CSF to pass out into the subarachnoid space.

The pons is the upward continuation of the brain stem from the medulla and is divisible into anterior and posterior parts. The anterior part contains the pontocerebellar fibres, which arise from the pontine nuclei. These pass through the middle cerebellar peduncle to the contralateral side of the cerebellum. In the rostral part of the pons, the lateral wall of the fourth ventricle is composed of paired superior cerebellar peduncles, with a thin superior medullary velum connecting them and forming its roof. These peduncles converge towards the midline as they pass into the midbrain, and contain cerebellar afferent and efferent fibres.

The brain stem contains many ascending and descending nerve tracts, some of which terminate in brain stem nuclei located in this region. Functionally, important centres controlling respiration, cardiovascular function and levels of consciousness, forming the reticular system, are located within the brain stem. The cranial nerves III–XII are attached to the brain stem, with their fibres either originating from, or terminating in, the appropriate cranial nerve nuclei.

On the dorsal surface of the midbrain can be identified four paired swellings called the superior and inferior colliculi, which form important parts of the visual and auditory systems.

THE MIDBRAIN

The midbrain is divided into anterior and posterior portions at the level of the cerebral aqueduct. The anterior portion is termed the 'tegmentum' and is bounded by the crus cerebri. The posterior portion is the tectum, made up from the inferior and superior colliculi (corpora quadrigemina). The inferior colliculus forms part of the ascending auditory pathway whereas the superior colliculus is part of the visual system.

The cerebral aqueduct traverses the length of the midbrain ventral to the colliculi, with the trochlear and oculomotor cranial nerve nuclei located adjacent to it within the periaqueductal grey matter. At the level of the inferior colliculus, the superior cerebellar peduncles are related to the central portion of the tegmentum. Ventrally, the midbrain tegmentum is the site of the substantia nigra, consisting in part of pigmented, melanin-containing neurons that synthesize dopamine as their transmitter. It is degeneration of the substantia nigra that is associated with Parkinson's disease and problems secondary to drug abuse. Anterior to the substantia nigra are the large crus cerebri, composed entirely of the descending cortical efferent fibres that have passed through the internal capsule after leaving the cerebral hemispheres. Sections through the brain stem are illustrated in Figure 1.9(a)–(e).

THE DIENCEPHALON

Above the midbrain lies the forebrain (termed the 'prosencephalon'). This comprises the paired cerebral hemispheres on each side of the centrally located tissue called the diencephalon. The diencephalon is the site of the epithalamus, thalamus, subthalamus and hypothalamus.

As the cerebral hemispheres overlie the midline structures within the diencephalon, only the anterior part of the hypothalamus can be viewed on the base of the brain. Both the thalamus and hypothalamus are nuclear groups forming the lateral wall of the centrally located third ventricle. The left and right thalami are often linked together across the third ventricle by an interthalamic adhesion.

The hypothalamus also contributes to the floor of the third ventricle. It is separated from the inferior aspect of the thalamus by a shallow hypothalamic sulcus and extends ventrally and medially to the subthalamus.

The epithalamus is a relatively small structure comprising the pineal gland and the habenular nuclei; it is located in the caudal and dorsal region of the diencephalon, immediately above the superior colliculus of the midbrain.

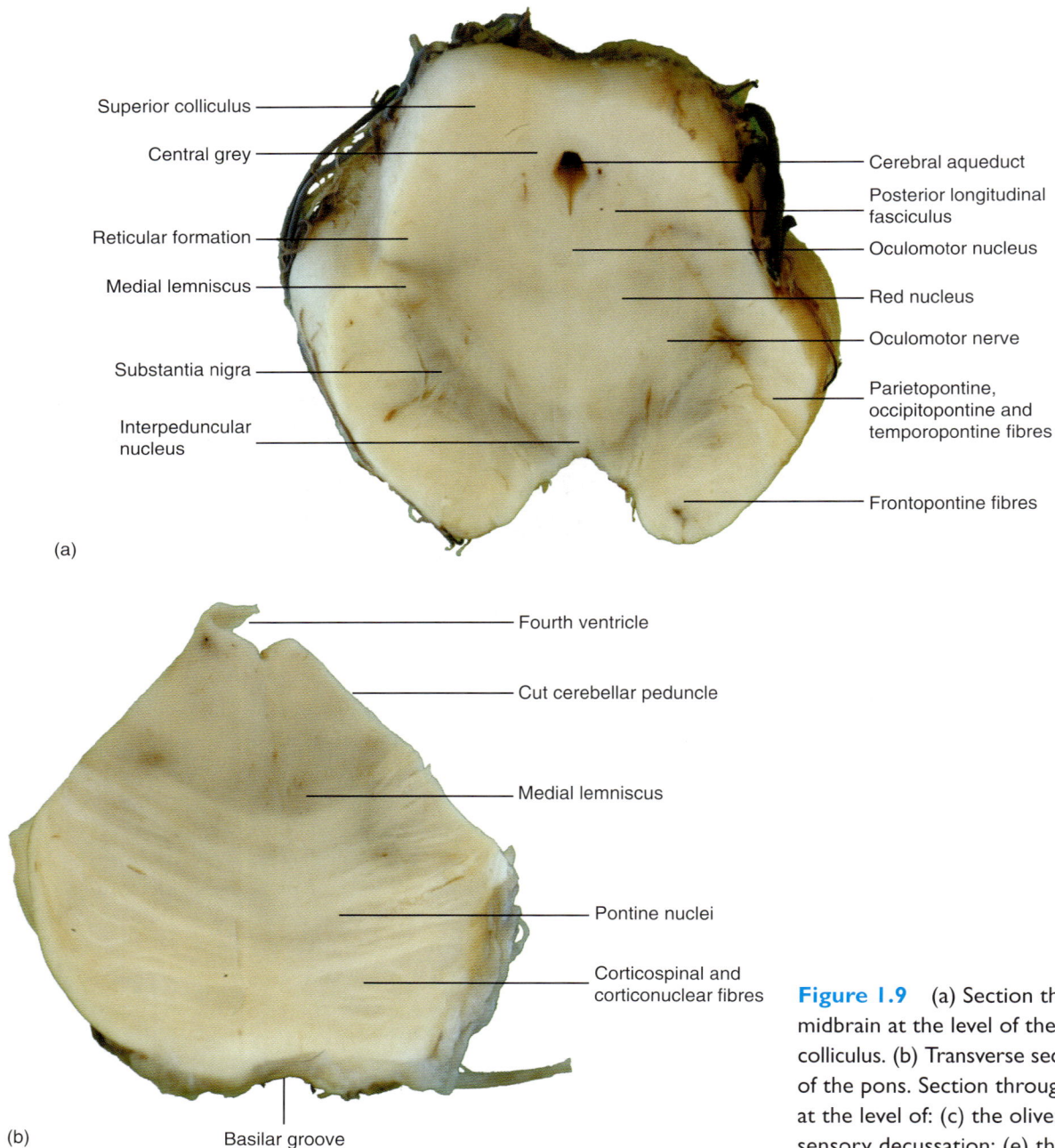

(a)

Superior colliculus
Central grey
Reticular formation
Medial lemniscus
Substantia nigra
Interpeduncular nucleus

Cerebral aqueduct
Posterior longitudinal fasciculus
Oculomotor nucleus
Red nucleus
Oculomotor nerve
Parietopontine, occipitopontine and temporopontine fibres
Frontopontine fibres

(b)

Fourth ventricle
Cut cerebellar peduncle
Medial lemniscus
Pontine nuclei
Corticospinal and corticonuclear fibres
Basilar groove

Figure 1.9   (a) Section through the midbrain at the level of the superior colliculus. (b) Transverse section at the level of the pons. Section through the medulla at the level of: (c) the olive; (d) the sensory decussation; (e) the pyramids.

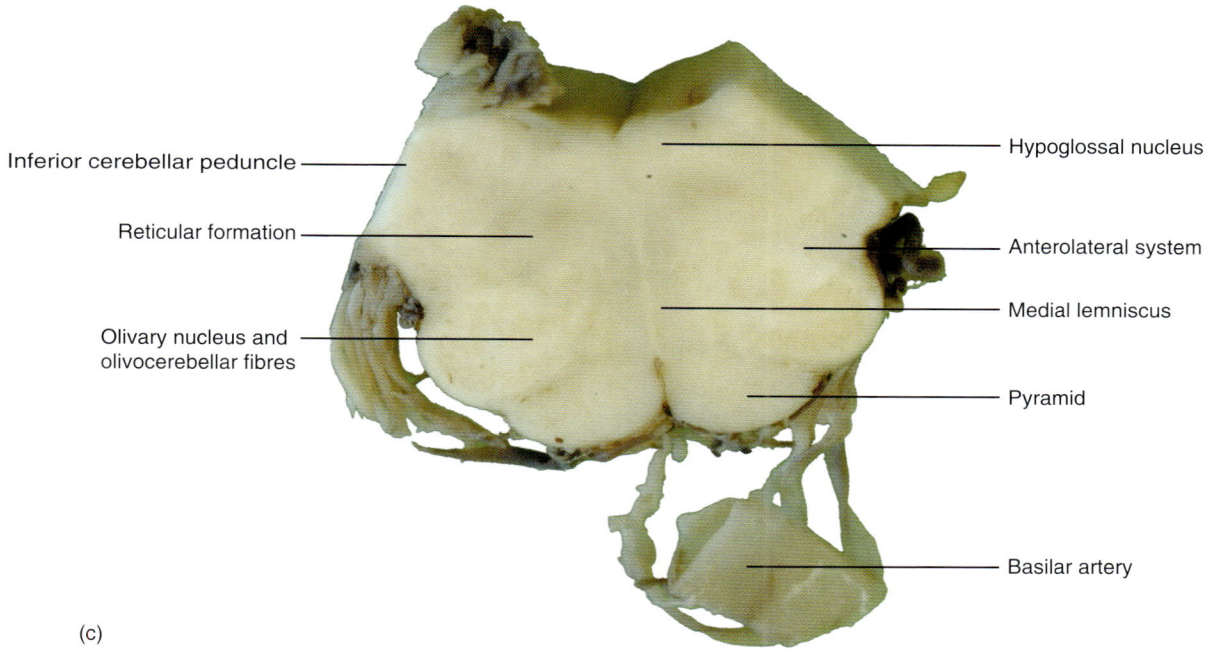

Inferior cerebellar peduncle

Reticular formation

Olivary nucleus and olivocerebellar fibres

Hypoglossal nucleus

Anterolateral system

Medial lemniscus

Pyramid

Basilar artery

(c)

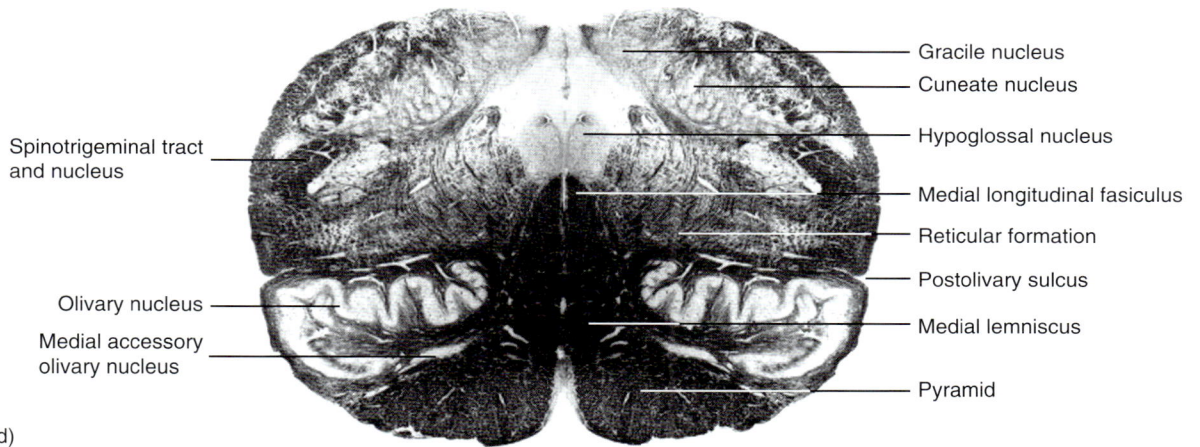

Spinotrigeminal tract and nucleus

Olivary nucleus

Medial accessory olivary nucleus

Gracile nucleus

Cuneate nucleus

Hypoglossal nucleus

Medial longitudinal fasiculus

Reticular formation

Postolivary sulcus

Medial lemniscus

Pyramid

(d)

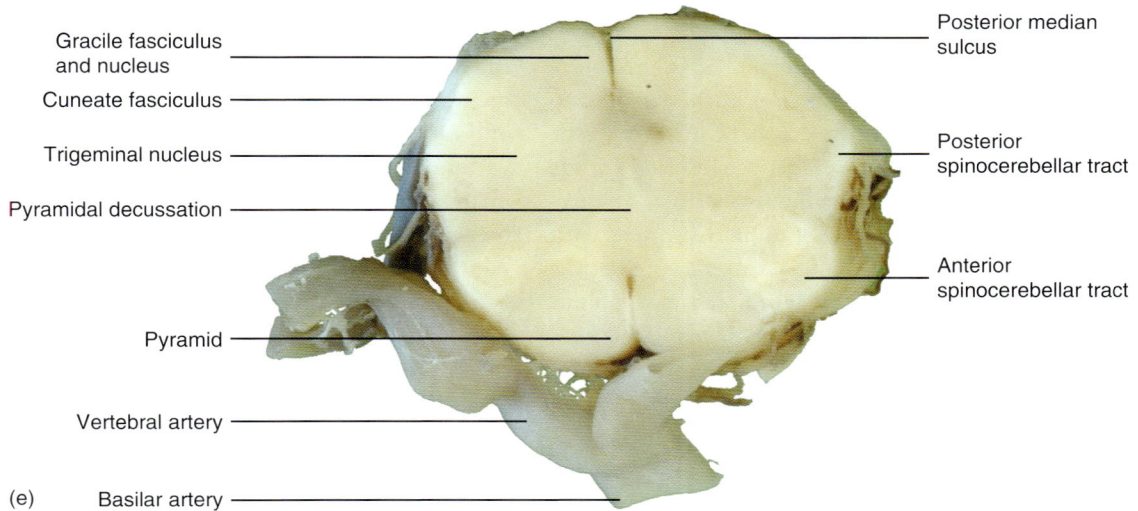

Gracile fasciculus and nucleus

Cuneate fasciculus

Trigeminal nucleus

Pyramidal decussation

Pyramid

Vertebral artery

Posterior median sulcus

Posterior spinocerebellar tract

Anterior spinocerebellar tract

(e) Basilar artery

Inferior to the thalamus and dorsolateral to the hypothalamus is the subthalamus, with its anterolateral aspect closely related to the internal capsule.

The internal capsule separates the thalamus from the lentiform nucleus and is a common site of strokes. It contains fibres running from the thalamus to the cortex and from the cortex to the brain stem and spinal cord. These fibres form the corona radiata that can be seen between the internal capsule and the cortex.

The internal capsule has a medially facing 'V' shape when examined in a horizontal section of the brain, with an anterior limb between the lentiform nucleus and caudate nucleus, a middle section known as the genu, and a posterior limb lying between the lentiform nucleus and the thalamus. Lateral to the thalamus, the capsule is known as the retrolentiform part.

The lentiform nucleus is itself lens shaped and is formed from the putamen and globus pallidus whereas the corpus striatum links the putamen and globus across the internal capsule.

Inferiorly, the diencephalon is related to the midline optic chiasma, caudal to which is a small midline elevation called the tuber cinereum. From its apex extends the infundibulum or pituitary stalk, which attaches to the pituitary gland. Further caudal to the tuber cinereum, a pair of rounded eminences, the mamillary bodies, is located on either side of the midline. These contain the mamillary nuclei of the hypothalamus.

THE CEREBELLUM

The cerebellum is part of the hindbrain that overlies the fourth ventricle on its dorsal aspect. It is connected to the brain stem by three stout pairs of fibre bundles called the inferior, middle and superior cerebellar peduncles. It comprises two laterally located hemispheres linked in the midline by the vermis, with its superior surface related to the deep surface of the tentorium cerebelli. Although the superior vermis forms a midline ridge, the inferior vermis lies in a deep groove between the hemispheres. The cerebellum has a highly folded and convoluted surface, with the folds oriented transversely. Fissures of varying depths lie between the folds. Some act as landmarks and subdivide the cerebellum into three lobes. The primary fissure is on the superior surface and separates the anterior lobe from the posterior lobe. On the inferior surface, a posterolateral fissure separates the flocculus and vermis, which together form the flocculonodular lobe.

Internally, the cerebellum consists of an outer layer of grey matter, the cerebellar cortex, and an inner core of white matter, composed of afferent and efferent fibres that run to and from the cortex. Irregular projections also extend up towards the cortex. Deep within the white matter are located four pairs of cerebellar nuclei, which have nerve connections with the cerebellar cortex and nuclei of the brain stem and thalamus. Sections through the cerebellar hemispheres are illustrated in Figure 1.10.

THE VERTEBRAL COLUMN

The vertebral column's function is to support the trunk and to protect the underlying spinal cord. Its structure undergoes progressive growth and developmental changes in the postnatal period. These continue in adulthood and lead to decline in senescence. The bony morphology is influenced internally by genetic, hormonal and metabolic factors, and externally by mechanical and environmental factors; this affects the column's ability to react to the dynamic forces of everyday life such as compression, traction and shear. The forces vary in magnitude and are influenced by occupation, locomotion and posture.

The column comprises 33 vertebral segments, each separated by a fibrocartilaginous intervertebral disc. Although the usual number of vertebrae is seven cervical, twelve thoracic, five lumber, five sacral and four coccygeal, this total is subject to frequent variations of between 32 and 35 bones.

In adults, the cervical region is curved convexly forwards and is the least marked; this is a lordosis. The thoracic curve is kyphotic, i.e. concave forwards and is a consequence of the increased posterior depth of the thoracic vertebral bodies. The lumbar curve is lordotic, i.e. convex forwards, having a greater curve in the female; the lordosis is caused by the greater anterior depth of the intervertebral discs and some anterior wedging of the vertebral bodies.

The vertebral canal follows these vertebral curves. Free mobility is present in the cervical and lumbar regions, and the canal is large with a triangular shape. Where movement is reduced in the thoracic region, the canal is small and circular. These canal size differences are matched by variations in the diameter of the spinal cord and its enlargements. The lumbar spinal canal has a gradual decrease in measurement between L1 and L5 with a greater relative width in the female.

Each vertebra has distinctive shapes to its body and spinous process as well as to the transverse processes and the intervertebral joints, although similarities between groups in the same region can aid identification of individual bones. In older people, it should be noted that, as bone undergoes age-related changes in its structure, this can lead to broadening and loss of height of the vertebral bodies. Such changes are more severe in women (Figure 1.11).

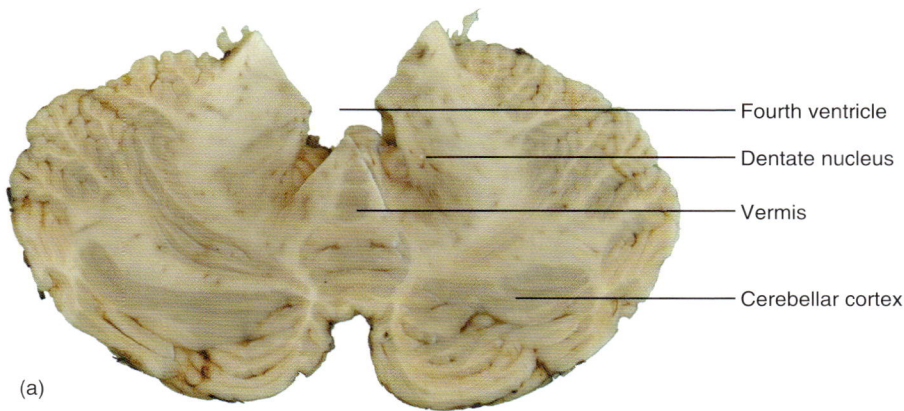

(a)

Fourth ventricle

Dentate nucleus

Vermis

Cerebellar cortex

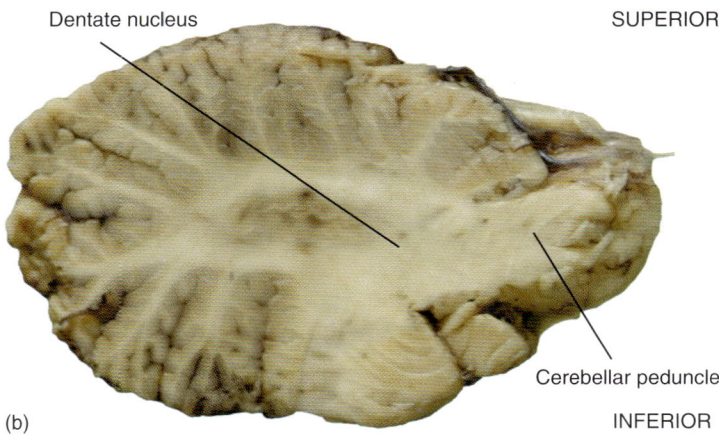

Dentate nucleus

SUPERIOR

Cerebellar peduncle

(b)

INFERIOR

Figure 1.10　Cerebellum (a) horizontal section; (b) sagittal section.

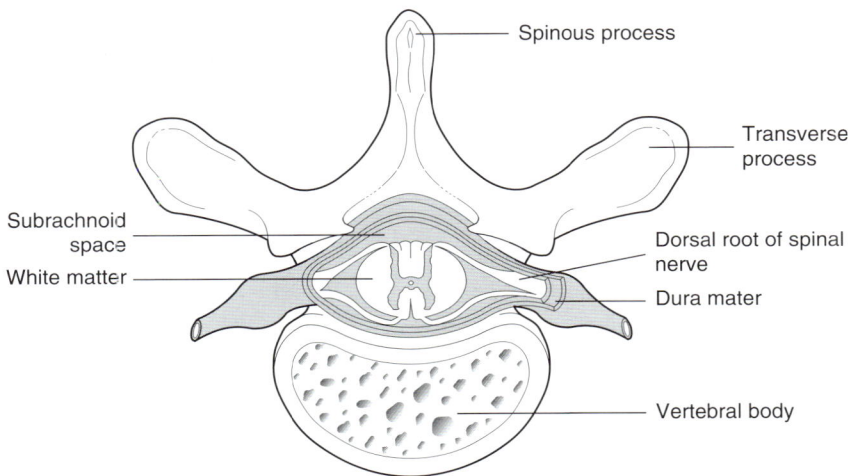

Spinous process

Transverse process

Subrachnoid space

White matter

Dorsal root of spinal nerve

Dura mater

Vertebral body

Figure 1.11　Spinal column.

THE SPINAL CORD

The spinal cord serves to link the brain stem to the appropriate segmental nerves of the peripheral nervous system as they emerge between each vertebra. In the adult, it extends from the lower end of the medulla oblongata at the level of the foramen magnum to the level of the upper border of the second lumbar vertebra. Below this level, it tapers and forms the conus medullaris. A band of pia mater called the filum terminale continues downwards from the end of the cord to the coccyx.

During early pregnancy, the cord in the fetus extends the entire length of the vertebral canal. As pregnancy progresses, the fetal vertebral column grows more rapidly in length than the spinal cord, with the result that it extends only to the level of the third lumbar vertebra at birth; it then slowly attains its adult level in childhood.

The spinal cord lies fully within the vertebral canal of the column and is thus normally protected from external trauma. Damage can thus only occur either as a consequence of massive external forces that compress this bony cage or by penetrating wounds, where an instrument or other penetrating agent passes between the vertebra in the narrow space occupied by the intervertebral disc.

The three meningeal layers, as described surrounding the brain, also enclose the spinal cord. The dura mater layer extends down to the level of the second sacral vertebra and is lined on its inner surface by the arachnoid mater. The pia mater layer firmly adheres to the cord's surface. Laterally, the cord is suspended within the dural sheath by serrated denticulate ligaments, which effectively create a shelf between the dorsal and ventral roots of the spinal nerves.

The CSF is enclosed by the subarachnoid space, which also extends down to the level of the second sacral vertebra.

External to the dura mater is the epidural space. This contains fat and the extensive vertebral venous plexus.

The cord is grooved on its anterior surface by the median fissure and on its posterior surface by the shallower median sulcus. Laterally, the posterolateral sulcus is present, to which the dorsal roots of the spinal nerves are attached.

The cord is divided into 31 spinal cord segments, from which arise the paired spinal nerves. Sensory fibres enter the cord via its dorsal (posterior) aspect, which also has a clearly defined dorsal root ganglion containing the cells of the nerve axons. Motor fibres leave the cord on the anterolateral aspect via the ventral (anterior) root.

The sensory and motor nerve roots fuse and merge into a single spinal nerve that leaves the vertebral canal via the intervertebral foramen. Leaving the foramen, the emerging nerves divide into anterior and posterior rami, each containing both motor and sensory fibres. The length of the spinal nerve within the vertebral canal increases progressively down the length of the cord, until after termination of the cord itself at the level of the second lumbar

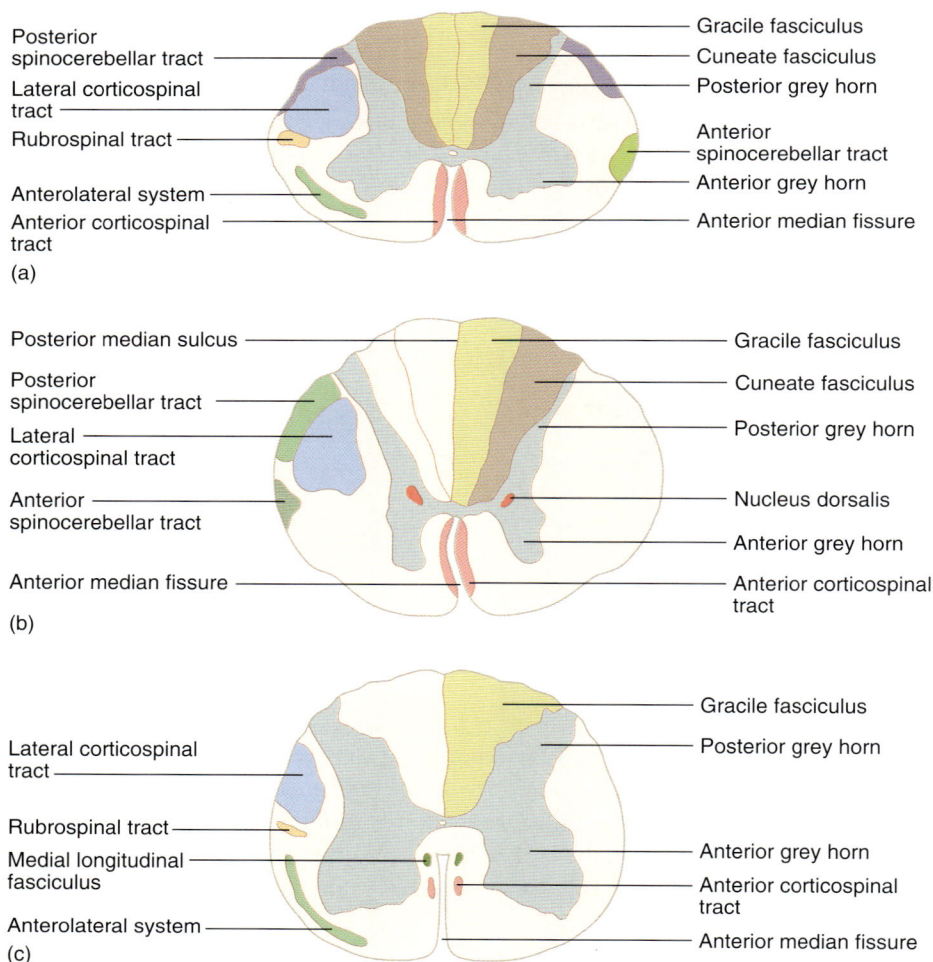

(a)

Posterior spinocerebellar tract
Lateral corticospinal tract
Rubrospinal tract
Anterolateral system
Anterior corticospinal tract

Gracile fasciculus
Cuneate fasciculus
Posterior grey horn
Anterior spinocerebellar tract
Anterior grey horn
Anterior median fissure

(b)

Posterior median sulcus
Posterior spinocerebellar tract
Lateral corticospinal tract
Anterior spinocerebellar tract
Anterior median fissure

Gracile fasciculus
Cuneate fasciculus
Posterior grey horn
Nucleus dorsalis
Anterior grey horn
Anterior corticospinal tract

(c)

Lateral corticospinal tract
Rubrospinal tract
Medial longitudinal fasciculus
Anterolateral system

Gracile fasciculus
Posterior grey horn
Anterior grey horn
Anterior corticospinal tract
Anterior median fissure

Figure 1.12 Section through the spinal cord at the level of: (a) C7; (b) T6; (c) L3.

vertebra. Below this level, the nerves alone form a bundle known as the cauda equina within the vertebral canal.

The blood supply of the spinal cord is maintained by the anterior and posterior spinal arteries. The anterior spinal artery is located within the anterior median fissure and is formed by the union of a branch vessel from each vertebral artery. The posterior spinal arteries are normally branches of the posteroinferior cerebellar arteries but can arise directly from the vertebral arteries. They are present on both sides of the posterior aspect of the cord. A series of radicular arteries also enters the vertebral canal through the intervertebral foramina. Surgery of aortic aneurysms may compromise these vessels.

Between the vertebrae and the dura mater is the epidural space, which contains the arteries supplying the spinal cord as well as the vertebral venous plexus. These are called Bateson's veins and are unusual because they contain no valves. Clinically, this is an important point because it allows metastases from malignant breast and prostate tumours to reach the vertebrae as the vertebral venous plexus are connected to the veins draining these organs.

INTERNAL STRUCTURE OF THE SPINAL CORD

A transverse section of the cord exhibits an H-shaped area of grey matter containing sensory and motor nerve cells. Within the grey matter, a central canal communicates superiorly with the fourth ventricle of the brain stem. The posterior horn of the grey matter contains the termination of the sensory nerve fibres whereas the larger anterior horn contains motor cells that supply fibres within anterior roots.

In the thoracic and upper lumbar regions are lateral horns containing the cells of origin of the preganglionic sympathetic fibres.

The grey matter of the cord is surrounded by white matter containing the ascending and descending nerve tracts (Figure 1.12a–c).

FURTHER READING

Atlases of anatomy and neuroanatomy

Abrahams, P.H., Marks, S.C. and Hutchings, R.T. 2003. *McMinn's Colour Atlas of Human Anatomy*, 5th edition. New York: Mosby.

Dean, D. and Herbener, T.E. 2000. *Cross-sectioned Human Anatomy*. Baltimore: Lippincott, Williams & Wilkins.

El-Khoury, G.Y., Bergman, R.A. and Montgomery, W.J. 1995. *Sectional Anatomy by MRI*, 2nd edition. Edinburgh: Churchill-Livingstone.

Ellis, H., Logan, B. and Dixon, A. 1999. *Human Sectional Anatomy*, 2nd edition. Oxford: Butterworth-Heinemann.

Logan, B.M., Hutchings, R.T. and Reynolds, P. 2004. *McMinn's Colour Atlas of Head and Neck Anatomy*, 3rd edition. New York: Mosby.

Textbooks of anatomy and neuroanatomy

Bear, M.F., Connors, B.W. and Paradison, M.A. 2002. *Neuroscience exploring the brain*. Baltimore: Lippincott, Williams & Wilkins.

Crossman, A.R. and Neary, D. 2000. *Neuroanatomy*, 2nd edition. Edinburgh: Churchill-Livingstone.

Fitzgerald, M.J.T. and Folan-Curran, J. 2002. *Clinical Neuroanatomy*, 4th edition. Edinburgh: WB Saunders.

Haines, D.E. 2004. *Neuroanatomy: An atlas of structures, sections and systems*, 6th edition. Baltimore: Lippincott, Williams & Wilkins.

Moore, K.L. and Dalley, A.F. 2006. *Clinically orientated anatomy*, 5th edition. Baltimore: Lippincott, Williams & Wilkins.

Osborn, A.G., Blaser, S. and Salzman, K.L. 2004. *Diagnostic imaging; brain*. Philadelphia: WB Saunders.

Standring, S.S. (ed.). 2005. *Gray's anatomy*, 39th edition. Edinburgh: Churchill-Livingstone.

2

Techniques

Helen L Whitwell

This chapter covers in outline the approach to postmortem examination with emphasis on the neuropathological aspects. It is not intended to be a fully comprehensive, detailed guide, but concentrates rather on specific points of note from a practical aspect in forensic neuropathological cases.

BEFORE THE POSTMORTEM EXAMINATION

In the UK and in many other jurisdictions postmortem examinations on head injury or other forensic neuropathological cases will be at the bequest of the legal authorities. It should be good practice that this authority is confirmed in writing, although, if a case is out of normal working hours, this may not be possible. In practice this will usually be a fax from the relevant office.

GENERAL POSTMORTEM EXAMINATION

A full postmortem examination should be performed in all cases. Recent guidelines have been published by the Royal College of Pathologists (2002); these form a base for good practice together with the recommendations of the Council of Europe (Brinkman and Mangin 2000). For forensic cases there is a joint Royal College of Pathologists (UK) and Home Office (England and Wales) document relating to Code of Practice and Performance Standards for forensic pathologists (Royal College of Pathologists 2004). There are also standard textbooks of postmortem practice (Ludwig 2002) and textbooks of neuropathology detailing the examination of the brain (Dawson and Neal 2002, Esiri 1996).

NEUROPATHOLOGICAL EXAMINATION

In many cases of head trauma there will be hospital records including radiological results. These, in particular, may aid in interpretation where death is delayed and secondary changes have taken place (Bauer 2004). Perusal of these is essential before the examination and particular attention should be paid to any medical procedures, including neurosurgery, that may have been undertaken. This avoids later confusion about whether or not an injury is genuine (Figure 2.1), e.g. insertion of a ventricular shunt or an intracranial pressure-monitoring device. Apart from the above, details of any known history of the circumstances relating to the case are essential. It is, however, recognized

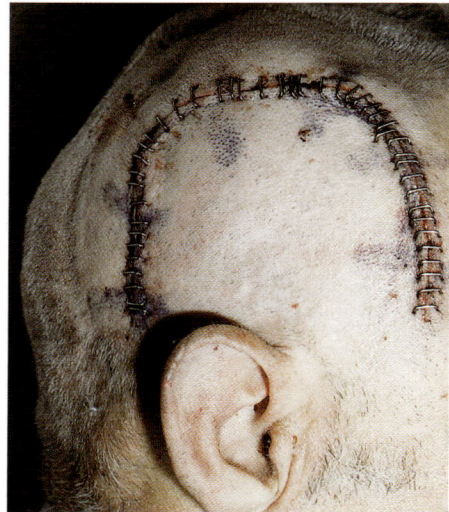

Figure 2.1 Bruising in association with a craniotomy wound.

that in the initial stages of any investigation these may be scanty and further review may be necessary at a later date, sometimes at a considerable time distance, reviewing relevant statements.

In paediatric cases, prior radiological examination should be performed. This will usually be plain radiography. However, it is likely that wider use will be made in future of other radiological techniques, including magnetic resonance imaging (MRI) and computed tomography (CT) depending on availability. It is essential that films are reviewed by an expert in paediatric radiology, with particular regard to injury. Radiological examination is also valuable in adult cases, in particular gunshot wounds, burnt bodies and decomposed bodies, and where there is extensive head/facial trauma.

RECORDING INFORMATION

Notes taken by the pathologist, including hand-written ones, and any diagrams, as well as tapes if dictation is used, should be retained as evidence. Photographs are ideal and essential in any case where criminal proceedings may take place. With the ease of digital photography these should be possible in the vast majority of cases. However, digital photography may not be recognized in some jurisdictions for legal proceedings. Video recording has value, particularly in the examination of suspected traumatic subarachnoid haemorrhage (see Chapter 6, p. 72). Diagrams of any injuries, together with depiction of any fractures, can also be undertaken. It may be necessary to produce body plans for court use – this is often preferred for juries rather than photographs. Computer technology has also aided in the production of such plans. Other investigations relating to the head before examination depend on circumstances but may include the following:

- cerebrospinal fluid (CSF) – virology, microbiology
- collection of hair – for toxicology, this should include roots if possible and be of pencil thickness from the crown – see also Chapter 14, p. 176. Hair may also be taken for forensic analysis relating to the scene or potential weapon.

EXTERNAL EXAMINATION

Part of the interpretation of any head injury case is the evaluation of external wounds. This is covered in detail in Chapter 3.

INTERNAL EXAMINATION

Subscalp bruising/haematomas

Following reflection of the scalp, all bruises should be documented with their location and size recorded, including involvement of muscles, e.g. temporalis muscle. Description should include an assessment of the age, i.e. recent, days, etc. Over-dogmatic interpretation about age should not be done because it is well recognized that only estimates can be given. In elderly people, bruising can persist for many weeks. Re-examination of the external aspect may be necessary to coordinate any findings. It is well recognized that scalp bruising may be much more extensive than the injuries noted externally. Kicking tends to produce marked deep bruising (Figure 2.2). Neurosurgical procedures may mask or cause bruising or soft tissue haemorrhage (Figure 2.3). Artefactual bruising may occur over the back of the head when an individual has been lying in a prolonged comatose state. Scalp petechiae are well recognized as an artefact in many situations and should not be taken as a sign of asphyxia (Figure 2.4).

Removal of the skull

This can be done either with an electric or hand saw. The latter method is indicated in suspected cases of Creutzfeldt–Jakob disease (CJD). It also tends (in experienced hands) to produce less in the way of damage to the underlying dura/brain. Care should be taken to avoid artefactual skull fracturing. Skull fractures are covered in Chapter 4. The dura should always be stripped. In

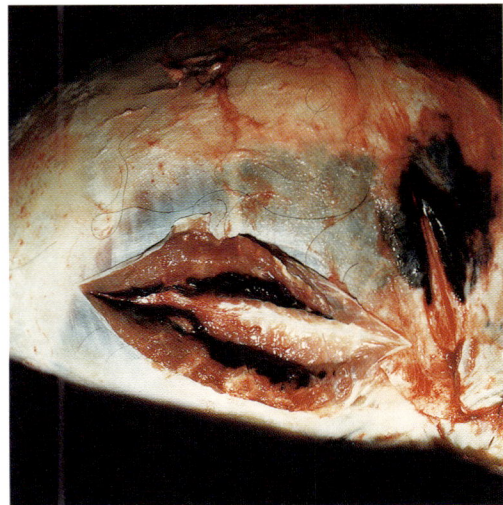

Figure 2.2 Haemorrhage into temporalis muscle caused by kicking.

addition, examination of the facial structures to assess bruising/fractures should be undertaken. Bruising to the jaw may be evident only on deep dissection, e.g. after a punch or slap to the face.

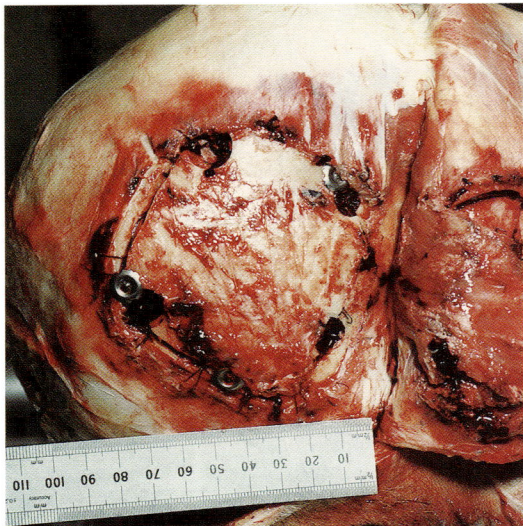

Figure 2.3 Bone flap with soft tissue haemorrhage caused by neurosurgery.

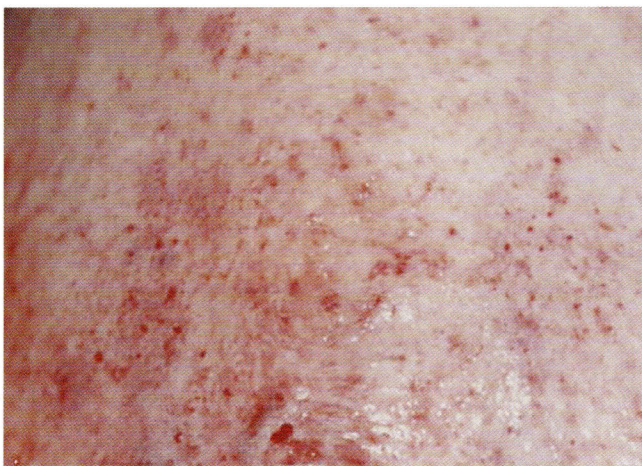

Figure 2.4 Scalp petechiae – of no diagnostic significance.

IMMEDIATE NEUROPATHOLOGICAL EXAMINATION

Immediate inspection on removal of the skull will assess the presence of intracranial haematomas – extradural and subdural – as well as subarachnoid haemorrhage (Figure 2.5). In the latter condition, if traumatic subarachnoid haemorrhage is suspected as a result of the history and

circumstances, the appearance of the brain may alert to the presence of this, with subarachnoid haemorrhage being just visible over the lower hemispheres (Figure 2.6). Other conditions including meningitis and brain swelling may also be immediately apparent. Documentation should be done before brain removal, which normally

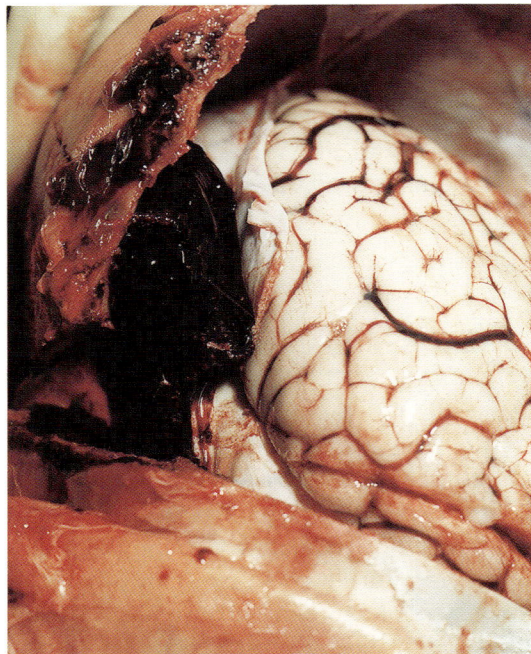

Figure 2.5 Acute extradural haematoma identified on removal of skull.

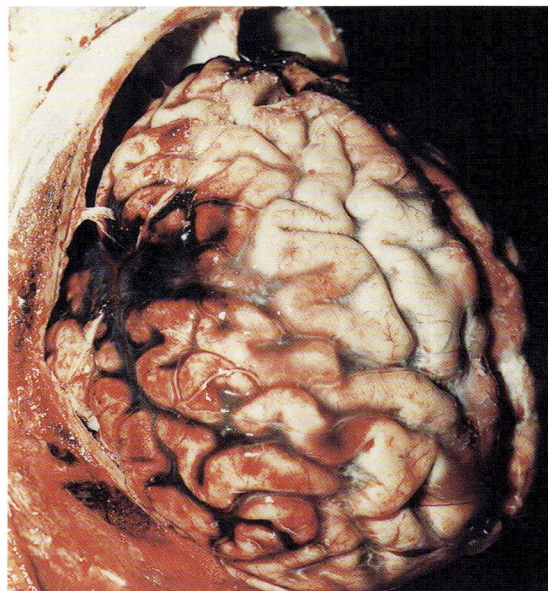

Figure 2.6 Appearances of traumatic subarachnoid haemorrhage visualized on removal of the skull.

should be carried out by the pathologist. Examination for vertebral/basilar tears may be undertaken as necessary. This is covered in detail in Chapter 6 under the heading Traumatic subarachnoid haemorrhage.

Once the brain has been removed, before fixation, other findings should be documented, including:

- assessment of the volume of any extradural haemorrhage (EDH) and/or subdural haemorrhage (SDH); most often only an approximate estimation of volume can be given
- lacerating and contusional injury, including pontomedullary tears
- sinus thrombosis
- identification of items associated with surgical intervention such as aneurysm clips or coils
- aneurysms: this should be done before fixation which makes identification at a later stage difficult, if not impossible; this may necessitate washing away blood clot and dissecting along the paths of the arteries
- infections, including meningitis, which may be predominantly basal
- intracerebral haematomas may also be identified and care should be taken that they do not disrupt artefactually before fixation.

The brain should be weighed fresh – standard tables are given in Appendices 1–3. It should be remembered, however, that these can be used only as a guide.

After removal of the brain, the dura should be stripped for detailed assessment of any bone pathology. Additional examinations include removal of the pituitary gland. Exposure of the middle and inner ears, orbit and air sinuses may be indicated in cases of infection.

Removal of the spinal cord again depends on clinical circumstances. This can be done via either the anterior or posterior approach using standard techniques. The posterior approach is preferred in exposure of the high cervical region/craniocervical junction. Removal of the spinal cord en bloc with the spinal column may be useful in some cases, e.g. compression fractures, vertebral artery trauma (see Chapter 6, p. 72).

REMOVAL OF THE BRAIN IN NEWBORNS AND YOUNG INFANTS

The skull should be opened along the sutures. This exposes the brain, the removal of which is aided if performed under water so that it can 'float'. Fixation can be accelerated using 20 per cent formalin in a 37°C oven for 2–3 days before section if return to the body is necessary. In cases of any posterior cranial fossa abnormality, a posterior approach is recommended. (See also Chapter 12, p. 139.) External examination of the brain should include examination for congenital abnormalities as well as external injury.

RETENTION OF THE BRAIN

It is advisable and considered good practice to retain the brain where there is any issue relating to the neuropathology. This is particularly so where medicolegal proceedings may take place. When a brain is retained for examination it is advised that full written documentation is prepared, including consent (if applicable). It may also be necessary to have signed continuity documentation regarding the location and possession not only of the brain but also blocks and slides. Although in some cases of head injury the relevance of formal neuropathology may not be immediately apparent (e.g. in cases where there may be other types of assault such as strangulation or stabbing), issues may arise at a later date that were unknown at the time of the original examination, which then cannot be answered to any degree of satisfaction. These include length of survival after head injury and underlying pre-existing disease. There have been considerable issues raised over organ retention in recent years in the UK and other countries, as a result of retention of organs after postmortem examination where the legal/consent issues were unclear (Burton and Wells 2001, Redfern et al. 2001). This has caused difficulties in retention of brains for formal neuropathology even in medicolegal cases. There is also considerable pressure for the return of the brain before burial of the body, which again will not allow review at a subsequent date. In England and Wales, samples of body tissue including toxicological samples may be formally exhibited, although it is unclear what legal protection this confers. The Royal College of Pathologists have issued guidelines relating to the retention of material in pathological cases where legal proceedings occur (Whitwell and Lowe 2002).

If there is no consent for prolonged retention of the brain, even a few days' fixation in formalin produces significantly better results than immediate sectioning and blocking. This is often possible and practicable because in many cases burial or other disposal of the body takes several days to organize. In potentially criminal cases this is often longer. It should be noted that information may be lost if a brain is not retained. This has been highlighted in paediatric cases where there has been a marked reduction in formal neuropathology with consequent loss of significant information (Walsh and Moore 2003).

Microwave fixation has been tried. This has some advantages where the brain needs to be returned to the body at the same time as the postmortem examination – the method is given in Appendix 4. However, with complex cases fixation is not optimal. In addition, histological sections prepared from such brains leads to a degree of artefactual change, which may make interpretation difficult against standard reference texts using established neuropathological techniques (personal experience).

The pathologist may, however, have to return the brain immediately. This is particularly so in cases where the cause of death is known and permission for brain retention has not been given. In these cases it may be possible to obtain permission for small samples. It is important that the pathologist documents the extent of examination and the reasons preventing formal detailed dissection in the report, in case medicolegal issues arise at a later date.

EXAMINATION AFTER FIXATION

After fixation of the brain, which in standard neuropathological practice is for at least 3–4 weeks, sectioning can be done. It may well be that at this time further details of circumstances and other information are available.

Further recording of findings on brain cutting is essential, of both the external appearance of the brain before cutting and subsequently on sectioning. Again photographs are ideal, including digital photography. In addition, drawings of the various brain slices can be used.

If the dura has been retained with the brain, this should be sampled, if necessary in multiple locations, particularly in the case of extra- or subdural haemorrhage. Further examination of the basal cerebral vessels should be undertaken before sectioning of the brain. The cerebellum and stem should be separated from the hemispheres. The stem should be sectioned at approximately 5-mm intervals and in some cases embedded in its entirety (head injury in infancy). Either the cerebellum can be cut in the horizontal plane or more detailed examination can be done by first dividing the hemispheres in the midline followed by sagittal cuts at approximately 5 mm through each hemisphere.

The standard method for the hemispheres is for coronal sections – the initial slice is through the mamillary bodies, then at 1-cm intervals through the hemispheres. Details of anatomy are given in Chapter 1, p. 10.

BLOCKING OF THE BRAIN

This depends on the circumstances of the individual case. Details of blocking for axonal injury, suspected non-accidental infant head injury and hypoxic brain damage are given in Appendix 5. Details for possible dementia cases are given in Appendix 6. Blocking schedules for sudden death in infancy and sudden unexpected death in epilepsy are given in Appendices 7 and 8. In other cases a systematic screening approach may be needed. The areas for a screening sampling are outlined in Appendix 9.

EXAMINATION OF THE SPINAL CORD

The dura should be examined externally for any pathology. The dural sheath can then be opened both anteriorly and posteriorly for examination. Sampling for histology should be guided by the circumstances. In traumatic injury, transverse sections from above and below the level of injury may be helpful on frozen sections for neutral lipids and myelin staining. This may be an aid to dating injury (Dawson and Neal 2002).

HISTOLOGICAL STAINS

In forensic neuropathology, stains should include haematoxylin and eosin, amyloid precursor protein (APP) and CD68 immunocytochemistry. In addition Perls' stain for iron is useful to assess haemosiderin. Additional stains of particular relevance may include Masson trichrome for collagen, EVG (elastic van Gieson) for blood vessels and Congo red or Aβ immunocytochemistry for amyloid angiopathy. Other stains required will depend on the clinical history, e.g. immunocytochemistry for neurodegenerative disorders, myelin stains for demyelinating disorders.

FINDINGS RELATING TO HOSPITAL THERAPY

- Scalp bruising: this may be extensive in association with surgery (see Figure 2.3).
- Herniation of the brain through craniotomy/burr holes: the severely swollen brain may herniate through a craniotomy site, producing areas of haemorrhagic pressure necrosis at the edge of the hernia. Swelling with oedema of brain tissue occurs followed by variable haemorrhagic or ischaemic necrosis. A burr hole may be associated with protusion of smaller amounts of brain tissue (Figure 2.7).

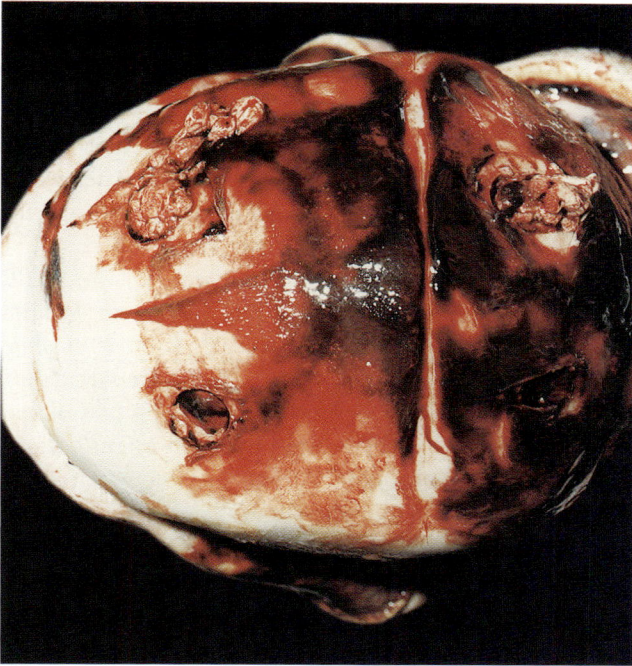

Figure 2.7 Burr hole with protrusion of brain tissue.

- Ventricular catheters: these may produce a variety of haemorrhages including cortical and white matter as well as injury to the corpus callosum.
- Non-perfused brain: this is covered in Chapter 9, p. 112.

ARTEFACTS

- Fixation artefacts may occur if care is not taken in suspending the brain. Flattening of the hemispheres with gross distortion may make subsequent interpretation difficult (Figure 2.8). If fresh slices are fixed, these may also become extremely distorted. Reforming the brain with stout absorbent paper between each slice, placing each half of the brain so that the mid-coronal slices are on the base of the bucket, will reduce this (Dawson and Neal 2002).
- Changes associated with decomposition: where there has been a delay in fixation postmortem bacterial activity produces a 'Swiss cheese' artefact (Figure 2.9a). Histology shows bacterial colonies (Figure 2.9b).
- Spinal cord artefact: excessive handling may cause localized softening/squashing with, if severe, the 'toothpaste' effect where there is distortion of internal architecture.
- Histological artefacts: neuronal shrinkage, particularly near the surface of the brain, may be seen with 'corkscrew' twisting of neurons and dark staining

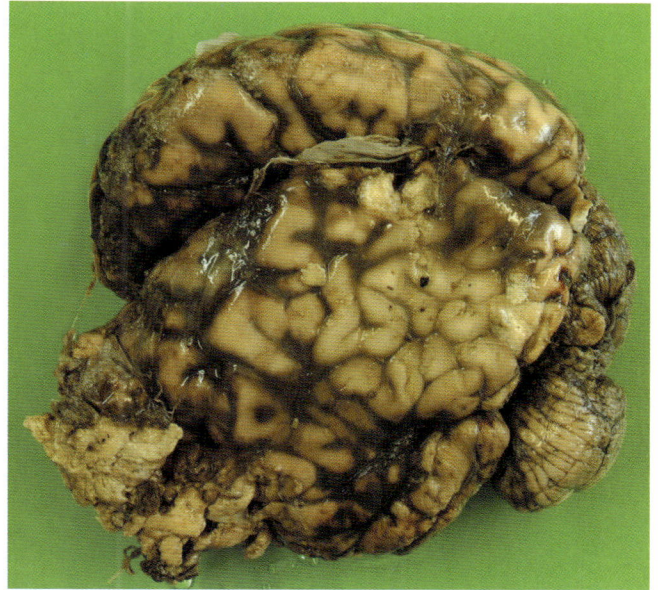

Figure 2.8 Distortion of brain as a result of poor suspension/fixation.

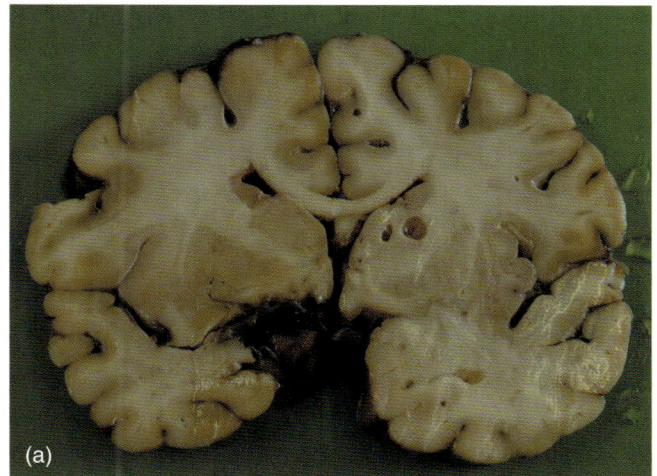

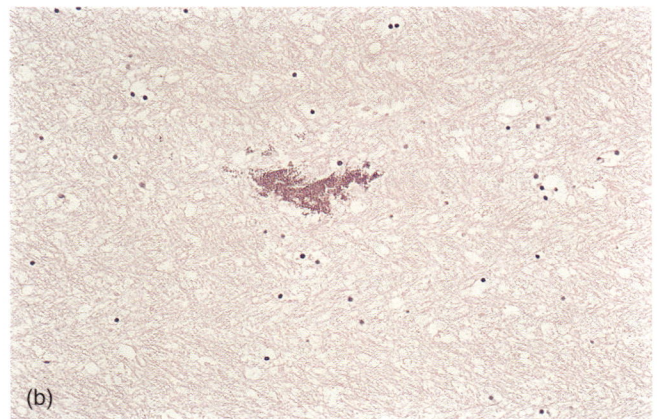

Figure 2.9 (a) Postmortem bacterial changes with cystic spaces seen in the right basal ganglia; (b) histological appearances with postmortem bacterial colonies.

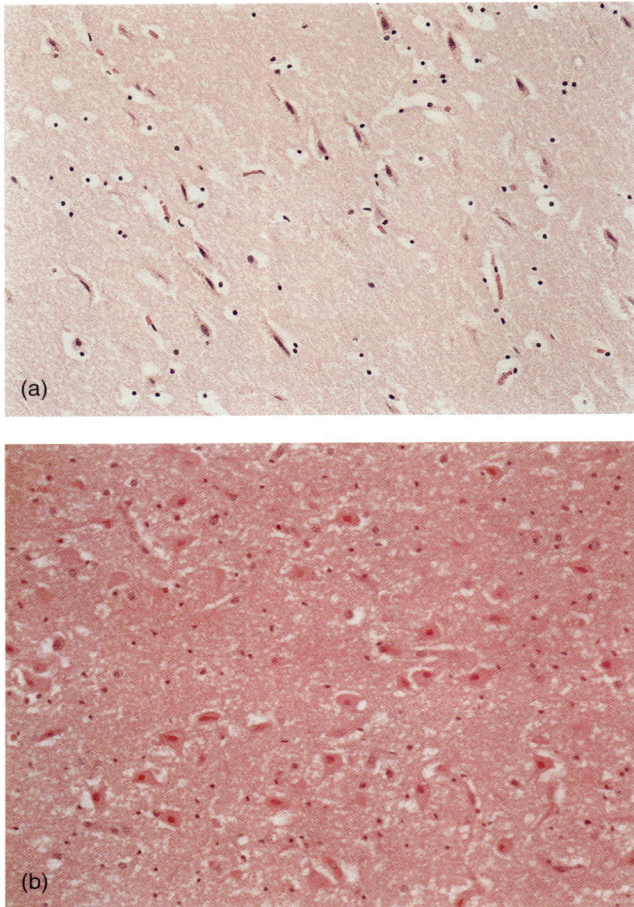

Figure 2.10 (a) Neuronal shrinkage with 'corkscrew' neurons not to be confused with (b) genuine hypoxic–ischaemic cell change.

nuclei. This may be confused with hypoxic–ischaemic cell damage (Figure 2.10).

OUTLINE OF ADDITIONAL PROCEDURES FOR PARTICULAR CLINICOPATHOLOGICAL ENTITIES

Sudden unexpected death in infancy (SUDI)

This type of death is particularly difficult and challenging. The term is used for deaths in infancy (the first year of life) where death is sudden and unexpected. In some, a cause such as congenital heart disease or infection will be determined. In others, despite full investigation, no definite cause is identified. There has been considerable variation in the terminology used in these cases, with terms such as sudden infant death syndrome (SIDS), unascertained, and sudden unexpected death in infancy being

used. The recent joint document (Royal College of Pathologists and Royal College of Paediatrics and Child Health 2004) has recommended standardization of terms between coroners and pathologists, with SIDS being used only where the death is shown to meet the international definition of sudden infant death syndrome after review of the histology, circumstances and postmortem examination to an agreed protocol (Willinger *et al.* 1991).

For sudden unexpected deaths in infancy the CESDI (Confidential Enquiry into Sudden Deaths in Infancy) protocol outlines in detail the samples for bacteriology, virology and metabolic studies (CESDI Report 2000). With regard to central nervous system (CNS) pathology a sample of CSF should be examined in all infant cases. A direct swab of the meninges for bacteriology should be undertaken. In addition a portion of cortex can be retained for virological examination. The joint Royal College of Pathologists and Royal College of Paediatrics and Child Health Publication, *Sudden unexpected death in infancy* (2004) includes an autopsy protocol for sudden unexpected death in infancy.

The details of blocking for neuropathological examination are given in Appendix 7.

Examination in cases of suspected stroke

In addition to the procedures outlined above, assessment of the following should be undertaken. Some of these come within the general examination and include a detailed assessment of the heart for left ventricular hypertrophy, myocardial infarction and mural or atrial thrombus, and assessment of the atheroma of the aortic arch to the carotid arteries with regard to severity of atheroma, together with evidence of other disease processes such as diabetes and hypertension. Specific neuropathological procedures related to the CNS include examination of the vertebral arteries (the procedure is the same as for examination in basal subarachnoid haemorrhages outlined in Chapter 6, p. 72). Examination of the carotid arteries will need to be from the origins of the carotid artery to the base of the skull, and subsequent examination of the intracranial portion of the carotid arteries including the cavernous sinus. This may necessitate fixation and decalcification (Langlois and Little 2003). Tissue for genetic analysis may be indicated in young people (snap-frozen brain, skeletal muscle and blood).

Infectious cases

All forensic cases are potentially high risk with a significant proportion from the intravenous drug abuser population.

With the increasing rise in drug abusers and communicable diseases associated with this, as well as the emergence of infections including tuberculosis (TB) and hepatitis viruses including C, and HIV, postmortem examinations may be performed in a high-risk mortuary, although with appropriate precautionary measures a standard mortuary may be sufficient. Most of the viruses, and in addition the infections that may be opportunistic in AIDS cases, are readily inactivated by formaldehyde. Additional investigations for virology, including polymerase chain reaction (PCR) analysis, may also be indicated.

Suspected or known cases of Creutzfeldt-Jakob disease, including the variant form, identified in young adults should be carried out in specialist mortuaries equipped to deal with the additional decontamination procedures required. Details of postmortem examination in suspected cases of Creutzfeldt-Jakob disease can be found in specialist neuropathological texts (for example, Dawson and Neal 2002).

Sudden death in epilepsy

Sudden death in epilepsy (SUDEP) is defined as 'sudden, unexpected, unwitnessed or witnessed, non-traumatic and non-drowning deaths in patients with epilepsy, with or without evidence for a seizure and excluding documented status epilepticus, where postmortem examination does not reveal a cause of death' (Thom 2003, Nashef 1997). In forensic practice, death may be related to an accident during a seizure (e.g. vehicular or drowning) or be a complication of an acute or older head injury.

Investigation of such cases at postmortem examination should include examination of bite marks to the tongue and oropharynx as well as examination for petechial haemorrhages, which may be an indication of an "asphyxial" death.

During the general postmortem examination particular attention should be paid to the heart, where changes such as contraction bond necrosis and sub-endocardial fibrosis may have been identified (Natleson *et al.* 1998). The respiratory tract may show evidence of inhalation of gastric contents, although this is not uncommonly seen in the larynx, trachea and bronchi where death is unrelated to epilepsy. Choking may occur on large pieces of food that occlude the upper airways.

In addition, full toxicology should be performed as for a number of SUDEP cases low levels of anti-convulsants have been identified. Alcohol and other drugs should also be screened for.

Full neuropathology of the fixed brain (ideally after fixation) should be undertaken. The blocking schedule is shown in Appendix 8.

MISCELLANEOUS

- CNS tissue may be useful in estimation of diatoms in cases of suspected drowning, although other tissues including long bones may also be used. Diatoms are aquatic unicellular plants with an extracellular coat (frustule) composed of silica (Figure 2.11a and 11b). There are many morphologically distinct varieties. They are found in naturally occurring water such as lakes, seas and rivers. The method for diatom analysis is given in Appendix 10.
- Toxicological analysis may be done for carbon monoxide and alcohol on any extracranial haematoma. In specialist toxicological hands, estimation of other substances may be possible depending on the circumstances. Expert toxicological advice should be sought in such cases.
- In cases of suspected excited delirium/restraint deaths, it is recommended that snap-frozen brain tissue be

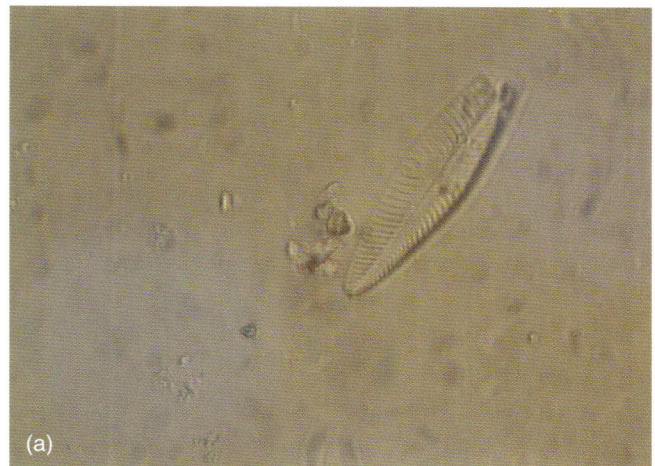

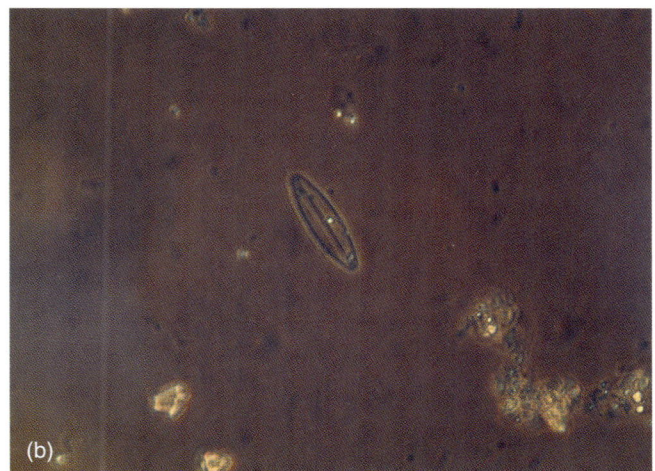

Figure 2.11 (a) Diatom (from lung); (b) diatom preparation after acid digestion of tissue (phase contrast illumination).

retained (Police Complaints Authority 2002). This should include the basal ganglia at the level of the anterior temporal poles with a block thickness of 0.5–1 cm snap-frozen in liquid nitrogen and stored at −70°C. This enables neurochemical examination to be undertaken, in particular for dopamine receptors (Mash and Staley 1999, Stephens *et al*. 2004).

REFERENCES

Bauer, M., Polzin, S. and Patzelt, D. 2004. The use of clinical CCT images in the forensic examination of closed head injuries. *J Clin For Med*, **11**, 65–70.

British Neuropathology Society. 2000. *Guidelines for Good Practice*. British Neuropathology Society: www.bns.org.uk

Burton, J.L. and Wells, M. 2001. The Alder Hey affair: implications for pathology practice. *J Clin Pathol*, **54**, 820–3.

CESDI. 2000. The CESDI SUDI studies. In: Fleming P., Blair P., Bacon, C. and Berry, P.J. (eds), *Sudden Unexpected Deaths in Infancy*. London: The Stationery Office.

Brinkman, B. and Mangin, P. 2000. Recommendation No. R (99) 3. The harmonisation of medico-legal postmortem examination rules and its explanatory memorandum. *Forensic Sci Int*, **111**, 5–29.

Dawson, T.P. and Neal, J.W. 2002. *Neuropathology Techniques*. London: Arnold.

Dekaban, A.S. and Sadowsky, D. 1978. Changes in brain weights during the span of human life: relation of brain weights to body heights and body weights. *Ann Neurol*, **4**, 345–56.

Esiri, M.M. (ed.) 1996. *Oppenheimer's Diagnostic Neuropathology: a practical manual*, 2nd edition. Oxford: Blackwell Scientific Publications Inc.

Geddes, J.F., Whitwell, H.L. and Graham, D.I. 2000. Traumatic axonal injury: practical issues for diagnosis in medicolegal issues. *Neuropathol Appl Neurobiol*, **26**, 105–16.

Langlois, E.I. and Little, D. 2003. A method for exposing the intraosseous portion of the carotid arteries and its application to forensic case work. *Am J Forensic Med Pathol*, **24**, 35–40.

Ludwig, J. 2002. *Handbook of Autopsy Practice*, 3rd edition. Totowa, NJ: Humana Press.

Mash, D.C. and Staley, J.K. 1999. D3 dopamine and kappa opioid receptor alterations in human brain of cocaine-overdose victims. *Ann NY Acad Sci*, **877**, 507–22.

Nashef, L. 1997. Sudden and unexpected death in epilepsy: terminology and definitions. Epilepsy and sudden death. Proceedings form an international workshop on sudden death in epilepsy. *Epilepsia*, **38**, s6–8.

Natleson, B.H., Suarez, R.V., Terrence, C.F. and Turizo, R. 1998. Patients who die suddenly with epilepsy have cardiac disease. *Arch Neurol*, **55**, 857–60.

Police Complaints Authority. 2002. *Policing Acute Behavioural Disturbance*, revised edn. Police Complaints Authority.

Redfern, M., Keeling, J.W. and Powell, E. 2001. *The Royal Liverpool Children's Inquiry Report*. London: HMSO.

Royal College of Pathologists. 2002. *Guidelines on postmortem examination Practice*. Report of the Working Group of The Royal College of Pathologists. London: Royal College of Pathologists.

Royal College of Pathologists. 2004. *Code of Practice and Performance Standards for Forensic Pathologists*. London: Home Office Policy Advisory Board for Forensic Pathology and The Royal College of Pathologists.

Royal College of Pathologists and Royal College of Paediatrics and Child Health. 2004. *Sudden unexpected death in infancy*. London: RCP and RCPCH.

Stephens, B.G., Jentzen, J.M., Karch, S. *et al*. 2004. Criteria for the interpretation of cocaine levels in human biological samples and their relation to the cause of death. *Am J Forensic Med Pathol*, **25**(1), 1–10.

Thom, M. 2003. *Post mortem examination in patients with epilepsy*, Version 1 – 31/8/03. UK: British Neuropathological Society Professional Affairs Committee. Retrieved from http://www.bns.org.uk/pdf/PMEpilepsy31803.pdf

Walsh, Z. and Moore, I.E. 2003. Changes in paediatric postmortem examination practice – are there benefits for parents? *J Pathol*, **201**(suppl 44A).

Whitwell, H.L. and Lowe, J. 2002. Guidance for retention of brain and spinal cord following post-mortem examination and where criminal proceedings are in prospect. *Bull R Coll Pathol*, **118**, 20–33.

Willinger, M., James, L.S. and Catz, C. 1991. Defining the sudden infant death syndrome (SIDS): deliberations of an expert panel convened by the National Institute of Child Health and Human Development. *Pediatr Pathol*, **11**, 677–84.

APPENDICES

Appendix 1 Normal adult brain weight with leptomeninges and dimensions

	Male	**Female**
Adult whole brain (g)	1400 (normal range 1100–1700)	1275 (normal range 1050–1550)
Adult cerebellum (g)	147 (normal range 136–149)	133 (normal range 127–137)
C:C ratio (%)	10–11	10–11
Sagittal diameter (cm)	16–17	15–16
Vertical diameter (cm)	12.5	12.5

Reprinted from Dawson, T. P. and Neal, J.W. 2002. Neuropathology Techniques. London: Arnold.

Appendix 2 Infant and child brain weights

Age (months)	Weight (g)	Age (months/years)	Weight (g)
Term	362	14	927
1	415	16	985
2	469	18	1021
3	520	20	1054
4	558	24	1098
5	619	36	1184
6	641	4 years	1228
7	680	5 years	1251
8	717	6 years	1267
9	758	7 years	1281
10	797	8 years	1292
11	830	9 years	1275
12	860	10 years	1314

Gender-dependent weight difference becomes evident at around 5 years. These are average values (metadata compiled from several sources).
Reprinted from Dawson, T. P. and Neal, J. W. 2002. Neuropathology Techniques. London: Arnold.

Appendix 3 Adult brain weights

Age (years)	Male (g)	Female (g)
10	1440	1260
15	1440	1280
20	1450	1310
30	1440	1290
40	1430	1290
50	1410	1280
60	1370	1240
70	1350	1230
80	1310	1171

Adapted with permission of Wiley-Liss Inc., a subsidiary of John Wiley & Sons, Inc. from Dekaban, A. S. and Sadowsky, D. 1978. Changes in brain weights during the span of human life: relation of brain weights to body heights and body weights. *Ann Neurol*, **4**, 345–56.

Appendix 4 Rapid microwave fixation method for brain fixation

- Remove brain as usual
- Place in plastic brain bucket in 10 per cent formalin, covering the brain
- Place lid on loosely
- MICROWAVE – normal domestic
- 20 min – reheat
- 20 min – low
- Leave to cool (about 20 min)
- Rinse in cold water
- Cut.

Note: infant brain times – 10 min – reheat, 10 min – low.

Care should be taken with this method as due to the toxicity of formalin fumes an appropriate extraction system should be used.

Appendix 5 Minimum sets of blocks and special stains recommended for determining the distribution and amount of microscopic brain damage in head injury

BLOCKS RECOMMENDED FOR AXONAL INJURY
See Figure 2.12
- Corpus callosum and parasagittal posterior frontal white matter
- Splenium of the corpus callosum
- Deep grey matter to include posterior limb of the internal capsule
- Cerebellar hemisphere
- Midbrain (to include decussation of superior cerebellar peduncle)
- Pons (to include superior or middle cerebellar peduncles).

ADDITIONAL BLOCKS RECOMMENDED IN SUSPECTED NON-ACCIDENTAL INFANT HEAD INJURY
- The entire pons and medulla
- All the upper cervical cord segments.

BLOCKS RECOMMENDED FOR HYPOXIC BRAIN DAMAGE
- Samples of all cortical areas (frontal, parietal, temporal, occipital) to include arterial watershed zones
- Deep grey matter

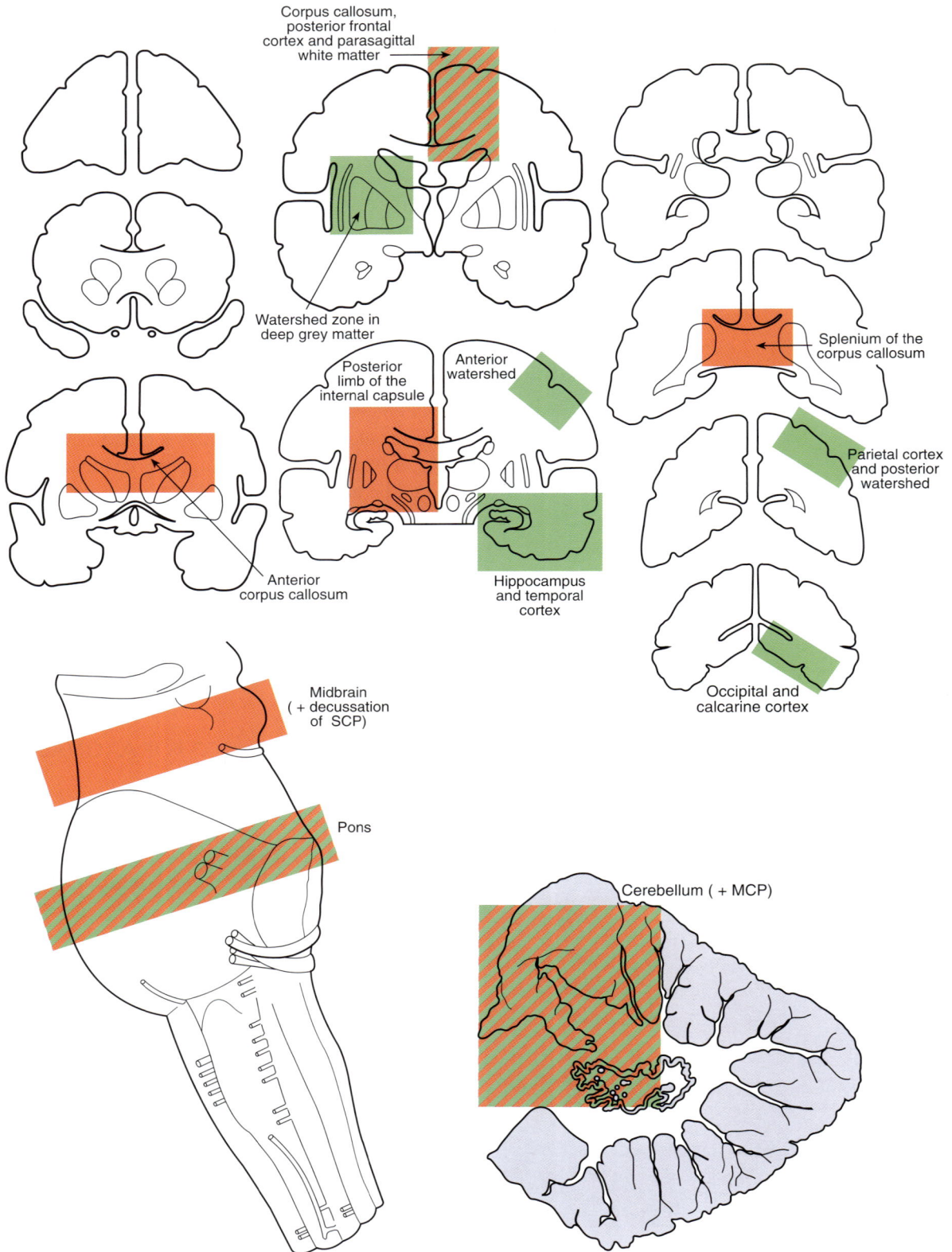

Figure 2.12 Suggested sampling protocol for microscopic brain injury in trauma (SCP, superior cerebellar peduncle; MCP, middle cerebellar peduncle). Blocks for axonal damage are shown in red; those in green are the blocks used to determine the extent of hypoxic brain damage, often an issue in trauma cases. Blocks coloured green and red should always be taken for both traumatic and hypoxic damage. Amended with permission from Dawson, T. P. and Neal, J. W. 2002. *Neuropathology Techniques*, London: Arnold.

- Hippocampus
- Cerebellar hemisphere
- Brain stem.

Note that if small (3 inch x 1 inch) slides are used, two samples of each of the above areas should be taken. Representative blocks should also be taken of any focal pathology seen, and any visible lesions should be sampled.

Modified from Geddes et al. 2000, by kind permission of Blackwell Publishing Ltd.

SPECIAL STAINS
- APP
- CD68
- Perls.

Note: Use of stains to be guided by circumstances.

Appendix 6 Blocks and special stains recommended for sampling the brain in cases of possible dementia

MINIMUM SET OF BLOCKS RECOMMENDED FOR THE ASSESSMENT OF DEMENTIA
- Cerebrum
- Hippocampal including hippocampus, entorrhinal cortex and inferior temporal gyrus
- Superior and middle temporal gyri
- Anterior tip of the temporal lobe
- Middle frontal gyrus
- Basal ganglia at the level of the mamillary bodies
- Thalamus
- Section of the cerebellum including cerebellar cortex and dentate nucleus
- Brain stem: midbrain including substantia nigra
- Pons including locus ceruleus (pigmented nucleus)
- Medulla.

In addition, sample any areas that macroscopically appear atrophic and any focal lesions (such as lacunar infarcts).

SPECIAL STAINS RECOMMENDED FOR THE ASSESSMENT OF DEMENTIA
Immunohistochemistry for the following antibodies should be undertaken.

Tau and b-amyloid
- Midbrain
- Pons
- Hippocampus

- Cerebellum
- Basal ganglia.

a-Synuclein
- Midbrain
- Pons
- Hippocampus
- Middle frontal gyrus
- Superior and middle temporal gyrus.

Ubiquitin
- Hippocampus
- Middle frontal gyrus
- superior and middle temporal gyrus.

Appendix 7 Minimum set of blocks recommended for SUDI cases

See Figure 2.13
- Mid frontal gyrus
- Mid corpus callosum
- Basal ganglia
- Parietal cortex
- Caudate
- Thalamus
- Hippocampi
- Calcinne sulucs
- Superior cerebellum
- Mid pons, whole medulla.

In addition, sample any visible lesions and if trauma is suspected, prepare blocks as described in Appendix 5.

Reproduced with permission from Dawson, T.P. and Neal, J.W. 2002. Neuropathology Techniques. London: Arnold.

Appendix 8 Minimum set of blocks recommended for SUDEP cases

See Figure 2.14
- Hippocampus (left and right)
- Cerebral cortex (including anterior and posterior watershed areas)
- Amygdala
- Basal ganglia and thalamus
- Insular cortex
- Superior cerebellum
- Brain stem: mid brain, pons and medulla

In addition, sample any macroscopically-identified lesions or abnormality.

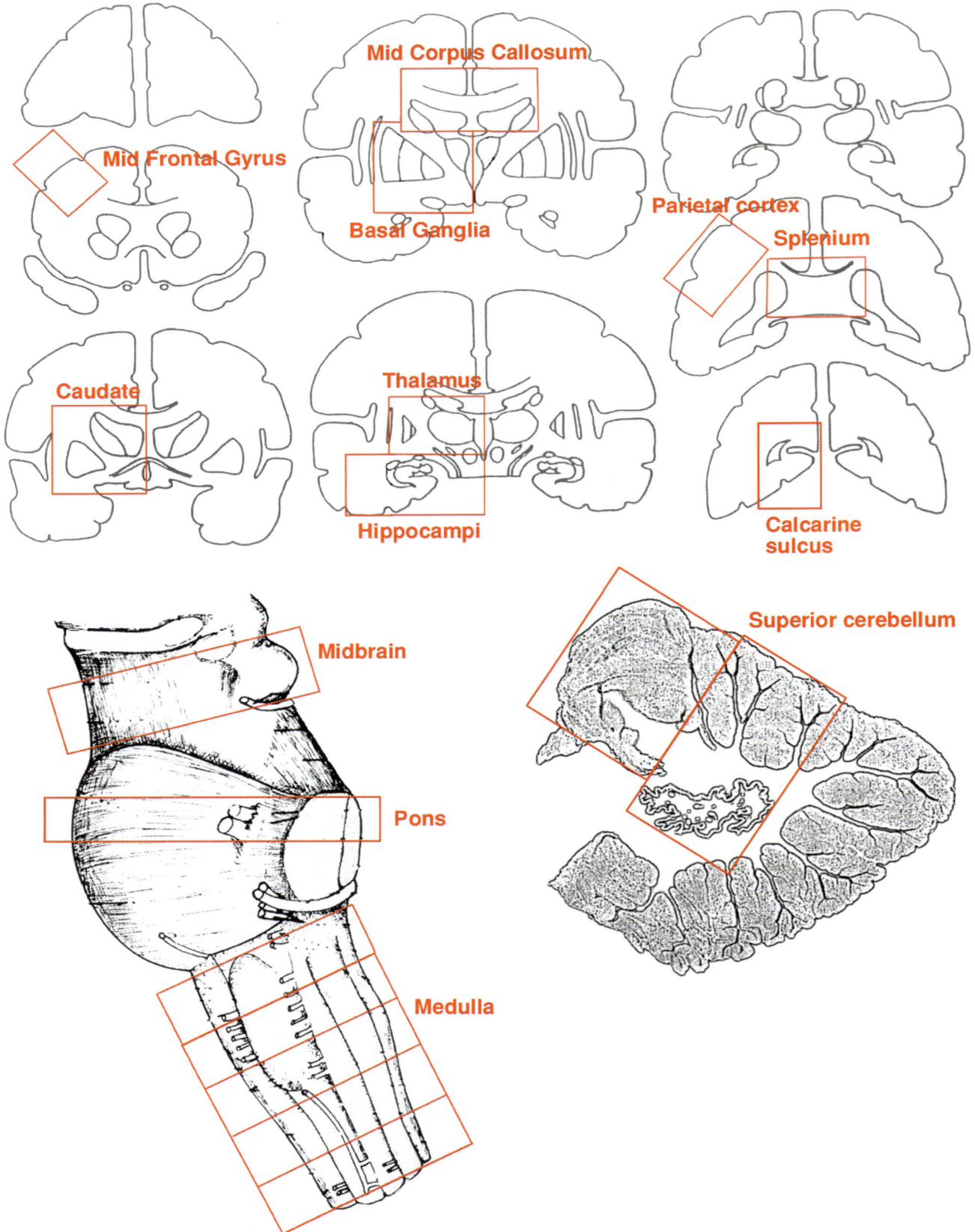

Figure 2.13 Suggested sampling protocols for sudden unexpected death in infancy. Reproduced with permission from Dawson, T. P. and Neal, J. W. 2002. *Neuropathology Techniques*. London: Arnold.

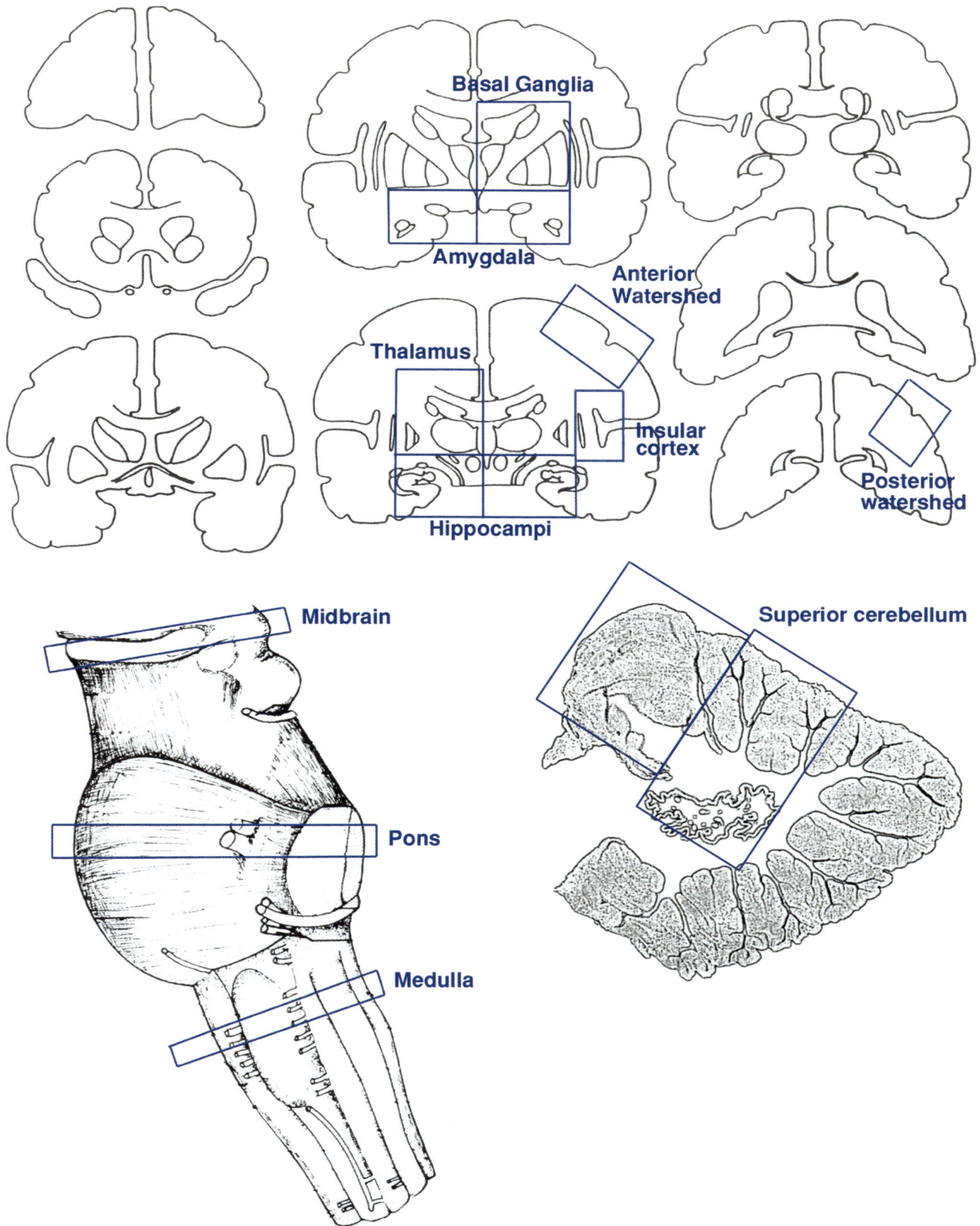

Figure 2.14 Suggested sampling protocols for sudden unexpected death in infancy. Reproduced with permission from Dawson, T. P. and Neal, J. W. 2002. *Neuropathology Techniques*. London: Arnold.

Appendix 9 Screening blocks

BRAIN-STEM SCREENING BLOCKS
- Pyramidal decussation
- Caudal end of olives
- Mid-olivary level
- Medullopontine junction
- Root of fifth cranial nerve including middle cerebellar peduncles
- Pontomesencephalic junction including decussation of superior cerebellar peduncles
- Midbrain through inferior colliculi
- Midbrain through superior colliculi to include the substantia nigra (usually the level of separation of the brain stem from the cerebrum)
- Any suspect lesions

CEREBELLUM SCREENING BLOCKS
- Both dentate nuclei with dorsal cerebellar cortex
- Hemispherical white matter with middle cerebellar peduncle
- Vermis (superior and inferior)
- Any suspect lesions

CEREBRAL HEMISPHERE SCREENING BLOCKS
- Gyrus rectus
- Middle frontal gyrus
- Tip of temporal lobe
- Anterior cingulate gyrus with corpus callosum and parasagittal white matter
- Head of caudate nucleus, putamen and nucleus accumbens
- Superior and middle temporal gyri
- Globus pallidus, putamen, claustrum, insula
- Anterior commissure, hypothalamus and substantia innominata
- Amygdala with uncus
- Mamillary bodies and hypothalamus
- Posterior limb of internal capsule and caudate nucleus
- Thalamus and subthalamic nuclei
- Superior parietal (motor and sensory) cortex
- Posterior cingulate gyrus, corpus callosum, parasagittal white matter
- Hippocampal formation at the level of the lateral geniculate body, with parahippocampal gyrus and collateral sulcus
- Angular gyrus (lateral parietal cortex)
- Occipital lobe with calcarine cortex
- Any suspect lesions

Modified from Dawson, T. P. and Neal, J.W. 2002. Neuropathology Techniques. London: Arnold.

Appendix 10 Procedure for diatom analysis

REQUIREMENTS
100 g fresh lung*, liver*, kidney, spleen or brain tissue. Preferred tissue is indicated with an asterisk.

N.B. It is mandatory to take a water sample from the scene because without this qualitative and quantitative analyses cannot be made.

METHODOLOGY
1. Observe the water sample using phase-contrast microscopy
2. Record the types and number of diatoms present diagrammatically and using digital photography
3. Extract possible diatoms from fresh tissue only if diatoms are present in the water (from the scene)
4. In a class 1 safety cabinet place the tissue sample in a clean, labelled flask using forceps rinsed in distilled water. If the sample is too large use a scalpel, rinsed in distilled water, to cut it into smaller sections
5. Transfer the flask/s to a fume cupboard and add $100 \, cm^3$ concentrated nitric acid
6. Cover the flask/s with clean loose lid/s and leave in a container in the fume cupboard for 48 hours.
7. When the tissue has dispersed into the acid, heat it slowly using a hotplate until the volume has reduced to $30 \, cm^3$. Allow to cool
8. Add $20 \, cm^3$ concentrated sulphuric acid and heat slowly until the solution turns black
9. Partly fill a separating funnel with concentrated nitric acid and set the tap to drip it slowly into the charred solution
10. While dripping in the nitric acid continue to heat the solution slowly until the solution turns a pale straw colour
11. If a large amount of residue is visible, hydrolyse it at 60°C in hydrochloric acid
12. Allow to cool, and then centrifuge aliquots of the solution at 3000 rpm for ten minutes, removing the supernatant via suction
13. Repeat until the flask is empty
14. Rinse the flask out with distilled water to remove any remaining residue, and centrifuge as before

If the pellet is from a sample of lung tissue
1. Remove the final supernatant and add $0.5 \, cm^3$ distilled water
2. Resuspend the pellet using a plastic disposable pipette and transfer it to a stoppered microfuge tube ($1 \, cm^3$ volume)

3. Transfer one drop to a clean slide, coverslip, and observe using phase contrast microscopy.

If pellet is from a sample of liver or other tissues
1. Reduce the residue to one large drop, and observe all of the material on the slide
2. Record the types and number of diatoms present both diagrammatically and using digital photography

3. Match the diatoms found in the water sample against any diatoms found in the tissue extracts

Courtesy of Ian Newsome, Department of Forensic Pathology, University of Sheffield.

3

Scalp, facial and gunshot injuries

Christopher M Milroy

Surface injuries are central to forensic medicine. The pattern of injury provides information on the mechanism of causation of both external and internal injuries. Although this chapter deals predominantly with injuries to the scalp and face, it should be remembered that clues to the causation of injury may be present on other parts of the body. A systematic approach should be made to recording injuries, including type, dimensions, shape and colour, with the anatomical position, which can be linked to appropriate anatomical landmarks. Medical descriptions of wounds and injuries can be at variance with how legal systems define wounds. Gunshot wounds are lacerations with specific patterns of injury and are dealt with under a separate section in this chapter. As the effects of penetrating injuries are intrinsically linked to the surface injuries, this chapter also includes the effects of penetrating trauma.

LEGAL DEFINITIONS

Black's Law Dictionary (Garner 1999) describes wounding as 'an injury', especially one involving rupture of the skin. Bodily injury is described as 'physical damage to person's body', but the two are not necessarily the same. In English law it is now established case law that a wound must involve the breaching of the full thickness of the skin, i.e. the epidermis and dermis (*Moriarty v Brookes* 1834). In the case of an airgun pellet that struck an eye causing internal haemorrhage but no breach of the surface, it was held not to be a wound (*C (a minor) v Eisenhower* 1984). Similarly an abrasion is not a wound in law (*M'Loughlin* 1838). A fracture of the clavicle without breach of the overlying skin was also not a wound (*Wood* 1830). Although the legal definition of a wound is restricted, alternative legal charges usually exist to take

account of this apparent anomaly. However, in describing a wound some knowledge of the legal definition (if any) in the jurisdiction in which the report is to be used is helpful. Terms such as 'cut', 'superficial laceration' and 'deep abrasion' are confusing and may result in legal challenges.

EXTERNAL EXAMINATION

For a clear demonstration of injuries on the head, the hair may need to be shaved during postmortem examination. It is surprising how frequently thick hair covers significant injuries and prevents identification.

Photographs of the injuries should be made in colour, preferably with an appropriate colour reference grid at the start of the film, although some jurisdictions still prefer black and white photographs, especially for presentation to juries. Injuries should be photographed with and without a measuring scale and at right angles, not obliquely. Where necessary, e.g. with bite marks and patterned bruising, a right-angled ruler should be used (Figure 3.1). Techniques such as use of ultraviolet light should be considered; they may enhance the features of an injury (Hempling 1981, Krauss and Warlen 1985). With some patterned injuries such as those caused by footwear, forensic scientists currently prefer black and white photographs, so that they can compare the pattern with footwear held on databases. Many practitioners make notes of injuries by drawing on prepared body plans.

Injuries may alter in appearance in the postmortem period. Abrasions may dry out and appear more prominent, although they will not significantly change in size, unlike bruises, which may increase in size and prominence. A second examination of the body may therefore reveal bruises that were not initially evident. However, caution must be

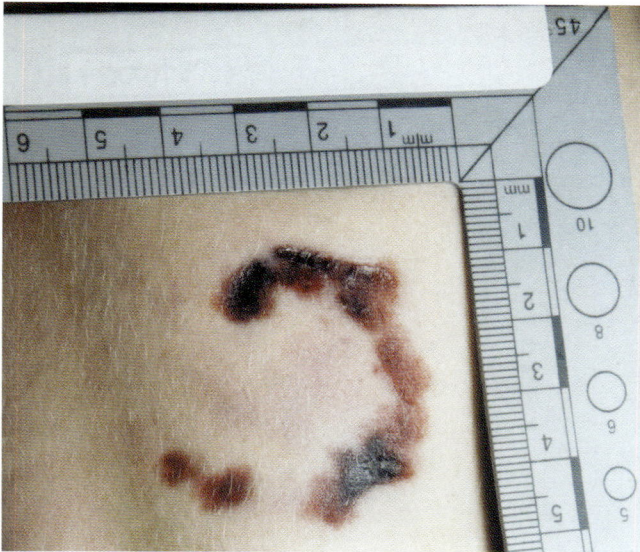

Figure 3.1   Bite mark: photographed with right-angled scale.

exercised that the injury is not an early change of putre-faction or an artefact of the initial postmortem examination. If necessary, microscopic examination of the skin can be undertaken. Microscopy can also be helpful in dating an injury, as detailed later.

CLASSIFICATION OF INJURIES

Abrasions

A pure abrasion is an injury involving only the epidermis. Abrasions heal without scarring. An abrasion may result from a tangential movement of skin in comparison with an abrading material, or by direct imprint. In the head and face abrasions commonly occur when a person falls. When a person falls forwards the prominent areas of the face will be injured if the person is unable to break the fall (Figure 3.2). Falls on to the back of the head may result in occipital or parieto-occipital region abrasions, along with bruising or lacerations (Figure 3.3). Scuffed kicks may abrade the skin. Where there is significant tangential movement of skin against an abrading surface, the direction of move-ment can be determined by the position of the skin tags in the abrasion. Imprint abrasions are seen when a patterned object impacts on to skin. A typical example is a ligature mark. Abrasions, as well as bruises and lacerations, may be seen when a spectacle wearer is struck in the face. If such injuries are seen on the bridge of the nose, or around the eyes, the possibility of the victim wearing spectacles should be considered, because this information is not always immediately available (Figure 3.4). Abrasions may be

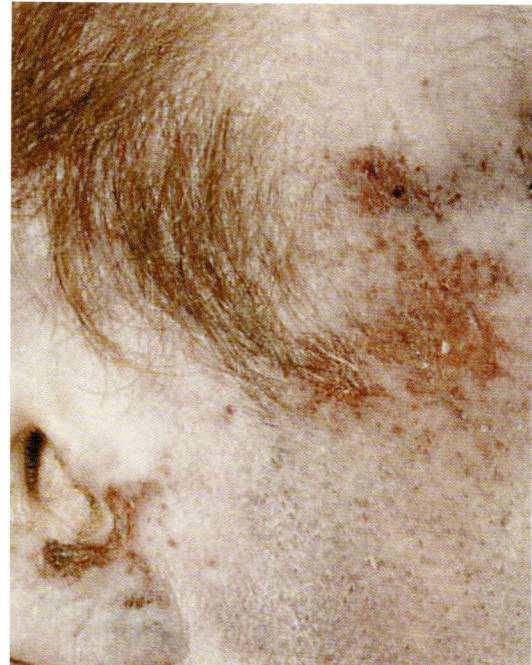

Figure 3.2   Abrasions over bony prominence of cheek.

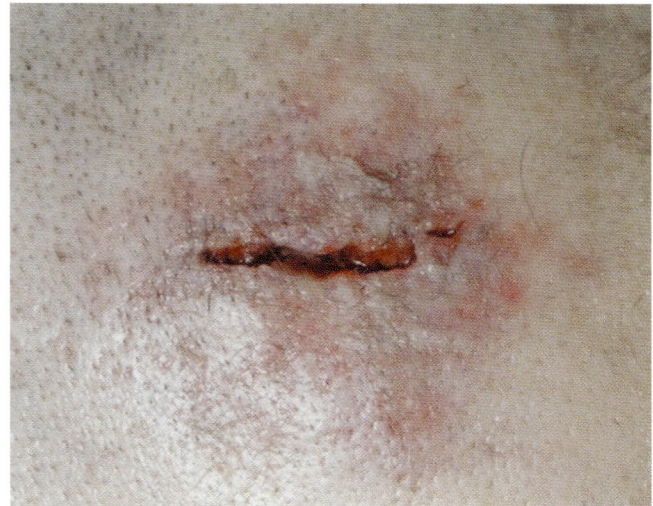

Figure 3.3   Linear laceration with surrounding abrasion and bruising.

inflicted after death, when they appear yellow, particularly during postmortem movement of the body. Abrading type injuries may be seen around the mouth when gastric acid from vomitus has spilt around it (Figure 3.5).

Bruises

Bruises are caused by trauma to the skin that ruptures blood vessels and blood escapes into the tissues. Particularly

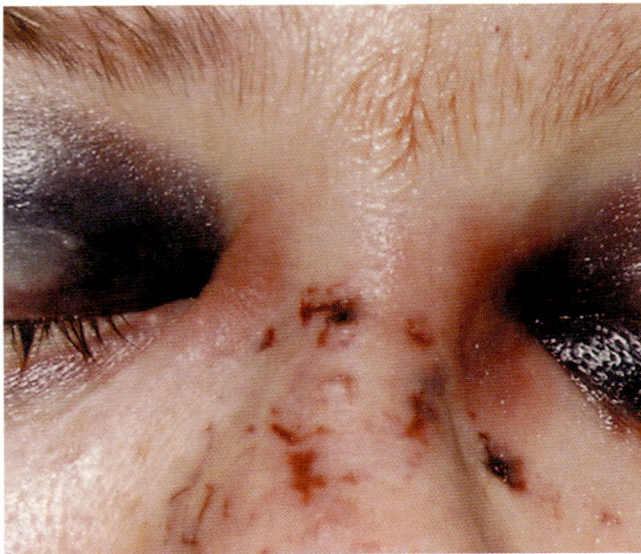

Figure 3.4 Abrasions on bridge and side of nose from spectacle damage. The black eyes are secondary to underlying orbital fractures.

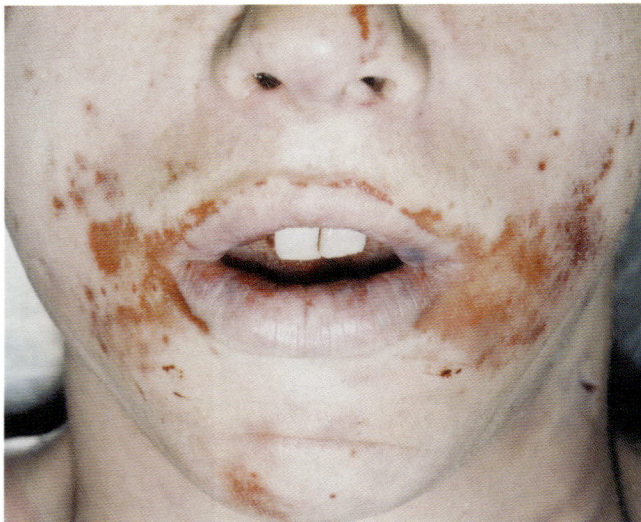

Figure 3.5 Skin damage caused by vomitus *post mortem*.

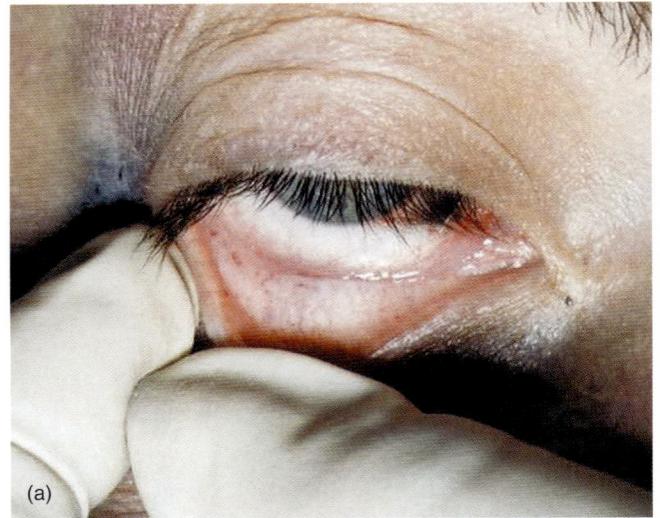

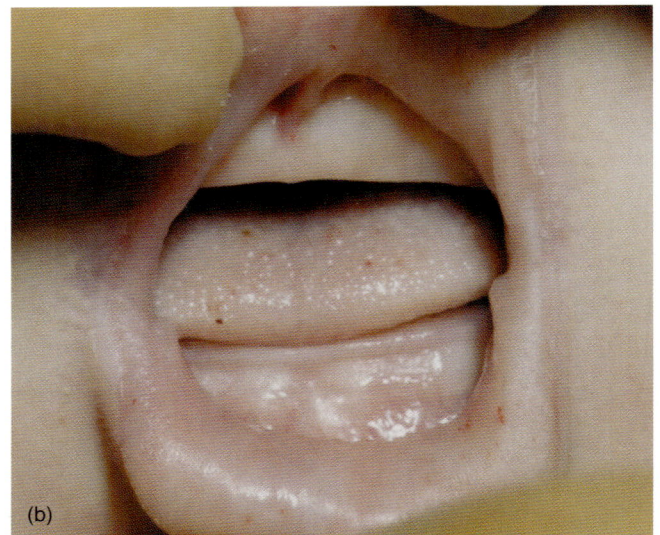

Figure 3.6 Petechiae (a) in the subconjunctiva; (b) in the mouth.

in deeper tissues, bruises are often called contusions. The older term 'ecchymosis' is also sometimes used. In forensic practice findings must be described in terminology that can be understood by non-medical lay people and if simple terminology can be used it is to be preferred. It is more likely that a jury will understand 'black eye' than periorbital haematoma.

The smallest pattern of bleeding is a petechial haemorrhage or a petechia. Petechiae are pinhead-sized areas of bleeding found in the skin and internal organs. Although often described as a sign of asphyxia, they are by no means specific and are seen in a number of different causes of death, as well as non-fatal cases. In the head they can be seen in the eyes (Figure 3.6a) and eyelids, on the face and behind the ears, as well as sometimes on mucosal surfaces (Figure 3.6b). Petechial-type haemorrhages on the undersurface of the scalp found following dissection are a well-recognized artefact, and should not be interpreted as arising *ante mortem* (see Chapter 2, p. 21). In the absence of other external petechiae such as may be seen in external pressure to the neck, they are meaningless in determining the mode of death.

A bruise may vary in shape and size depending on how, where and when it was inflicted. Bruises are well recognized to undergo colour changes after infliction. These colour changes have been used in an attempt to date the time of infliction. Forensic pathologists try to differentiate fresh bruises from older injuries, although studies have shown that considerable caution must be exercised in attempting to date a specific bruise. Indeed an analysis of standard forensic pathology textbooks showed a surprising variation in stated age of a bruise as assessed by colour (Wilson 1997). Bruises that are fresh are often blue or red–purple. They go through a series of changes through brown and green to yellow. This sequence, however, will vary and depend on a number of factors such as size and site, age of victim and medical condition present, as well as idiosyncratic factors in the victim. Although a red or blue swollen bruise is likely to be fresh, often defined as less than 24–36 h old, red and blue coloration may be seen in older bruises. A brown bruise may be fresh or old. In a study of bruising, Langlois and Gresham (1991) found that the only reliable finding was that a yellow bruise was at least 18 h old. Stephenson and Bialis (1996) examining children found that a yellow bruise was at least 24 h old. Caution must therefore be exercised in determining the age of a bruise, although as a rule bruises of a similar size and site but of significantly differing colour are likely to have been inflicted at different times.

A large collection of blood under the skin is known as a haematoma. Haematomas are an important finding in head-injured patients, particularly young children. In these cases bruising on the external surface of the scalp may not be discernible, whereas reflection of the scalp at postmortem examination may reveal significant areas of bruising. The risk of an underlying skull fracture and intracranial injury has been shown to increase with the size of scalp haematoma and haematoma location in infants, where temporal and occipitoparietal haematomas were strongly associated with skull fracture, although frontal haematomas were not. Skull fracture without associated haematoma was common in infants aged under 3 months (Greenes and Schutzman 2001).

Bruises may sometimes be patterned, replicating the weapon or implement that inflicted them. A relatively common example on the head is where the victim has been stamped on (Figure 3.7). Careful documentation of these injuries is important because the type of shoe may be identified by comparison with databases. Another characteristic pattern is that caused by a cylindrical shaped object such as a baton, baseball bat or similar shaped object. These weapons often cause a tramline bruise, where there is a central zone of pallor with a band of bruising either side (Figure 3.8). Another type of patterned bruise that may be seen on the face is a bite mark. In such marks there may be abrasion and laceration (see Figure 3.1). When such a mark is suspected expert forensic odontological advice should be sought. 'Love' bites, commonly seen on the head

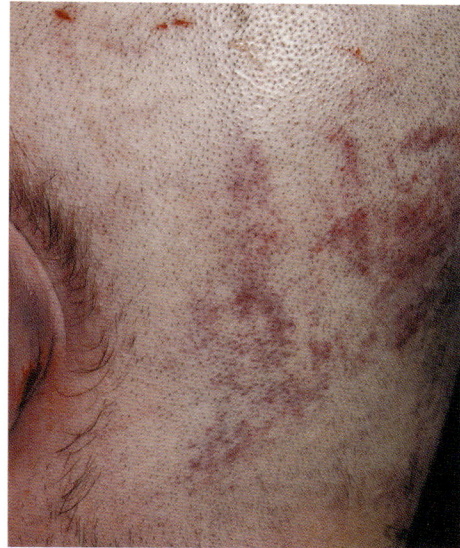

Figure 3.7 Patterned bruising caused by stamping with a shod foot.

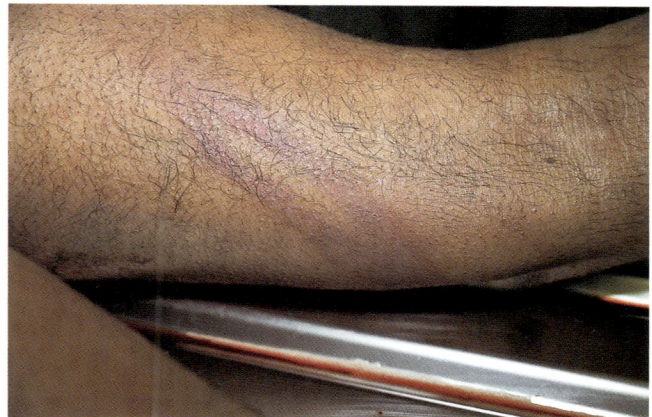

Figure 3.8 Tramline bruising caused by blows with a baseball bat.

and neck, are caused by suction, rather than biting. It is important to look inside the mouth, because this is a common site for bruising after a blow to the area, often associated with damage to teeth (Figure 3.9).

Bruises may track through the tissues, so a bruise on the forehead may settle under gravity around the eyes. Similarly, bleeding from a fractured nose may result in bilateral black eyes. It is generally accepted that bruises are ante-mortem injuries. However, postmortem bruising can be produced but requires much more force than when a circulation is present. Polson *et al.* (1985) produced postmortem bruises of the occiput by blows with a mallet. Caution should therefore be exercised in ascribing a bruise to the back of the head as being caused by ante-mortem trauma if a person has been lying face up for days with a

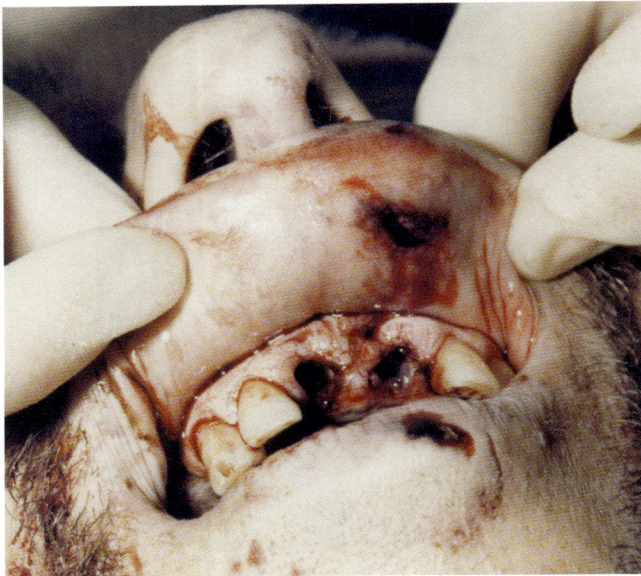

Figure 3.9   Bruising and laceration to inside of lips with associated damage to the teeth caused by a blow to the mouth.

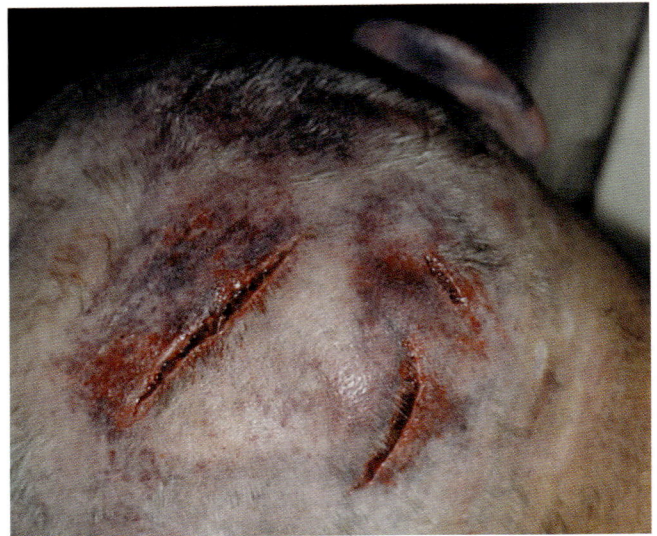

Figure 3.10   Multiple linear lacerations with associated bruising caused by blows with a blunt implement.

head injury, or where the rough handling of a body *post mortem* cannot be excluded.

Lacerations

Many doctors as well as non-medically qualified people call all wounds lacerations, not differentiating a laceration from an incised wound (see below). A laceration is a splitting or tearing of the skin caused by blunt trauma. The head is a particularly common area for lacerations to occur because the skin is closely associated with underlying bone in this region and therefore, when blunt force is impacted on the skin, the skin is rapidly compressed against the bone. In comparison, considerably more force is required to inflict a laceration in the chest or abdomen. Lacerations on the face may result from a punch, but a punch would not be expected to cause a laceration on the scalp, unless a ring, knuckleduster or similar object was worn. On the face, lacerations occur over bony prominences, including the orbital ring, and over the maxillae. With severe damage to the underlying skeleton, the bones may break through the skin, lacerating it in the process. An unusual example of this mechanism was described by Ainsworth and Hunt (1993) where a blow to the nose, which pushed the nose upwards, resulted in the nasal bone penetrating the skin, before returning to its original position. This resulted in a curved laceration of the tip of the nose. Punches to the lower jaw do not typically cause a laceration of the skin, even with a fractured jaw, although the assailant's hand may be lacerated from the blow, especially if he strikes the teeth of the victim.

Lacerations may be linear, or slightly curved on the scalp where the blow is from a linear object (Figure 3.10). Falls on to the head may also result in a linear laceration (see Figure 3.3). Unlike an incised wound a laceration has irregular edges, with bruising and abrasion of the margins. Tissue may bridge the wound and hair across the wound may be uncut, unlike an incisional wound. Dirt may have been forced into the wound. Lacerations, bruises and abrasions to relatively inaccessible areas where a blunt agent is normally used, such as under the chin or beneath the eyebrow, are often the result of a blow with a shod foot (Figures 3.11a and 3.11b), also causing a black eye (Teare 1961). Blows with large implements, such as a lump hammer or sledgehammer, may produce complex lacerations (Figure 3.12).

The shape of a laceration may reflect the shape of an implement used in an attack. A characteristic example is a blow by a hammer that produces a crescent-shaped laceration (Figure 3.13). Lacerations on the back of the head are a common finding after a fall and there may be associated abrasion and bruising (see Figure 3.3). More than one laceration on the scalp raises concerns that the lacerations are the result of an attack, but does not prove an attack. An intoxicated person, for example, may fall, get up and fall again, sustaining more than one laceration (see Chapter 11, p. 129).

Blows with a relatively sharp object, such as an axe, may produce a linear wound that is difficult to differentiate into a laceration or incised wound. Help in identifying the

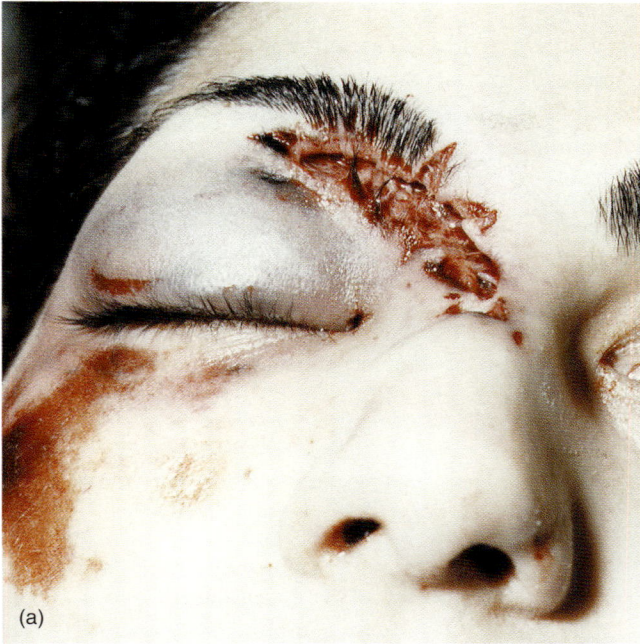

(a)

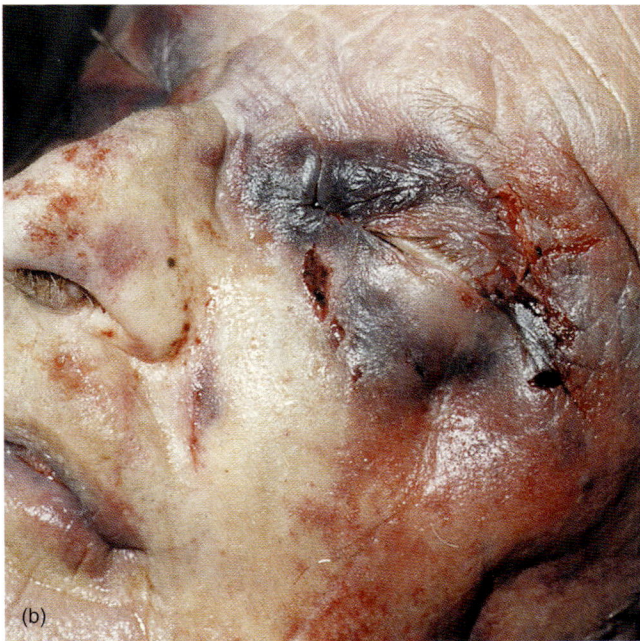

(b)

Figure 3.11   (a) Laceration under the eyebrow caused by kicking; (b) with more superficial lacerations and bruising.

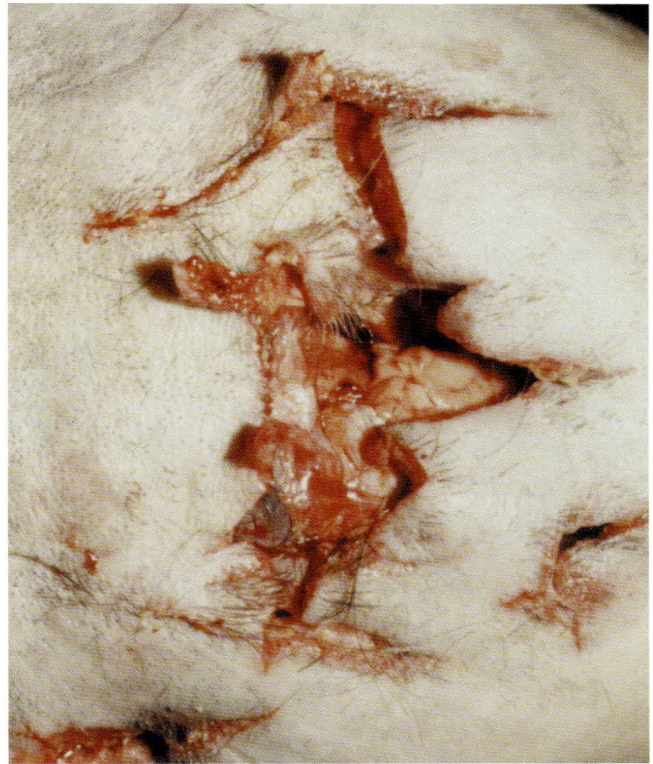

Figure 3.12   Complex laceration caused by hammer blows.

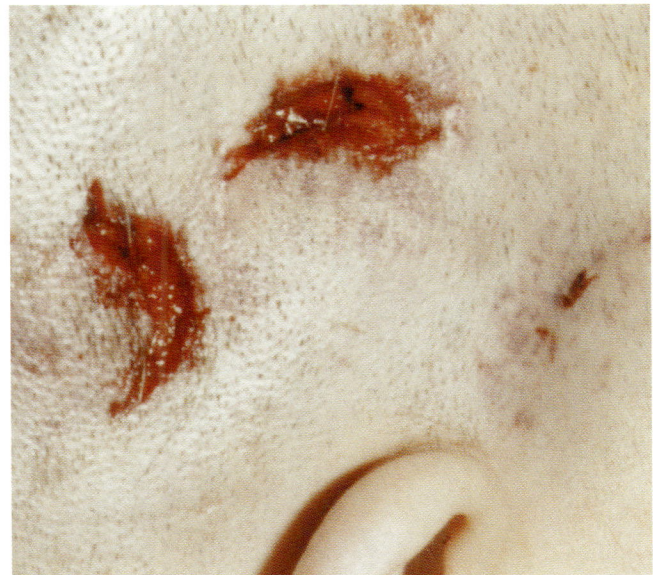

Figure 3.13   Crescent-shaped lacerations caused by blows with a ball-pein hammer.

cause of the wound may be provided by examining the underlying skull, where an axe is likely to cause damage to the bone, as is a machete.

Punches to the mouth can cause bruising and laceration to the inside of the lips, when no external injury is visible (see Figure 3.9). There may be associated damage to the teeth. A common pattern of external injury in accelerated falls, producing a significant head injury, is bruising with

or without laceration of the inside of the lip from a punch, associated with bruising, laceration and abrasion to the occiput. A torn frenulum is an important injury in child abuse (see Chapter 12, p. 138).

Incised wounds and stab wounds

Injuries caused by sharp-edged objects produce a clean-edged wound. The paradigm of an incised wound is the surgical wound produced by a scalpel. However, any sharp-edged object may produce an incised wound, and facial injuries inflicted by broken glass are an important cause of morbidity. Although incised wounds to the head are frequently seen, they are rarely fatal, although, as with lacerations, occasional deaths from haemorrhage occur.

Self-inflicted incised wounds to the head may be suicidal or inflicted for personal gain. Where they are inflicted for personal gain they are typically superficial and do not involve sensitive areas of the face such as the eyes or lips.

Most fatal stab wounds are inflicted to the trunk, particularly the chest, but penetrating wounds to the head are occasionally encountered (Figure 3.14). Stab wounds should be recorded by site, external dimensions, direction of travel and depth. Although some knives leave injuries that indicate whether they are single or double edged, this is often not possible to determine. Movement of the knife in relation to the victim may cause variation in the size of the surface wound. Some weapons have serrated edges, but this is often not possible to determine from the wound. Stab wounds aimed at the head may penetrate the skull, but if they do not the wound may be significantly affected by movement against the skull. Some stabbing weapons, notably scissors, may produce more irregular wounds. More blunt stabbing weapons, such as screwdrivers, may also produce penetrating wounds. In these cases, the screwdriver can produce a small slit-like laceration, belying the degree of damage caused by the passage of the weapon into the brain (Figures 3.15 and 3.16).

Burns

Burns may be the result of the application of dry or moist heat, or of the contact of a chemical with the body. Burns can be divided by depth of injury into three groups: superficial

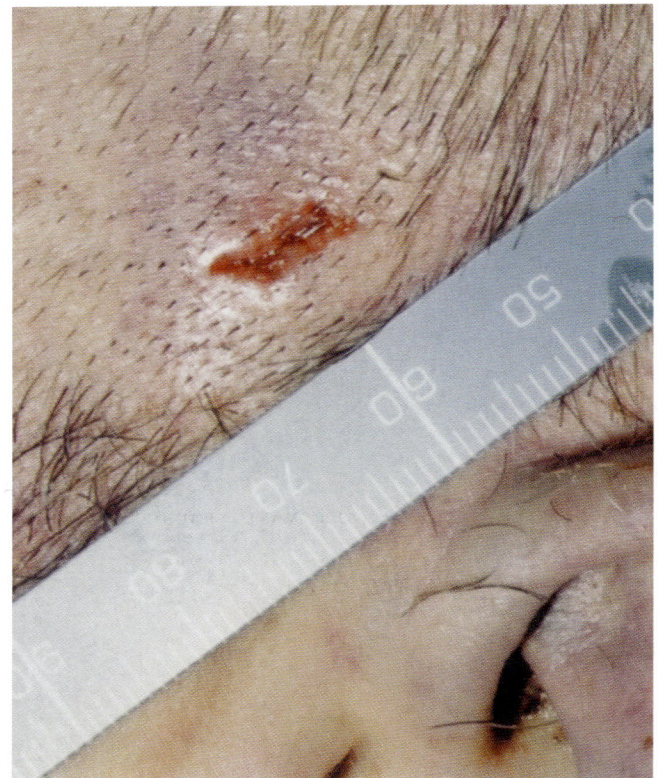

Figure 3.15　Stab wound caused by a screwdriver.

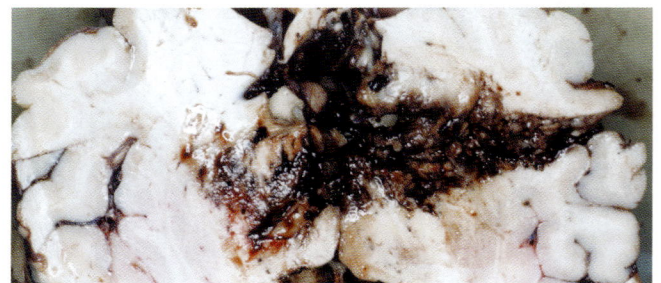

Figure 3.16　Cerebral damage caused by penetration with a screwdriver.

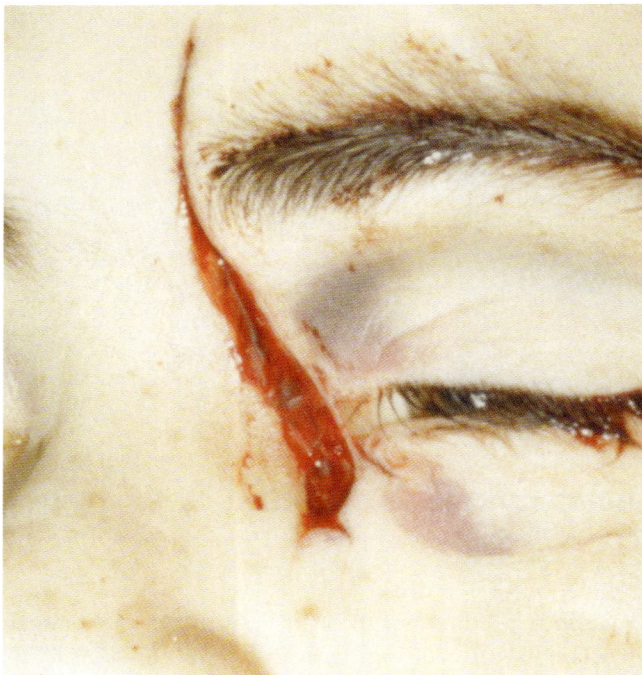

Figure 3.14　Stab wound penetrating the eye.

(first degree), deep dermal (second degree) and full thickness (third degree). Moist heat characteristically causes scalds to the skin (first degree), whereas dry heat may cause any degree of damage. With extensive burning, as is commonly encountered in house fires, or after deliberate attempts to destroy the body, extensive destruction involving the body may be seen with exposure of muscle and bone. However, complete destruction of the body is rare and requires prolonged high temperatures. With extensive burning of the skin of the head, artefactual splits may occur and these must not be confused with ante-mortem injury. Bodies subject to significant burning of the head frequently have artefactual heat fractures (Figure 3.17) and extradural heat haematoma (see Chapter 5, p. 64).

Electrical injury causes damage to the skin, which can vary from a small, localized injury that may be difficult to identify to extensive destruction with high-tension current, such as when victims have climbed electricity pylons, the burns often involving the head, as well as other areas of the body.

Deliberately inflicted burns, where a weapon such as a cigarette is pressed against the skin, may be encountered, particularly in child abuse, where such injuries are pathognomonic of abuse.

Firearms injuries

Firearms are the most frequent cause of penetrating wounds of the head. There are two principal categories of firearms: rifled weapons and smooth bored shotguns. Rifled weapons may be short or long barrelled and have a series of grooves made into the barrel. These grooves are spiralled, with varying numbers, running in a clockwise or anticlockwise direction. The metal ridges left between the grooves are called lands. This rifling imparts gyroscopic stability on the bullet in flight. Examination of the wound characteristics in firearms injuries provides information on type of weapon used and distance discharged from. Accurate determination of range requires test firing of the gun with the appropriate ammunition. Before starting a postmortem examination in suspected firearm's deaths radiographs of the body should be taken (Figure 3.18) (DiMaio 1999, Besant-Matthews 2000).

RIFLED WEAPON INJURIES

Rifled weapons usually discharge a single bullet, unlike shotguns. Hand-held weapons are of two main types: revolvers and self-loading pistols. Revolvers usually have a muzzle velocity of around 200 m/s (700 feet/s). The self-loading pistol loads bullets in a spring-loaded magazine. Each time a bullet is fired, a replacement bullet is moved into the firing chamber until the bullets are all discharged. The spent cartridge case is expelled, unlike the revolver, where the cartridge cases are retained until emptied by hand. In a semi-automatic setting, a bullet is advanced into the firing chamber singly each time the trigger is fired, whereas in fully automatic mode bullets are continually discharged while the trigger is depressed until all are used up or pressure is released. Self-loading pistols typically have a higher muzzle velocity than revolvers, being around 300–360 m/s (1000–1200 feet/s).

All .22 rifles usually have lower muzzle velocities, similar to those of handguns. Other long barrelled rifled weapons

Figure 3.17 Fire death with artefactual heat fractures of the skull.

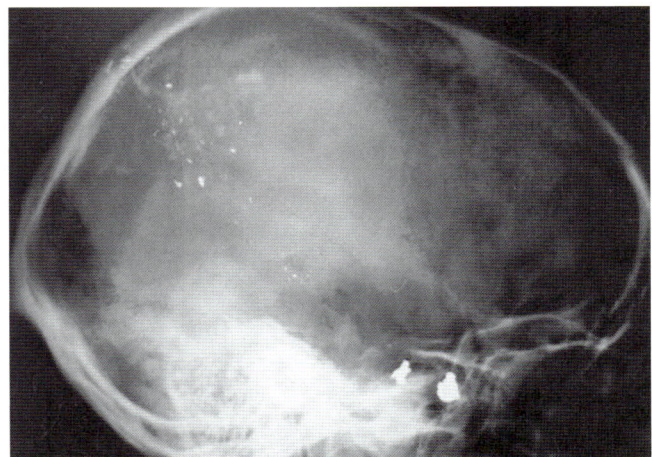

Figure 3.18 Skull radiograph of .22 gunshot wound to head (contact).

usually have a high muzzle velocity, over 300 m/s (1000 feet/s). High-powered rifled weapons, such as military rifles, may have muzzle velocities of 600–1200 m/s (2000–4000 feet/s), although all these quoted velocities will be modified depending on the type of cartridge used.

Contact wounds

In the head the contact is often firm, the skin being in close approximation to the bone. Muzzle gases enter the wound, along with other ejecta, including primer and propellant residues as well as the bullet. Externally, little soot and powder burning may be evident, and may require microscopic examination to confirm that it is a contact wound. As muzzle gases contain a high percentage of carbon monoxide, the tissues may appear pink, and chemical analysis reveals the presence of carboxyhaemoglobin and carboxymyoglobin.

When the hard contact is over bone, most characteristically seen in the head, the muzzle gases pass into the skin and the pressure causes splitting of the skin as they exit, producing a stellate or cruciate laceration (Figure 3.19). Close examination of the edges of the wound will reveal powder burns and soot. The amount of splitting of the skin will depend on the amount of gas formed after firing of the gun. Bullets with lower charges will produce less splitting, if any. Splitting of the skin may be seen in intermediate and distance shots, when the bullet strikes a bony prominence. When the muzzle is in hard contact with skin, the muzzle may become imprinted on to skin as a muzzle abrasion (Figure 3.20). When the contact is not hard, a muzzle abrasion is less likely, but a ring of soot and powder will occur around the entrance, because a small gap is present between muzzle and skin during discharge. In angulated contact wounds the gun is pushed into the skin so that the muzzle is only partially in contact with the skin. Muzzle gases and powder will escape between the gap and give an eccentric pattern of staining on the skin. If the gap is sufficiently large, unburnt powder can escape and powder tattooing may be seen on the opposite side to where the barrel was in contact with the skin.

Near contact and intermediate wounds

Near contact wounds occur when the muzzle of the gun is close to, but not in contact with, the skin. The characteristic appearance is of an entrance wound surrounded by soot with blackened, heated skin. Unburnt powder may be present. If the barrel is angled to the skin, the distribution of the soot will be eccentric to the entrance wound. No muzzle abrasion will be present, because the muzzle has not been in contact with the skin.

When the muzzle of the gun is held away from the skin, unburnt powder produces powder tattooing of the skin

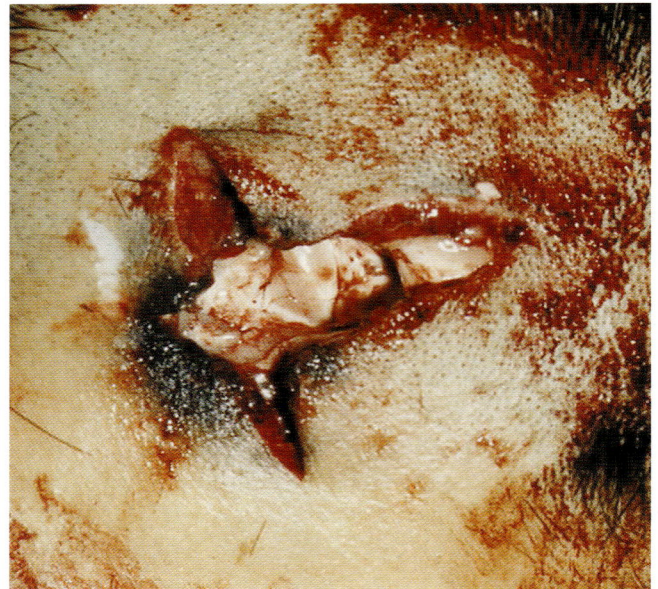

Figure 3.19 Contact gunshot wound of head with stellate entrance wound.

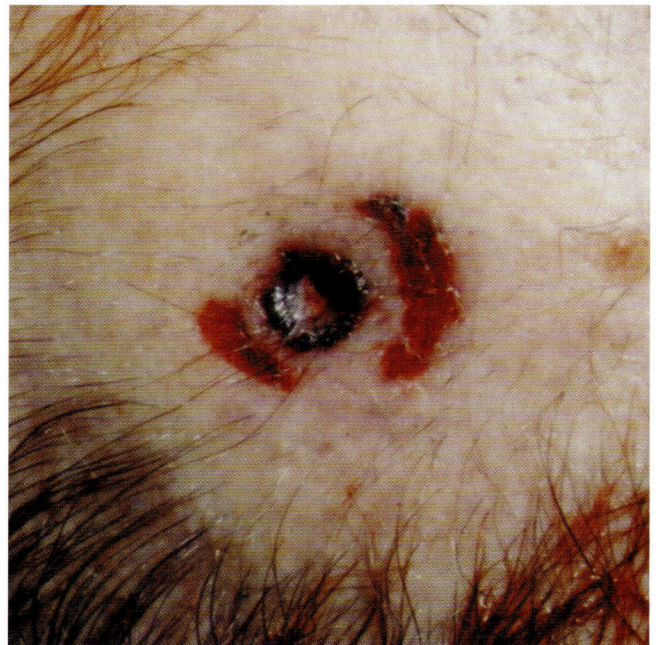

Figure 3.20 Contact wound with muzzle abrasion.

(Figure 3.21). As well as powder grains, soot will be discharged from the barrel and deposited on the skin. The spread of the pattern of tattooing and soot staining will depend on a number of factors including type of weapon and ammunition, length of barrel and distance fired from the body. With longer barrels, soot deposition occurs for a greater distance from the target than a shorter barrel. In general soot and powder deposition can be seen up to a

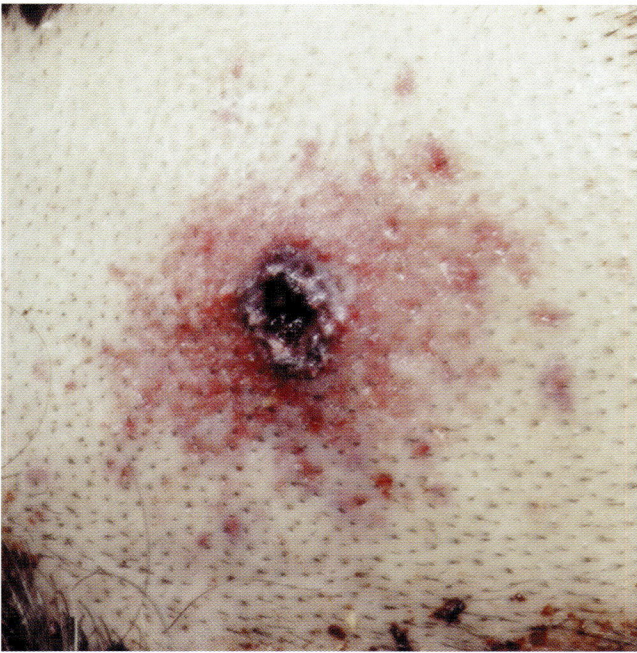

Figure 3.21 Entrance gunshot wound with powder tattooing.

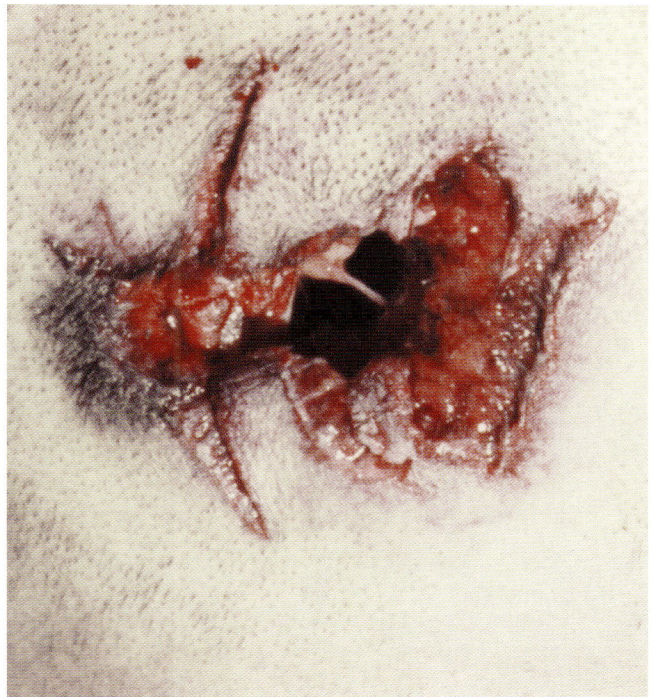

Figure 3.22 Atypical entrance gunshot wound of occiput.

range of 20–30 cm with handguns, although soot may occur at a greater range depending on the powder used in the cartridge. Ball powder travels farther than modern flake powder. As the range increases, the powder tattoo pattern will become less dense and more widely dispersed. Shorter barrelled weapons will produce a more widely dispersed pattern than a longer barrelled weapon for the same ammunition and range. Clothing will of course interfere with these patterns.

Distant wounds

Distant wounds are characterized by the absence of soot deposition and powder tattooing. With handguns this is of the order of 60–100 cm, although, as with all firearms injuries, test firing will be required to determine any range with accuracy. However, if the shot has been fired from a range beyond which powder tattooing and soot deposition would occur, no maximum range can be given because the distance may be a few metres or hundreds of metres. Clothing may modify the picture. If clothing is present and the person shot from intermediate range, all soot and powder grains may be absorbed, giving the entrance wound the appearance of a distance shot. The wound will consist of an entrance hole with or without an abrasion collar, but no powder tattooing or soot deposition. Small tears radiating from the entrance hole may be seen. Other intermediate targets such as glass may modify the wound appearance, as will ricocheting bullets causing instability in the bullet and an irregular entrance wound (Burke and Rowe 1992).

Tangential wounds

If a bullet hits the skin at a shallow angle, a number of different wounds may result. In tangential entrance gunshot wounds, sometimes called atypical entrance gunshot wounds, the bullet passes superficially through skin causing multiple splits. The direction of the bullet may not be obvious and a failure to appreciate that a bullet can cause such an injury may result in the wound being interpreted as a non-firearm injury. If the firearm is discharged close to the skin, powder and soot staining may be seen at one end of the wound (Figure 3.22). If the bullet passes very superficially across the skin, an abrading injury or superficial splitting of the skin may occur.

Skull damage

Bullets penetrating the skull will characteristically produce a bevelled wound with the inner table hole larger than the external table – so-called internal bevelling (Figure 3.23). The opposite effect is seen with exit wounds, where there is typically external bevelling (Figure 3.24). However, on occasions external bevelling may be seen in some entrance wounds. This most commonly involves contact wounds, but distance shots may also produce this phenomenon.

Tangential wounds of the skull have been called 'gutter wounds'. There may be various degrees of penetration of the skull. In first-degree gutter wounds the outer table is damaged. In second-degree wounds pressure waves fracture the inner table. In third-degree wounds the outer table

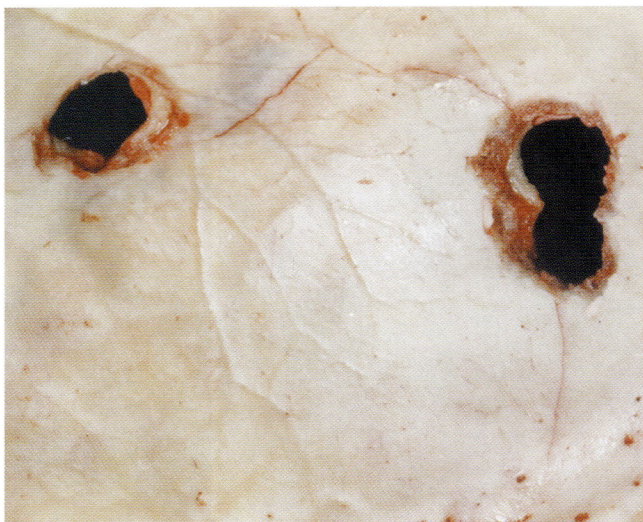

Figure 3.23 Inside of skull showing internal bevelling caused by buckshot.

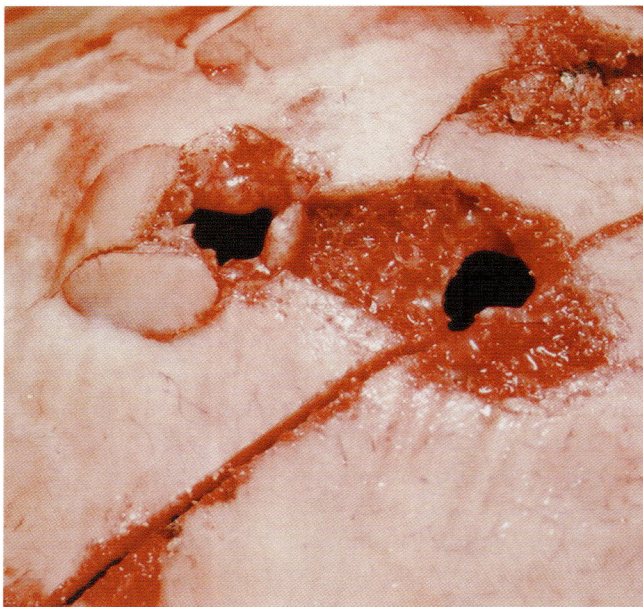

Figure 3.24 Outside of skull showing external bevelling caused by buckshot.

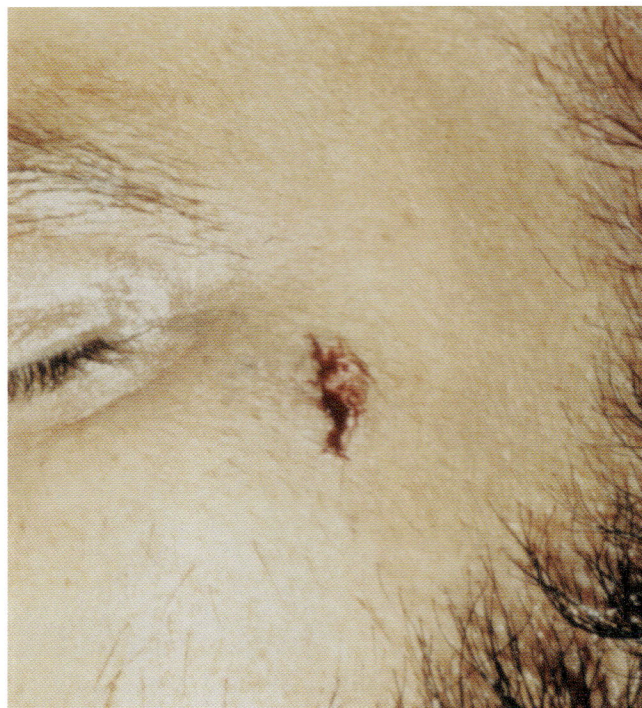

Figure 3.25 Slit-like exit wound.

Fracture lines may be seen radiating from the entrance hole. Secondary fractures occur as a result of raised intracranial pressure. As with other head injuries these are most commonly seen in the orbital plates. Temporary cavitation will also produce secondary fractures. When the skull has been damaged after death, reconstruction of the skull with Superglue will allow identification of an entrance wound.

Exit wounds

Bullets passing through tissue tend to become unstable. As they exit they will therefore tend to produce an irregular wound in comparison with the entrance wound (Figure 3.25), although this is not always the case. As the bullet exits through the skin, it will be pushed outwards, splitting the skin. Exit wounds may be stellate or slit like. They may on superficial examination mimic a wound caused by another weapon.

SHOTGUN INJURIES

When a shotgun is fired the contents of the cartridge, i.e. the shot and wadding, will be discharged from the barrel, along with muzzle gases, burning and unburnt powder, soot, detonator and cartridge material. These all affect the characteristics of the wound, depending on the specific contents of the cartridge and the distance from which the gun is fired (Breitenecker 1969). The shot may vary from a single slug to multiple small pieces of shot, typically made of lead. The site of the injury also has considerable importance in determining the characteristics of the wound.

is fragmented and the inner table fractured. Fragments of bone may be driven into the brain. When the bullet superficially perforates the skull, there may be an entrance and exit wound in the skull close together. When the bullet strikes the skull at a shallow angle a keyhole wound may be produced. The bullet most commonly breaks as it penetrates bone so that part of the bullet travels under the scalp, with the bulk of the bullet penetrating through the skull. This process produces a keyhole-shaped wound with one side of the wound being a typical entrance wound and the other end having external bevelling.

Contact wounds

Contact shotgun wounds to the head are characteristically self-inflicted and the head is a common site for suicidal wounds. The gun may be placed under the chin or in the mouth or to the temple region. Occasionally the gun may be placed behind the ear or on the cheek. These wounds to the head have a devastating effect and are often associated with major disruption of the skull and brain, because of the effects of both the shot and the muzzle gases produced. The brain may be extruded through the resulting disrupted scalp and skull (Harruff 1995).

When the mouth is the entrance wound, laceration of the lips and bruising may be seen. Examination of the inside of the mouth will typically reveal an entrance wound in the hard palate with soot and powder blackening. If the forehead is chosen as the entrance wound a tangential wound may be the result, with either an exit wound slightly higher on the forehead or even a wedge-shaped wound formed by the coalescence of both entrance and exit wounds.

An entrance wound of the occiput is very suspicious of being inflicted by another party, although self-inflicted wounds to this site are encountered. When the muzzle of the gun is held in contact with the skin a muzzle abrasion will result. When disruption of the skin is extensive and heavily blood stained, it may be difficult to identify features of a contact wound. Suturing together of the wound may allow the features to be observed after the reconstruction.

Near contact and intermediate wounds

When a shotgun is discharged beyond 1–2 cm from the body powder, tattooing will occur. As with rifled weapons the presence or absence of powder tattooing depends on the type of ammunition and length of the barrel. The wadding may cause additional damage. With plastic wadding, after firing it opens into a 'petal' formation that may be replicated on skin, depending on the range (Figure 3.26). At close range the wadding will enter the wound and no identifiable wadding injury will be present. All .410 ammunition has three petals and the resulting damage can be seen at a range of 7.5–12.5 cm (3–5 inches). 12- to 20-bore shotgun ammunition has four petals, marks from which can be seen between 30 and 90 cm (1–3 feet). Wadding may produce damage up to 6 m (20 feet).

As the distance between the gun and the target increases, the presence of soot and powder tattooing decrease until they are no longer present and shot will begin to widen, producing a larger entrance hole. The shot fired from a .410 will begin to widen at a shorter distance than larger bores. At a distance of about 1 m the entrance wound from a 12 bore will be of the order of 2.5–4 cm. This will depend partly on the choke of the barrel as the more choked a barrel the more together the shot will remain. By 1.3 m the

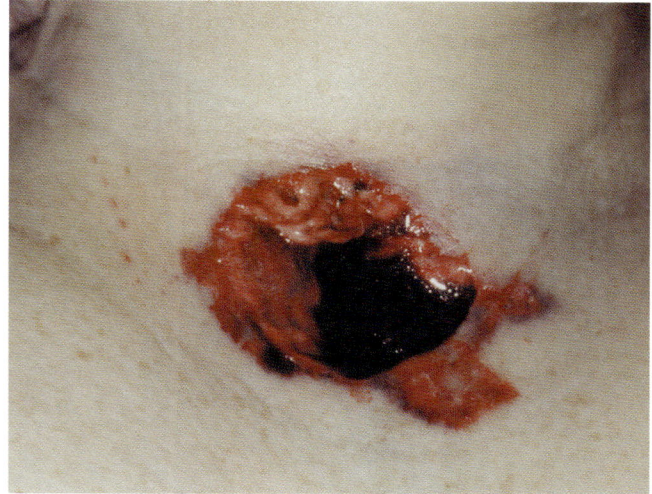

Figure 3.26 Shotgun wound to neck showing petal-shaped damage from wadding.

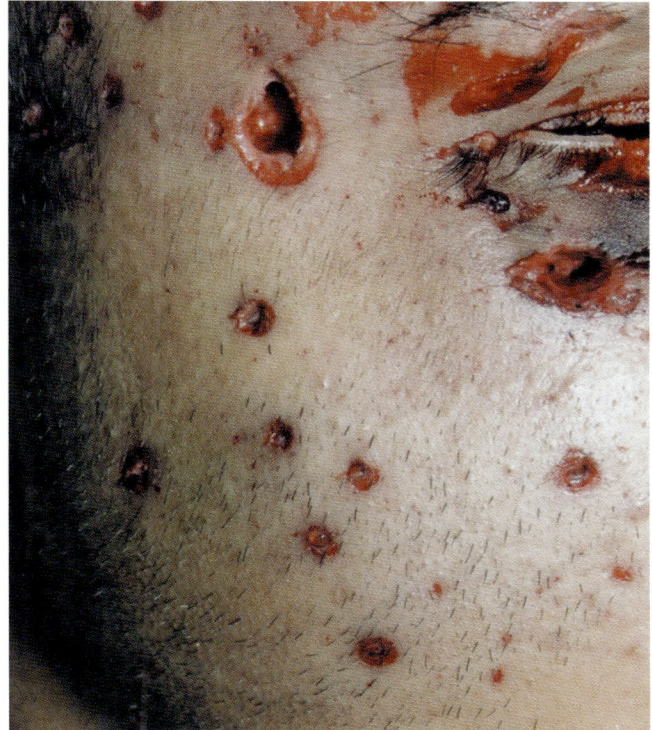

Figure 3.27 Shotgun wound with spread shot including buckshot.

entrance wound is likely to have a ragged appearance and satellite pellets will penetrate the skin around the wound.

Distant wounds

Shot will disperse as the distance increases and an increasing peppering of shot will surround the central entrance wound, until there is no central entrance wound, but a spread of shot (Figures 3.27 and 3.28). At these distances,

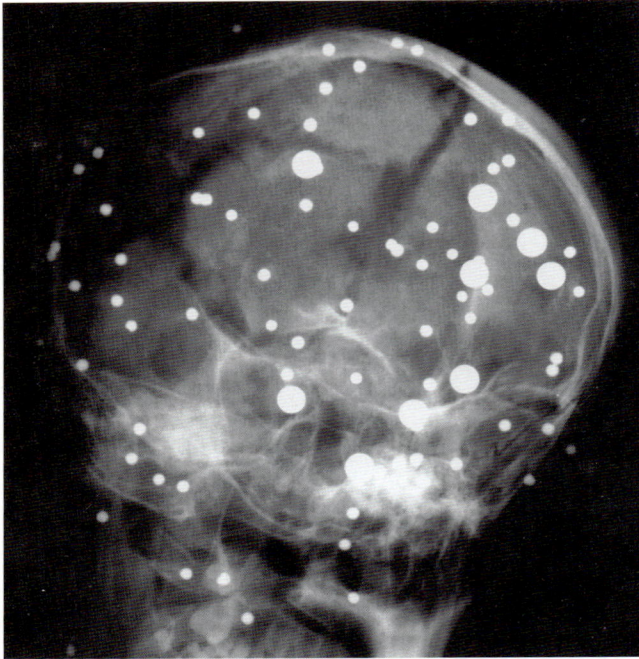

Figure 3.28 Skull radiograph of Figure 3.27, showing different size of shot (home-made ammunition).

smaller shot is less likely to penetrate, unless it strikes a vulnerable area, such as the orbit.

Wound ballistics

Traditionally, wounds have been described as high- or low-velocity wounds, principally based on the speed of the bullet. Bullets travelling at above 330 m/s (1100 feet/s), the approximate speed of sound in air, are high-velocity bullets, those below this figure, low velocity. The energy imparted is determined by the equation

$$E = \tfrac{1}{2}MV^2$$

The energy that a bullet releases having passed through the body is therefore determined by the formula:

$$E = \tfrac{1}{2}M(V1 - V2)^2$$

where $V1$ is the initial velocity on entering the body and $V2$ is the velocity of the bullet on exiting the body. Although the velocity of the bullet is an important component of its injury potential, the type of damage done to the body also depends on the construction of the bullet, how intact the bullet remains and on what tissues are involved by the wound tract.

Wounds are now referred to as low-energy transfer and high-energy transfer wounds. Low-energy transfer wounds pass through the skin, lacerating and crushing the skin and the other tissue in its path, but only those tissues that come into immediate contact with the bullet's path are damaged. No significant damage occurs by energy transfer to adjacent tissues outside the bullet path (Ryan *et al.* 1997)

In high-energy transfer wounds, as well as the damage caused by the immediate track of the bullet, there are two other methods of damage: shock waves and cavitation injury. With high-energy transfer, as the bullet passes through body tissues, the bullet compresses the tissue in front, and a spherical shock wave is transmitted before the bullet. This shock wave travels at a speed of around 450 m/s (1500 feet/s), the approximate speed of sound in water. Solid tissues such as muscle, brain and liver are particularly susceptible to damage by these shock waves. The shock waves pass along blood vessels, thus transmitting the damage some distance from the track.

The third cause of damage in high-energy transfer wounds is temporary cavitation. As the missile passes through the body, the energy is released into local tissues, which are displaced forwards and outwards, forming a large cavity. The cavity exists only for a few milliseconds and is formed under negative pressure and clothing, so bacteria and other external materials can be sucked into the cavity. The brain, liver and muscle, as solid organs, are particularly susceptible to cavitation injury, whereas the lungs, being less dense, are less damaged.

At speeds over 760 m/s (2500 feet/s), the tissue loses it elastic quality and the cavitation process becomes truly explosive, because the tissues can no longer recoil after the expansion. Tissue destruction therefore becomes much more severe with bullets above this speed.

A stable bullet that passes right through a limb may impart only 10–20 per cent of its energy. An unstable bullet may impart 60–70 per cent of its energy. A bullet that does not exit will impart 100 per cent of its energy to the body.

The pattern formation of a temporary cavity depends on the type of bullet. Experiments using material such as soap and gelatine blocks provide information on when cavity formation occurs. With full metal-jacketed bullets such as the standard NATO 7.62mm military round, a narrow channel is produced, before cavitation begins. In the case of the NATO 7.62mm this is of the order of 15 cm (Sellier and Kneubuehl 1994). A full metal-jacketed bullet can pass through a limb before the cavitation phenomenon has developed. With soft-point bullets, such as a soft-point 7.62mm used for hunting, cavitation will start immediately after penetration of the skin. The first soft-point bullets, produced from the British factory at Dum Dum in India in the 1890s, caused significant damage to the opposing armies. As a result of their controversial nature, such bullets were banned for use in warfare under the Hague Convention of 1899.

Appearance of the brain

The wound track in the brain may be associated with contusional damage at the site of entry into the brain. Contusions may also be seen at the site of exit. Raised intracranial pressure, evidenced by gyral flattening and cerebellar tonsillar grooving, can be seen (Kirkpatrick and DiMaio 1978). Three zones of macroscopic damage may be seen in the track of a bullet (Figure 3.29). The centre of the track is a blood-filled permanent cavity, outside which is an area of haemorrhagic necrotic tissue, with a marginal zone of pinkish-grey tissue. In addition more remote areas of haemorrhage may occur and subdural and extradural haematoma formation can develop.

Microscopic examination of the brains of people who have survived for more than 90 min has revealed axonal damage with β-amyloid precursor protein (APP) positivity throughout the cerebral hemispheres and the brain stem (Koszyca et al. 1998). An examination of 20 brains for the effects of temporary cavitation was performed on victims who survived for less than 90 min. They had been shot with bullets with a muzzle energy of less than 500 J, 17 being bullets from handguns and three from low-velocity rifles. Microscopic examination revealed a zone of astrocyte damage in an area of haemorrhagic extravasation surrounding the permanent track. Further from the permanent track the damage was lessened, with damage seen for a distance of 18 mm (Oehmichen et al. 2000).

After passing through the brain, a bullet may exit the skull. If it does not, it may ricochet back into the brain, or more commonly travel around the skull, parallel to the inner table, resulting in a shallow gutter wound track in the cortex of the brain (DiMaio 1999). Any bullet recovered should be carefully handled for subsequent ballistics examination. Metal forceps should not be used because they may mark the bullet, so plastic forceps should be used.

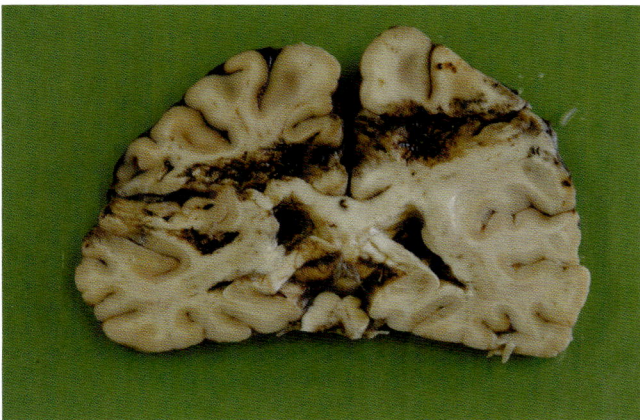

Figure 3.29 Coronal section of cerebral hemispheres showing gunshot wound track.

UNUSUAL WEAPONS

Air-powered weapons

Deaths from air weapons occasionally occur. The head is the most common part of the body penetrated. The entrance site may be through the eye, the temple or the forehead. The wound will be the approximate diameter of the pellet. There may be a small rim of abrasion, but powder burning will not be present, as it would be in a rimfire .22 rifle fired at near contact or contact range (Figure 3.30). No skin splitting is seen. If the weapon has been placed against the skin, an impression of the muzzle of the gun may be left. The muzzle impression may vary but, if the gun is available, it should be compared with the wound appearance (Milroy et al. 1998). Air-powered weapons may be modified illegally to become firearms, as may other weapons. These home-made firearms are sometimes referred to as 'country' guns or 'zip guns' (Book and Botha 1995).

Rubber bullets

With the increasing problem of urban unrest and the requirement to pacify demonstrators with non-lethal force, various weapons have been developed to hurt, but not significantly injure. In 1970 the British security forces in Northern Ireland introduced the rubber baton (often called the rubber bullet) as a method of riot control. The baton consisted of firm rubber constructed into a cylindrical mass 15 cm long with

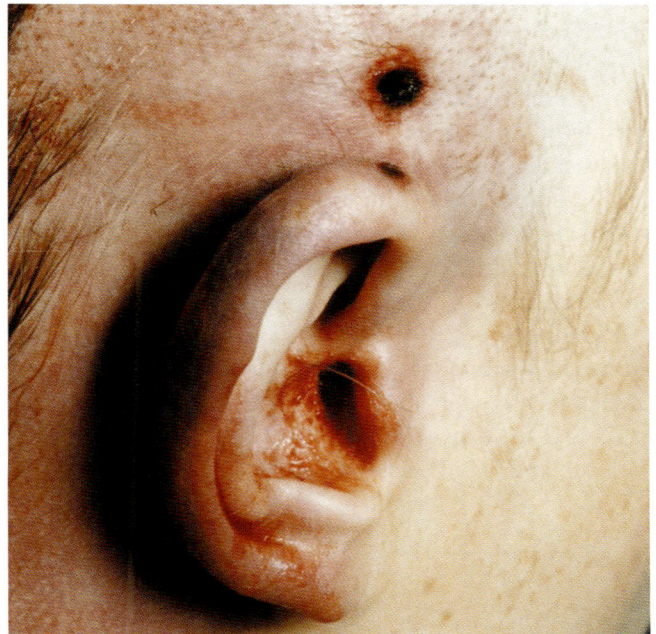

Figure 3.30 Entrance wound from a .22 air rifle pellet.

a rounded head, base 3.5 cm in diameter, and weighing approximately 140 g. The baton was designed to be fired at a range of 20–40 m. The object was to strike the lower limbs of the rioters, but the weapons were not accurate. Between 1970 and 1975, 55 000 rubber batons were fired. Three fatalities were attributed to the baton during this period.

Rubber batons were replaced as a riot method by plastic baton rounds. These rounds are made of Teflon, constructed as a cylindrical round without tapering, 10 cm long with a flat head and base 3.7 cm in diameter. The round weighs 135 g and has a muzzle velocity of 71 m/s and energy of 325.1 J. Since its first use in 1975 over 60 000 rounds have been discharged in Northern Ireland. Despite its greater stability, the plastic baton has been associated with 14 deaths in Northern Ireland. Its stability tends to result in the baton striking end-on and may also cause damage by ricocheting. Ten victims were hit in the head and four in the chest. Typically these wounds cause an annular abrasion with surrounding bruising. This pattern has apparently resulted in allegations that a 'D' or 'LR20' battery has been used as a projectile (Crane 2000).

The Israeli military have used both plastic and rubber bullets that are of a different construction to those used by the British. Hiss and colleagues have reported on these weapons (Hiss and Kahana 1994, Hiss et al. 1997). The plastic bullet weighs 0.85 g in a 5.56 × 45 mm cartridge. The bullet is made of polyvinyl chloride and is discharged with a muzzle velocity of 1250 m/s and energy of 663.7 J. Rubber bullets have also been used. There are four types: two spherical and two cylindrical. The spherical bullets are 1.8 cm in diameter. One is wholly made of rubber, as is one of the cylindrical bullets. The other two types are made of steel with a rubber shell. The rubber bullets weigh 8.3 g and have a muzzle velocity of 75–100 m/s and a kinetic energy of 23.3–41.5 J. The rubber and steel bullets weigh 15.4 g with a muzzle velocity of 100 m/s and energy of 77 J. Hiss and colleagues reported ten deaths from these rubber bullets and seven from plastic bullets. The rubber was separated from the steel in nine cases. Fatal penetrating injuries were to the head and chest, but in three cases there was no penetration, the victims being struck in the head and thoracic spine.

Mahajina and colleagues (2002) reported on the use of rubber and plastic bullets by Israeli forces against Arab Israelis in October 2000. Of 152 injured patients, three died, two from penetrating ocular injuries. Fractures of the skull and facial skeleton were recorded.

Humane killers

Humane killers are used by veterinary surgeons and in abattoirs to kill larger animals. They are of two main types, either firing a captive bolt or a bullet with a low velocity. Typically these weapons are used in suicides, and show features of a contact injury, including contact abrasion. The captive bolt may leave an injury similar to penetration by a non-firearm-pointed weapon (Hunt and Kon 1962).

Crossbow injuries

A number of reports have appeared detailing injuries from crossbows (Gresham 1977, Claydon 1993, Downs et al. 1994, Byard et al. 1999). Crossbows can produce fatal injuries that may simulate firearms injuries, especially if the bolt is removed. They also have the ability to produce fatal injuries at considerable distances. Crossbows may fire an arrow at 61 m/s (200 feet/s) with a range of around 270 m.

Crossbow arrows may be straight bolts or the arrowhead may have a number of blades attached to it. These two types of arrowhead give different wound appearances. With bolts with blades in the arrow tips, the wound shape can be characteristic, depending on the number of blades present, which may be two, three or four. These blades form incised wounds, which with edges opposed will indicate the nature of the arrow. Crossbow bolts with a circular head may produce circular entrance wounds that mimic a bullet entry wound. If the bolt is removed, considerable difficulty may be encountered, with an entrance wound resembling a bullet wound.

DEFENCE INJURIES

Defence injuries are wounds to the hands, arms and less commonly legs, produced when the victim tries to ward off blows with a weapon or with fists or feet. Firearms may also be associated with defence wounds. The identification of these wounds is important because they indicate that the victim was able to defend him- or herself for at least part of the attack. These injuries are often inflicted when the hands and arms are used to cover the head, and may indicate that more blows were aimed at the head than is apparent from the number of injuries found there.

HISTOLOGY OF SKIN WOUNDS

Considerable effort has been invested into using microscopic examination to date injuries (Raekallio 1980, Betz 2003). After an initial open injury the first change seen is margination of neutrophils, which can occur within 30 min or so and continues for a few hours. By 4 h neutrophils begin to infiltrate tissue, followed by mononuclear cells. For the first 24 h neutrophils predominate, but over the next 24–28 h mononuclear cells increase. Fibroblast proliferation begins after 48–72 h. Fibrin is seen, which at 16 h will stain red with Martius scarlet blue; before then it stains yellow. At 2–4 days a wound will have fibroblasts at the wound periphery and new capillary growth is seen at about 4–5 days. The epithelialization of an abrasion or small wound may be complete in 3 days. Over the next few days, fibroblasts decrease as collagen is laid down. The epithelium decreases in thickness by 5–7 days. Healing is completed by decrease in vascularity, cellular regression and restoration of collagen. The epithelium regains a basement membrane at 14 days.

Bruises typically have a neutrophil inflammatory response after about 4 h. Red blood cells may be seen intact in wounds at this and later stages, so the presence of intact red blood cells should not be interpreted as a fresh injury. Haemosiderin begins to be seen in wounds on day 3, and this can be a useful indicator of age, although previous injury at the same site should be considered, when haemosiderin might still be present. Haematoidin is occasionally found, but does not appear until around day 9.

A considerable amount of work has been published on the histochemical and immunohistochemical findings in wound healing. Antibodies to a variety of antigens have been used. Among these are fibronectin, which can be seen in a wound as early as 20 min, but routinely after 4 h, the interstitial cell adhesion molecule (ICAM-1) after 90 min and selectins, which can be positive at 1 h. Antigens marking later stages, such as fibroblast proliferation and collagen formation, can be used. Grellner (2002) examined the role of proinflammatory cytokines in human skin wounds. Interleukin-1β (IL-1β) and IL-6 showed enhanced expression after 15–20 min with increase in epidermal activity, with a marked increase after 30–60 min with IL-1β and 60–90 min with IL-6. Leukocytes reacting to these interleukins were seen after 2 h. Tumour necrosis factor alpha (TNF-α) showed changes at 15 min with marked enhancement at 90 min. These techniques remain essentially research tools rather than of practical value for casework, although they provide potential for more accurate delineation of timing. For a detailed discussion the reader should consult the original works (Betz 2003).

REFERENCES

Ainsworth, R.W. and Hunt, A.C. 1993. The enemy within. *Med Sci Law*, **33**, 358.

Besant-Matthews, P.E. 2000. Examination and interpretation of rifled firearm injuries. In Mason, J. and Purdue, B.N. (eds), *The Pathology of Trauma*, 3rd edn. London: Arnold.

Betz, P. 2003. Pathophysiology of wound healing. In Payne-James, J., Busuttil, A. and Smock, W. (eds), *Forensic Medicine. Clinical and pathological aspects*. London: Greenwich Medical Media, 83–9.

Book, R.G. and Botha, B.C. 1995. Zulu zip-guns and an unusual murder. *Am J Forensic Med Pathol*, **15**, 319–24.

Breitnecker, R. 1969. Shotgun wound patterns. *Am J Forensic Pathol*, **52**, 269–85.

Burke, T.W. and Rowe, W.F. 1992. Bullet ricochet: A comprehensive review. *J Forensic Sci*, **37**, 1254–60.

Byard, R.W., Koszyca, B. and James, R. 1999. Crossbow suicide: Mechanisms of injury and neuropathologic findings. *Am J Forensic Med Pathol*, **20**, 347–53.

Claydon, S.M. 1993. A bolt from the blue. *Med Sci Law*, **33**, 349–50.

Crane, J. 2000. Violence associated with civil disorder. In Mason, J. and Purdue, B.N. (eds), *The Pathology of Trauma*, 3rd edn. London: Arnold.

DiMaio, V.J.M. 1999. *Gunshot Wounds. Practical aspects of firearms, ballistics and forensic techniques*, 2nd edn. London: CRC.

Downs, J.C.U., Nichols, C.A., Scala-Barnett, D. and Lifeschultz, B.D. 1994. Handling and interpretation of crossbows injuries. *J Forensic Sci*, **39**, 428–45.

Garner, B.A. (ed-in-chief) 1999. *Black's Law Dictionary*, 7th edn. St Paul, MN: West Group.

Grellner, W. 2002. Time-dependent immunohistochemical detection of proinflammatory cytokines (IL-1β, IL-6, TNF-α) in human skin wounds. *Forensic Sci Int*, **130**, 90–6.

Greenes, D.S. and Scutzman, S.A. 2001. Clinical significance of scalp abnormalities in asymptomatic head-injured infants. *Pediatr Emerg Care*, **17**, 88–92.

Gresham, G.A. 1977. Arrows of outrageous fortune. *Med Sci Law*, **17**, 239–40.

Harruff, R.C. 1995. Comparison of contact shotgun wounds of the head produced by different gauge shotguns. *J Forensic Sci*, **40**, 801–4.

Hempling, S.M. 1981. The application of ultraviolet photography in clinical forensic medicine. *Med Sci Law*, **21**, 215–22.

Hiss, J. and Kahana, T. 1994. The fatalities of the Intifada (uprising): the first five years. *J Forensic Sci Soc*, **34**, 225–9.

Hiss, J., Hellman, F.H. and Kahana, T. 1997. Rubber and plastic ammunition lethal injuries: the Israeli experience. *Med Sci Law*, **37**, 139–44.

Hunt, A.C. and Kon, V.M. 1962. The pattern of injury in humane killers. *Med Sci Law*, **2**, 197–202.

Kirkpatrick, J.B. and DiMaio, V.J.M. 1978. Civilian gunshot wounds of the brain. *J Neurosurg*, **49**, 185–98.

Koszyca, B., Klombergs, P.C., Manavis, J. *et al.* 1998. Widespread axonal injury in gunshot wounds to the head using amyloid precursor protein as a marker. *J Neurotrauma*, **15**, 675–83.

Krauss, T.C. and Warlen, S.C. 1985. The forensic science use of reflective ultraviolet light. *J Forensic Sci*, **30**, 262–8.

Kury, G., Weiner, J. and Duval, J. 2000. Multiple self-inflicted gunshot wounds to the head: report of a case and review of the literature. *Am J Forensic Med Pathol*, **21**, 32–5.

Langlois, N.E.I. and Gresham, G.A. 1991. The ageing of bruises. A review and study of the colour changes with time. *Forensic Sci Int*, **50**, 227–38.

Mahanja, A., Abdoud, N., Harbaji, I. *et al*. 2002. Blunt penetrating injuries caused by rubber bullets during the Israeli-Arab conflict in October 2000: a retrospective study. *Lancet* **359**, 1795–800.

Milroy, C.M., Clark, J.C., Carter, N., Rutty, G. and Rooney, N. 1998. Air weapon fatalities. *J Clin Pathol*, **51**, 525–9.

Polson, C.J., Gee, D.J. and Knight, B. 1985. The Essentials of Forensic Medicine, 4th edn. Oxford: Pergamon Press, 102–4.

Oehmichen, M., Meissner, C. and Konig, H.G. 2000. Brain injury after gunshot wounding: morphometric analysis of cell destruction caused by temporary cavity. *J Neurotrauma*, **17**, 155–62.

Raekallio, J. 1980. Histological estimation of the age of injuries. In Perper, J.A. and Wecht, C.H. (eds), *Microscopic Diagnosis in Forensic Pathology*. Springfield, IL: Charles C. Thomas.

Ryan, J.M., Rich, N.M., Burris, D.G. and Ochsner, M.G. 1997. Biophysics and pathophysiology of penetrating injury. In Ryan, J.M., Rich, N.M., Dale, R.F., Morgans, B.T. and Cooper, G.J. (eds), *Ballistic Trauma. Clinical relevance in peace and war*. London: Arnold, 31–46.

Sellier, K.G. and Kneubuehl, B. 1994. *Wound Ballistics and the Scientific Background*. London: Elsevier.

Stephenson, T. and Bialas, Y. 1996. Estimation of the age of bruising. *Arch Dis Child*, **74**, 53–5.

Teare, R.D. 1961. Blows with the shod foot. *Med Sci Law*, **1**, 429–36.

Wilson, E.F. 1977. Estimation of the age of cutaneous contusions in child abuse. *Pediatrics*, **60**, 750–2.

TABLE OF CASES

4

Adult skull fractures

Philip D Lumb and Helen L Whitwell

A fracture can be defined as an abnormal break in the continuity of a structure such as a bone produced by stress and strain. Despite a 'lay view', skull fractures do not generally cause death directly but are an indicator of force applied. It is the underlying brain injury that causes death. External herniation of the brain through a fractured skull may decrease morbidity in some circumstances. Skull fractures are most commonly caused by direct force and less commonly by indirect force, such as a ring fracture to the posterior fossa caused by a fall from a height.

The biomechanics of skull fractures is complex. Deformation of the skull occurs with force and, if the limit of the elasticity of the skull is exceeded, fractures will occur. A focal impact leads to a depressed skull fracture and a broader impact causes linear fracturing. As a generalization, the more severe the fracture the greater the force and the greater the injury to the underlying brain (Leestma 1988, Payne-James *et al.* 2003).

DOCUMENTATION OF SKULL FRACTURES

Precise documentation is essential. When considering skull fractures, there are several features that must be carefully observed. Scene information and a scene visit with the body *in situ*, and interpretation of the assault/accident scenario, with identification of weapons or sites of potential impact, may greatly assist. Assessment of clothing, in particular items worn on the head, which may alter the appearances of scalp/skull injuries, should also be done at this point.

If considered clinically appropriate, e.g. in firearms injuries, skull stab wounds or 'hit-and-run' fatalities, radiological examination should be undertaken to identify foreign bodies and evidence of bony injury. In those individuals with survival in hospital, review of medical records, including radiology, should be done.

As with all injuries, whenever possible fractures should be photographed, from both an extracranial and an intracranial viewpoint. Preparation for photography by the pathologist is paramount in order that, when reviewing the case, the precise nature of the fracture can be reassessed accurately. After photography of all external injuries associated with both the head and elsewhere in the body, the scalp should be reflected both anteriorly and posteriorly. Before the temporalis muscles are dissected and the periosteum scraped, the skull should be photographed from four quadrants, including two left and two right, front and back images from both sides. This enables documentation of deep bruising associated with fracturing. Preparation of the fracture sites can then be made for more detailed close-up photography. The periosteum should be scraped away with a sturdy postmortem knife (such as a PM40) and cleaned of blood. Temporalis muscles should be removed in order to allow fracture extensions beneath this muscle to be readily identified.

Where necessary, the scalp should be reflected as far as is possible to reveal the course of the fracture. This may involve removal of the facial skin to show continuity with facial fractures or further reflection of the occipital scalp into the neck. It may be difficult to photograph skull fractures that extend into the occipital region and this is best done when the body is laid on its front. All photography should be done with and without measurement scales. After adequate photographic documentation, it is the author's preference to measure and record the fractures either diagrammatically or using computer graphics. Where possible the fracture lines should be avoided and, when cut through with the saw, this should be noted. This facilitates review of the body with other experts. During the removal of the skull cap, there may be collapse of certain areas and fall-out of bone segments, which should all be documented. The skull should be drilled in such a manner that the dura mater is kept intact. This involves skilful control of the drilling blade. Once this skull cap has been created, before its removal it is important to ensure that the photographer is prepared, because haematomas tend to slip away on to the mortuary bench. Therefore a quick photograph straight after the skull cap is removed can identify the haematoma in its original position.

If the meninges are still intact, the location of a haematoma can be readily assessed and compared with the location of the fracture site and underlying brain injury. Any meningeal defect should be noted along with its associated fracture. The brain can now be carefully removed along with the meninges. The inside of the skull cap should be examined and any injuries documented. After removal of the brain the meninges around the base of the brain and inside the remaining calvarium should be removed carefully. In younger individuals traction with soft tissue paper can do most of the meningeal stripping. In elderly individuals the dura mater tends to adhere more to the skull and dura strippers may be employed. Once all the dura has been removed, the inside of the cranium should be carefully examined and again the fracture lines documented diagnostically and photographed. It is important to attempt to separate the sides of the skull from one another because many fractures are not apparent with just the naked eye unless opened slightly. Gentle traction on the skull will open up the fracture line and make them more readily visible. Photography of these distracted segments should also be taken.

After documentation, measurements of skull thickness may be taken, although the value of these measurements is questionable. The adult cranium varies in thickness with thin/weaker areas supported by the buttresses, for example the occipital protuberance and saggital ridge. Most pathologists take measurements from the frontal and parietal bones (6–10 mm), temporal bones (4 mm) and occipital bones (15 mm) (Saukko and Knight 2004). The thicknesses can be documented in the postmortem report along with normal variables. The skull thickness can also be photographed with a scale.

In complex cases of skull fracture, reconstruction may aid interpretation. The method for this is given in the Appendix to this chapter and is illustrated both prior to and following reconstruction.

LINEAR SKULL FRACTURES

Linear fractures occur with a broad impact such as ground contact or from a flat object and account for around 70 per cent of skull fractures. Following impact there is inbending of the skull at the site of impact, with outbending peripheral to the impact site. A linear fracture is initiated by outbending of the skull and then usually extends to the impact point and, in the opposite direction, towards areas of structural weakness (in particular the base of the skull). Gurdjian and colleagues (1950) studied deformation patterns following blunt impact using a stresscoat method, which utilised a strain sensitive lacquer applied to the skull that cracks as a result of tension stress. He identified that the first area of outbending to occur was the area of primary stress level and found that with increasing force of impact additional linear fractures may occur at secondary and tertiary stress areas.

Occipital

Simple linear fractures of the occipital skull are most commonly seen in falls that may be accelerated or unprotected, for example when an individual is intoxicated or unconscious from a direct blow to the front of the face. Individuals in the elderly age group, who may be at risk of syncope, may take an unprotected fall on to the back of the head. In forensic practice, the acceleration is usually provided by means of a blow to the front of the face such as a punch. The point of contact of the blow should be investigated by examination of the injuries to the face and also deep dissection of the facial tissues. Facial fractures may occasionally be identified as a result of this direct blow, in association with a variety of other features of blunt force trauma including lacerations and bruising. Acceleration to a fall is provided by the delivery of the blow to the face. Acceleration may also occur as a result of individuals pulling on an object, such as a handbag, with loss of grip on the object providing sudden acceleration. Typically the head is accelerated on to a firm surface such as the pavement or a tarmac road. An elderly individual is also more at risk of developing intracranial haematomas, particularly subdural, caused most probably by an age-related decrease in cerebral volume (Leestma 1988, Hartshorne et al. 1997).

Externally the scalp may show evidence of blunt force impact with combinations of laceration, bruising and abrasion. A variety of linear fractures may be produced in the occipital bone – vertical fractures extending into the occipital suture line or horizontal fractures curving towards the base of the brain (Figures 4.1 and 4.2). The fractures tend to avoid the strengthened buttresses of the internal occipital protuberance (Saukko and Knight 2004). The linear fracture may continue into the base of the skull into the posterior fossa, where it often follows a complex path to the anterior fossa with dissemination of the energy of the impact away from the primary impact site. When springing of the sutures is identified, further fracturing may be seen extending into the posterior fossa away from the primary impact site. Typically in such accelerated falls

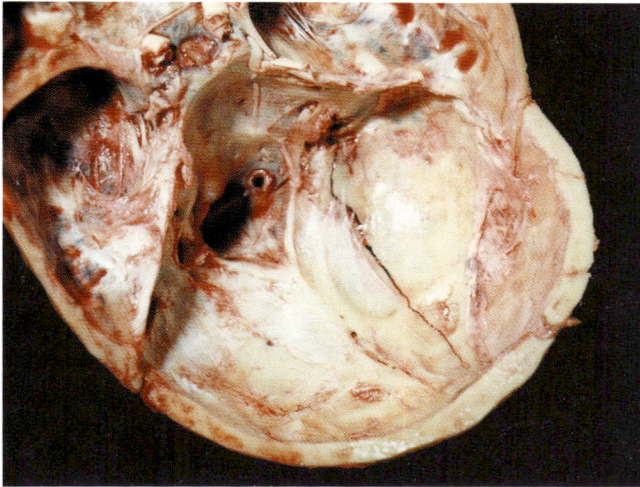

Figure 4.1 Occipital fracture after a fall on the back of the head.

the brain shows a contre-coup pattern of contusion and laceration (see Chapter 7, p. 85). Contre-coup orbital plate fractures are not infrequently seen in this scenario (see below).

Linear squamous temporal fractures

The squamous temporal bone is particularly vulnerable to fracturing as a result of its thin, unsupported structure. This most often occurs as a result of direct trauma such as a blow by an object (golf ball, bat) or a fall. Fractures to this region may injure the middle meningeal artery, which courses within the cranial cavity on the deep aspect of this bone (Figure 4.3). Injury to this artery causes extradural haematoma formation (see Chapter 5, p. 62).

Linear frontal fractures

Linear frontal fractures are also encountered in the context of accelerated falls to the front of the head as well as with direct trauma such as kicking. As with occipital fractures, they are commonly associated with external injuries such as lacerations and bruising. Frontal fractures are most commonly vertically oriented and extend towards the face inferiorly and the parietal skull posteriorly. Most fractures remain ipsilateral to the point of contact with

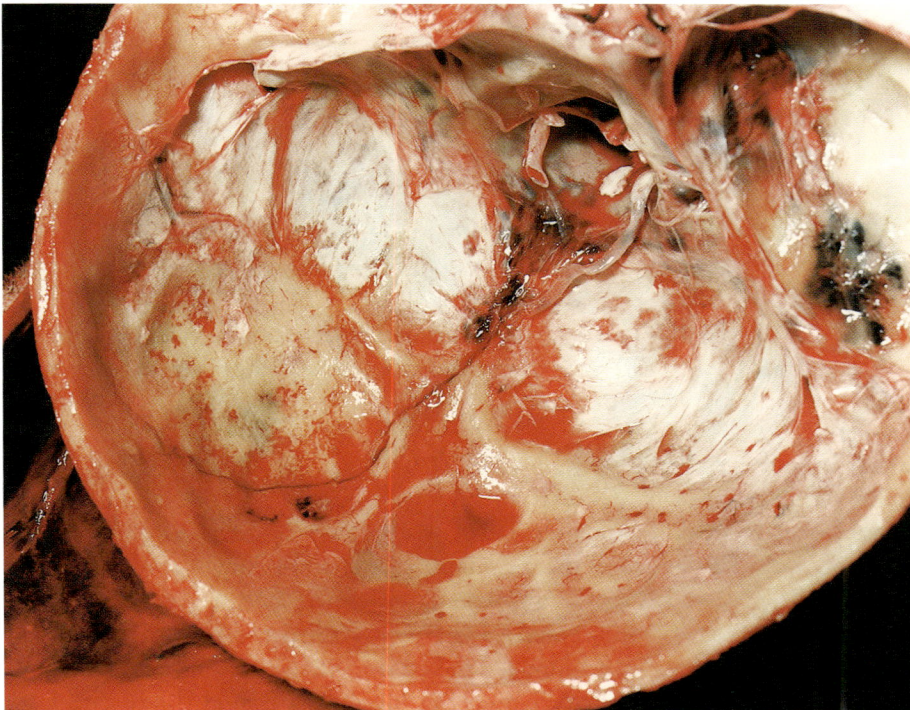

Figure 4.2 Curved occipital fracture involving the internal occipital protuberance.

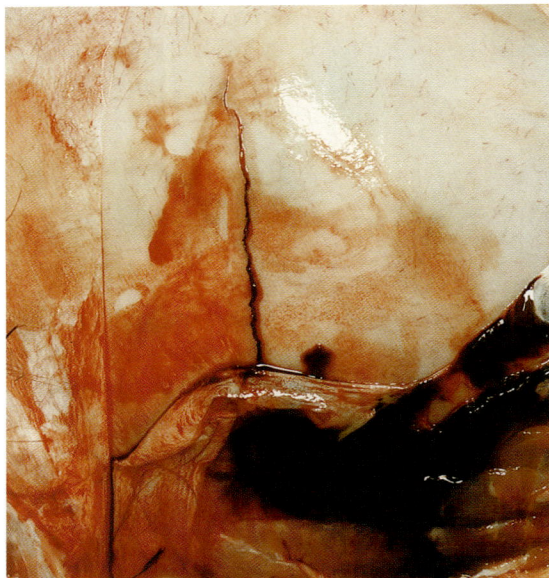

Figure 4.3 Linear sqamous temporal fracture in association with underlying acute extradural haematoma.

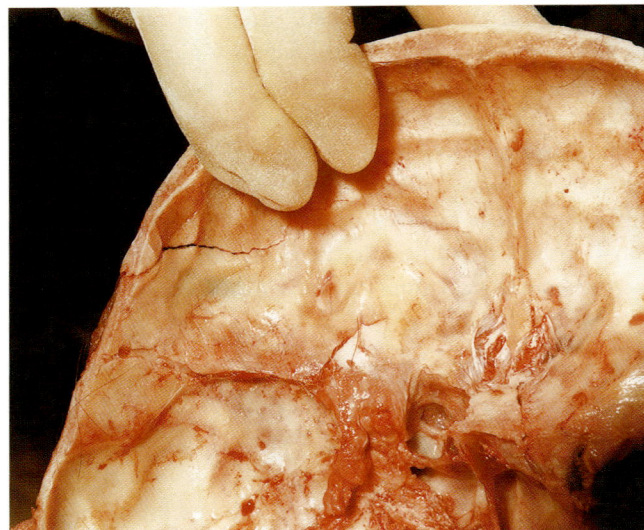

Figure 4.4 Frontal fracture extending into the anterior cranial fossa.

the ground although some curve posteriorly and cross the midline away from the region of impact. They may also terminate within the sutures, causing these sutures to spring (coronal or sagittal). With linear fractures of the frontal bone shorter emanating fractures may be identified. Inferiorly the fracture may extend in a variety of directions: into the roof of the orbit through the supraorbital ridge, posteriorly across the anterior cranial fossa (Figure 4.4). The fracture may also extend down the inside of the orbit across its medial wall, through the lacrimal bone and down into the maxilla (Figure 4.5). These fractures may also involve the hard palate. Again, as with occipital fractures, it is important to establish the point of contact for the initial blunt force trauma. Occasionally a frontal fracture may represent a secondary impact, after injury to the posterior scalp. These accelerating forces may be associated with a depressed fracture which may be comminuted.

Front to back hinge fracture

A frontal fracture may form part of an injury complex creating a 'front-to-back hinge fracture'. This occurs when there is a high-energy impact over the occipital skull, which drives the front of the head rapidly into a firm surface. This may occur in pedestrian road traffic incidents. If a pedestrian is hit by a car bumper or side mirror travelling at speed, the primary impact will create a large comminuted, depressed or mosaic fracture (see below). The energy from

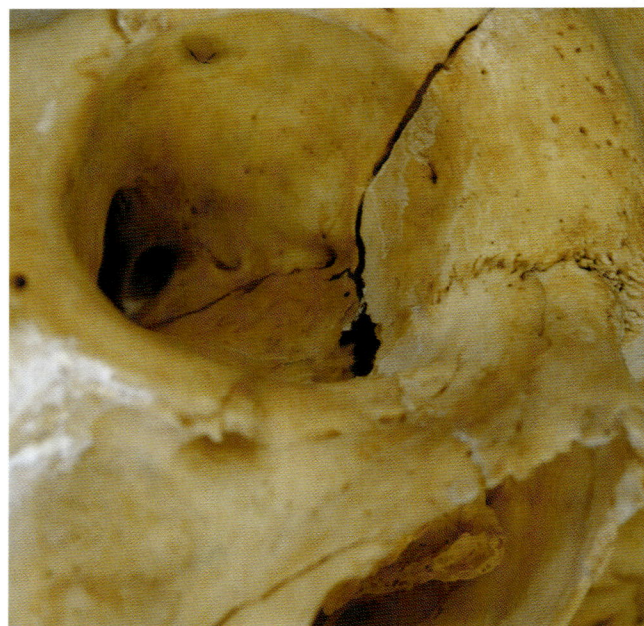

Figure 4.5 Involvement of facial bones, with a fracture extending from a frontal fracture into the orbit continuing to the maxilla.

this primary impact may drive the front of the face with considerable velocity into the pavement or roadside. This causes the secondary impact and creates a frontal fracture. With the high energies involved, both the primary impact fracture and the secondary impact (frontal) fracture may extend towards the base of the skull. Both fractures may terminate within the foramen magnum. Removal of the

Figure 4.6 Puppe's rule: the second later fracture (vertical) terminates in the first earlier fracture.

skull cap *post mortem* allows the two halves of the skull to form a 'front-to-back' hinge fracture. The primary and secondary impact sites may also communicate within the calvarium as the posterior extension of the frontal fracture (secondary impact) communicates with the anterior extensions of the primary fracture complex. When there is a primary and secondary impact site, such as in the situation described above, the secondary impact site with the frontal fracture is usually on the contralateral side to the primary impact site, because the primary impact site offset from the midline drives the skull away from the region in which the forces apply.

PUPPE'S RULE

Puppe's rule can be used to assess the order in which fracturing occurred. Evidence for two separate episodes of blunt force trauma must be corroborated by external features. After injury the first fracture is formed. Subsequent fractures caused by additional blunt force trauma will terminate and not cross in this first fracture line (Saukko and Knight 2004). Figure 4.6 gives an example of such an injury.

Depressed skull fractures

Depressed skull fractures are caused by focal impact to the skull. Indentation of the skull may occur without fracturing, and with increasing degrees of force the inner table may fracture, then both the inner and outer tables, and with greater force still both comminuted fractures and depressed fractures may occur (Gordon *et al.* 1987). The bone fracture may rupture the underlying meninges and force may be transmitted directly to the brain with cerebral contusion or laceration (Figure 4.7). The force is delivered over a small well-defined area and may mirror the shape of the impacting object although with tangible blows they may appear wedge-shaped. The pattern of external injury to the scalp may also be useful in defining the properties of the object used. Blunt implements such

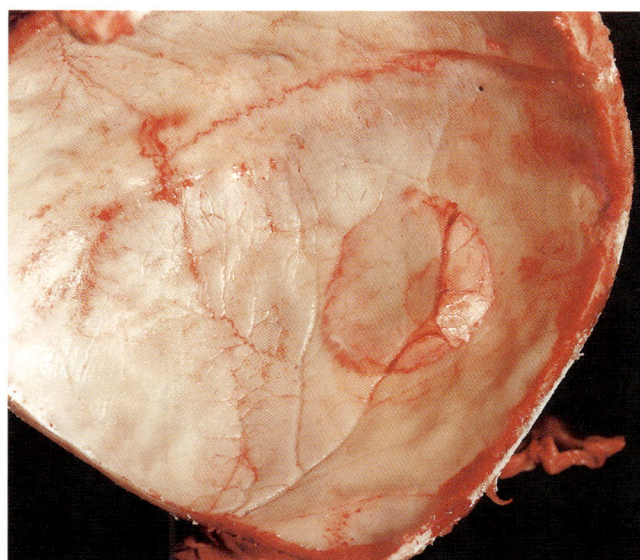

Figure 4.7 Depressed skull fracture (hammer).

as hammers, rocks and baseball bats are among the more common weapons. Depressed fractures may also be seen in vehicular incidents, either where there is contact with a protruding object in the vehicle, or where a pedestrian is struck by a part of a vehicle.

Pond fracture

A pond fracture is a shallow depressed fracture forming a 'concave pond'. It is more common in the infant's pliable skull where true fracturing may not actually occur and the skull shows a concave depression (Saukko and Knight 2004).

Mosaic fracture

A mosaic fracture comprises a comminuted, depressed area of fracturing with linear fractures radiating from an apparent central region and with intercepting fractures between the radiating arms of the fracture, which give the complex appearance described as a spider's web. The degree of depression may be minimal or absent. Plates of free-floating bone are created by the intersection of radiating fractures and intersecting fractures. At the centre of the region of impact these intersections are closer together, producing smaller free-floating areas. Further away from the epicentre of impact the free-floating segments are somewhat larger and eventually disappear as the radiating arms of the fracture extend outwards. Radiating arms of the fracture may be extensive and extend into the base of the skull and may also communicate with secondary fractures caused by secondary impacts.

Fractures of this nature are caused by severe impact with objects causing focal and general skull deformation. This may include vehicle parts such as bumpers, wing mirrors or other presenting parts of a vehicle impact. This fracture may also be identified as the primary impact site in a fall from a height. It is not usually encountered as a consequence of an accelerated fall such as that caused by a punch to the face, unless the head contacts with such an object as described above.

As a result of the broad nature of the impact, the scalp injury associated with these extensive fractures may be non-specific and include small lacerations and abrasions. However, on the deep aspect of the scalp more extensive bruising is usually present.

Ring fractures

Ring fractures are encountered in the context of falls from a height. They represent the delivery of energy to the base of the brain through the upper cervical vertebrae. The fracture encompasses the foramen magnum, occasionally making this free floating although most fractures are partial. Examination of the spinal column may reveal fracturing to the vertebral bodies and it is essential to examine the lower limbs and pelvis thoroughly to assess (if possible) the site of the primary impact (such as the feet or buttocks). Ring fractures may also be encountered in those individuals who sustain impact to the vertex of the skull. They are encountered occasionally in the context of road traffic incidents, particularly involving motor cyclists. However, they are usually combined with a hinge fracture in this context (Mason and Purdue 2000).

Isolated ring fractures may also be caused by hyperflexion/hyperextension of the cervical spine. The ligamentous attachments about the base of the skull are biomechanically stronger than the base alone. Violent swinging of the skull can cause the base of the brain to fracture, forming a ring around the foramen magnum. The mechanism of fracture may be enhanced by a heavy helmet, such as that worn by motorcyclists (Konrad et al. 1996).

Hinge fractures

The rare 'front-to-back' hinge fracture is described above. The classic hinge fracture involves an injury that extends across the base of the skull from one side to the other (Figure 4.8). After removal of the calvarium the anterior and posterior parts of the skull may be parted to form two

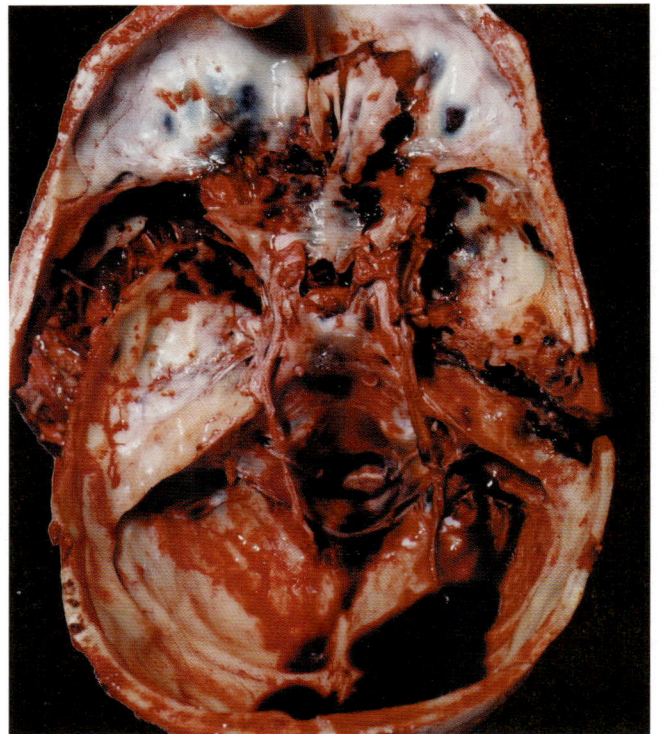

Figure 4.8   Hinge fracture (road traffic accident).

halves. Typically the hinge fracture extends through the petrous temporal bones, across the pituitary fossa and into the opposite petrous temporal bone. These fractures are most commonly encountered in road traffic incidents including injury caused by airbags (Perez and Palmatier 1996) (see also Chapter 11, p. 127). They may also be encountered within the context of falls from a height with primary impact about the vertex or in the context of blunt force trauma by an object or implement driven into the skull.

Transmitted fractures/contre-coup fractures

Transmitted fractures or contre-coup fractures are found away from the primary impact site and comprise isolated linear or occasionally complex areas of fracturing (Figure 4.9). These are most commonly encountered in the cribriform plate or orbital roofs (Hein and Schultz 1990, Asano et al. 1995). Other sites include the thin bones over pneumatized spaces, such as the lacrimal, sphenoid and frontal bones. They should not be confused with direct trauma to this area. They are most frequently encountered in falls on to the back of the head, and are associated with a fracture at the primary site of impact. Intracranial pressure produced by missiles such as bullets travelling across the skull can also produce the contre-coup fracture.

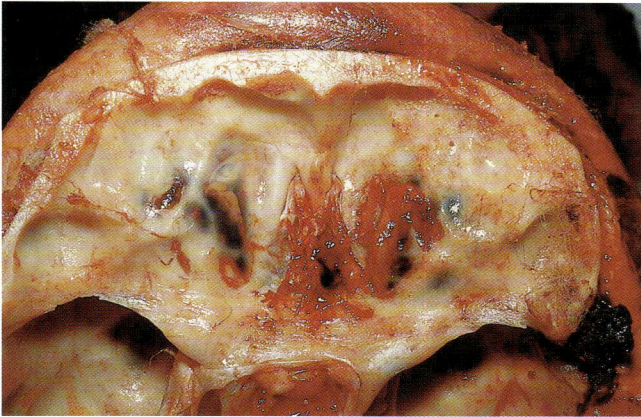

Figure 4.9   Contre-coup fractures as a result of a fall on to the occiput.

The differential diagnosis of the isolated anterior cranial fossa fracture includes penetrating injuries through the orbital roof as well as direct trauma. Fragments of glass from, for example, a road traffic accident may penetrate the orbit and orbital roof. Radiological examination may identify glass fragments, but one should be aware that not all glass fragments are identified by this means.

HISTOLOGY OF SKULL FRACTURES

The healing of skull fractures proceeds differently to that of long bone fractures. Little information is currently available in the medical literature on the histology of skull bone healing. The internal periosteum is replaced with dura mater, which contains few osteoblasts. No significant external callus forms in the healing of skull fractures. This may be an evolutionary adaptation, because a large callus may create a space-occupying lesion, displacing the cerebral contents. Dating the age of a fracture can be difficult. Healing of fractures may be influenced by a variety of factors including infection, hypoxia, fracture stability, interposition of tissue and severity of injury. Dating the age of the injury may require examination of associated injuries such as underlying haematomas or scalp bruising/laceration (Figure 4.10).

After a fracture, bleeding occurs immediately, from the injured marrow cavity, torn dura, external periosteum and cortical vessels. Bone necrosis occurs about the edges of the fracture within a few hours of the fracture occurring. The haematoma organizes (Figure 4.11) and early fibrosis occurs. The haematoma may disappear within a week to 10 days (Robbins *et al.* 1994, McCarthy and Fraasica 1998). Callus develops, which comprises a mixture of fibrosis, osteoblasts and osteoclasts. Active bone remodelling,

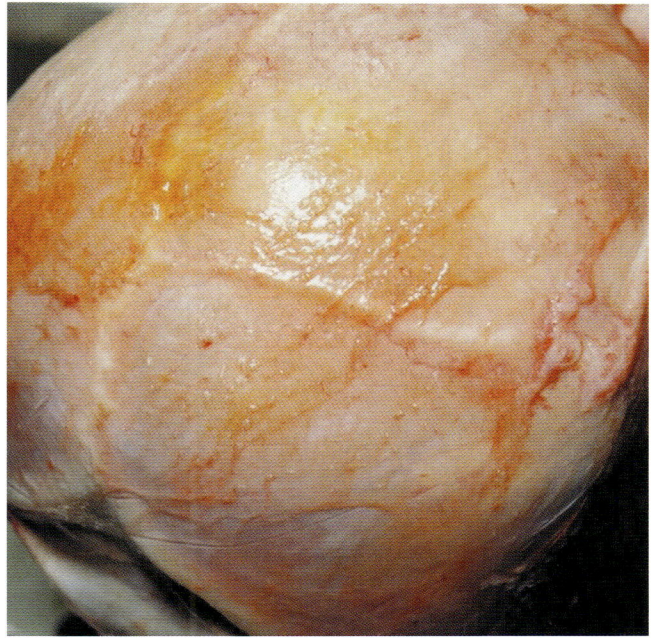

Figure 4.10   Haemosiderin staining over healing fracture – may be more useful in dating the injury.

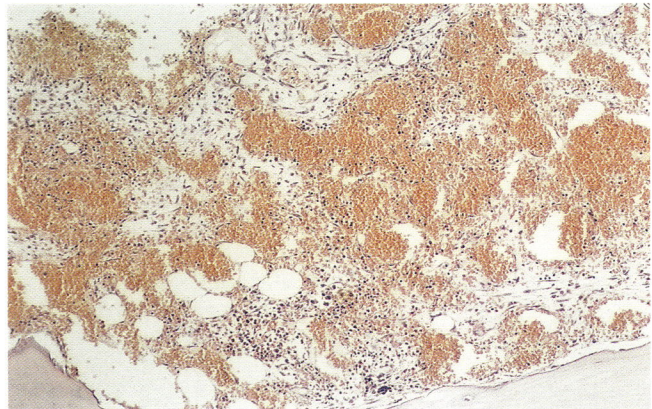

Figure 4.11   Haematoma within a skull fracture showing early organization.

with the presence of bone resorption and osteoblastic synthesis of bone matrix, can be identified. New bone may be visible at approximately 1–2 weeks. After bone remodelling, it may not be possible to identify a healed fracture site.

MEDICOLEGAL ISSUES

Bruising

Fractures to the base of the brain are commonly associated with additional bruising away from the scalp impact sites.

This includes bruising about the mastoid process (battle sign) and also bilateral periorbital bruising. Bleeding may also be identified emanating from the nose and both ears. This should not be mistaken for other primary impact sites. If there is doubt that bruising about the eye is caused by direct trauma and if clinically appropriate (such as survival for a period of time), the eye should be formally examined for evidence of direct trauma. Features include commotio retinae and traumatic changes to the lens.

Post traumatic epilepsy

Epilepsy occurs particularly with depressed fractures and may be delayed in onset for several months or longer. Detailed neurological review may be necessary to link a head injury with the epilepsy, which may cause sudden death at some time distant to the injury. General complications of head injury are covered in Chapter 15.

Infections

Infections, including meningitis (most commonly pneumococcus or staphylococcus) and cerebral abscess, may occur either as a direct result of contamination from an external wound such as a penetrating gunshot wound (in particular high-velocity wounds, which cause extensive tissue damage), via fractures of the anterior cranial fossa (which may be contre-coup fractures), or from air sinuses. Infection is less common with simple linear fractures than in compound fractures, including compound depressed fractures. In some cases presentation may be delayed for several weeks or even longer.

REFERENCES

Asano, T., Ohno, K., Takada, Y. *et al*. 1995. Fractures of the floor of the anterior cranial fossa. *J Trauma*, **39**, 702–6.

Gordon, I., Shapiro, H.A. and Berson, S.D. 1987. *Forensic Medicine: a guide to principles*, 3rd edition, Edinburgh: Churchill Livingstone.

Gurdjian, E.S., Webster, J.E. and Lissner, H.R. 1950. The mechanism of skull fracture. *J Neurosurg*, **7**, 106–14.

Hartshorne, N.J., Harruff, R.C. and Alvord, E.C. 1997. Fatal head injuries in ground-level falls. *Am J Forensic Med Pathol*, **18**, 258–64.

Hein, P.M. and Schultz, E. 1990. Contrecoup fractures of the anterior cranial fossae as a consequence of blunt force caused by a fall. *Acta Neurochir*, **105**, 24–9.

Konrad, C.J., Fieber, T.S., Schuepfer, G.K. and Gerber, H. 1996. Are fractures to the base of the skull influenced by the

mass of the protective helmet? A retrospective study in fatally injured motorcyclists. *J Trauma*, **41**, 854–8.

Leestma, J.E. 1988. *Forensic Neuropathology*. New York: Raven Press.

McCarthy, E.F. and Frassica, F.J. 1998. *Pathology of Bone and Joint Disorders with Clinical and Radiographic Correlation*. Philadelphia: WB Saunders.

Mason, J.K. and Purdue, B.N. 2000. *The Pathology of Trauma*. London: Arnold.

Payne-James, J., Busuttil, A. and Smock, W. 2003. *Forensic Medicine. Clinical and pathological aspects*. London: Greenwich Medical Media.

Perez, J. and Palmatier, T. 1996. Air bag-related fatality in a short, forward positioned driver. *Ann Emerg Med*, **28**, 722–4.

Robbins, S.L., Cotran, R.S. and Kumar, V. 1994. *Pathologic Basis of Disease*, 5th edition. Philadelphia: WB Saunders.

Saukko, P. and Knight, B. 2004. *Knight's Forensic Pathology*, 3rd edn. London: Arnold.

APPENDIX

Reconstruction of skull

1. Fix all fragments of bone plus any scalp, dura or blood clot in ten per cent formalin
2. Rinse carefully in running water, taking care not to lose small fragments
3. Carefully scrape all fragments clean using a scalpel blade. Palpate all tissues for possible bone fragments
4. Wash in water
5. Immerse in dilute hydrogen peroxide (approximately 5 per cent, or 20 volume strength) until bone is bleached white: on average the process will take up to one week
6. Wash in water
7. Immerse in absolute alcohol for two to four hours to dehydrate
8. Dry on tissue and place in chloroform to degrease. Change chloroform regularly and check bone for degreasing. The time needed for degreasing varies depending on amount of fat present: on average the process will take up to three weeks
9. Dry bone with tissue and place in a 60°C incubator for one hour. Wet areas indicate presence of fat
10. When degreased and dried, remove any remaining fragments of tissue, blood, etc.
11. Begin to reassemble skull initially using tape to tack pieces together
12. The pieces can, if necessary, be joined using minute amounts of cyanoacrylate glue, but only spot join pieces with these adhesives, or the fractures cannot

be filled in later on. Make use of blood vessel tracks, suture lines or bone thickness to aid reassembly

13. Use self-hardening model plastic to join the fragments and show the fractures. Make up a watery solution and introduce it into the fractures and joints using a Pasteur pipette. When set (in a few minutes), scrape the excess away

14. When the skull has been reconstructed, rub it down with sand paper

15. Finally, give the skull several coats of clear matt polyurethane varnish.

A fume cupboard with appropriate extraction should be used as necessary.

Figures 4.12a and b show the skull prior to and following reconstruction. In severe cases of fragmentation additional aids can be used, such as drawing a map plan of the pieces or marking the interior of each piece with a reference mark, letter or number. (Courtesy of Professor J. Crane.)

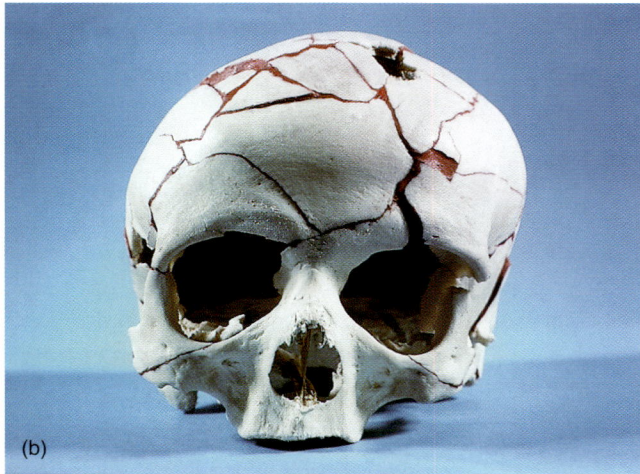

Figure 4.12 Complex skull fracture as a result as a of a gunshot wound to the head (a) prior to reconstruction; and (b) following reconstruction (courtesy of Professor J Crane).

5

Intracranial haematomas: extradural and subdural

Helen L Whitwell

This chapter covers the various types of intracranial haematomas including extradural haematoma (EDH) and subdural haematoma (SDH). Subarachnoid haemorrhage is covered in Chapter 6, intracerebral traumatic and non-traumatic haemorrhages in Chapter 7, and forensic anatomy in Chapter 1.

EXTRADURAL HAEMATOMA

This is also known as an epidural haematoma. Bleeding in this condition takes place between the skull and the dura. This type of haemorrhage is found in between 5 and 15 per cent of fatal head injuries (Graham *et al.* 2002). Most commonly it occurs with a laceration or tear of the middle meningeal artery or branch beneath the squamous part of the temporal bone in association with a skull fracture. Less commonly bleeding may be venous in origin (see Chapter 15, p. 183). In children, skull fracture is less common radiologically, although at subsequent operative treatment one may be found. It is thought that this relates to the flexibility of the child's skull, causing deformation rather than a fracture, with consequent evulsion of the dura from its under-surface and tearing of the blood vessel (Mealey 1960, Mazza *et al.* 1982).

A significant minority of cases of EDH (between 20 and 30 per cent) occur elsewhere, including the frontal region and the posterior fossa. Children have a higher incidence of EDH in these locations (Koc *et al.* 1998). With increased use of neuroradiology it is now recognized that a 'pure' extradural haematoma is less common than previously supposed. Imaging studies have shown that around 50 per cent of cases of EDH are associated with SDH and/or

contusional brain injury (Phonprasert *et al.* 1980, Servedai *et al.* 1995a). The morbidity/mortality is affected not only by the presence of other intracranial pathology – in particular contusional injury and/or traumatic axonal injury – but also by delay in diagnosis/operative intervention (Servedai *et al.* 1995b; see later).

Circumstances and clinical features

Extradural haematoma occurs as a result of focal impact to the head. The most frequent causes are falls, road traffic deaths (in which there is a higher incidence of other brain injury) and increasingly assaults. The classic clinical presentation is a history of head injury followed immediately by a period of concussion (this may be brief). This is followed by recovery – a lucid interval. It is during this period that blood accumulates within the cranial cavity to give rise to the features of a space-occupying lesion, with raised intracranial pressure and eventual lapse into coma. The time course for the lucid interval is variable, ranging from 30 minutes to several hours and may, very rarely, be of several days duration (see also Chapter 15). The presence of any other brain injury will also modify the clinical picture.

Death occurs in EDH primarily as a result of increasing pressure within the cranial cavity and brain-stem compression. It is, however, becoming increasingly recognized that the morbidity and mortality are affected not only by the presence of other intracranial pathology (see above) but also by delay in diagnosis and surgical intervention (Bricolo and Pasut 1984). Early scanning immediately after head injury may fail to pick up the development of an EDH (Mandavia and Vilagomez 2001).

External injury

This can be variable depending on the nature of the impact. The hair may both hide and alter the external injury. Abrasions are classically seen with ground contact such as in a fall. Lacerating injury may be seen where injury has occurred with an implement such as a stick, hammer or other weapon. Unless there are specific characteristics relating to the implement in question, it is usually impossible to be dogmatic in linking a particular weapon to the injury from the wound appearances alone. It may also be impossible to differentiate ground contact and a single weapon impact if the wound is a simple bruise or laceration.

Pathological appearances

As this results from a focal impact, the external injury should correlate well with bruising of the scalp tissues and any fracture beneath.

Examination of the skull and brain most commonly shows a fracture in the squamous part of the temporal bone. The skull is thin in this area and fracturing occurs relatively easily. In around 15 per cent per cent of cases no associated fracture is present. Fractures are less commonly seen in other locations. The usual site for EDH is the temporoparietal area. The dura is stripped from the inner aspect of the skull and becomes separated by the haematoma. As the haematoma enlarges it forms an ovoid 'bun'-shaped mass; this progressively indents the adjacent brain, leaving, on removal, a concave depression (Figure 5.1). The volume of the clot varies: in fatal cases it

Figure 5.1 Acute extradural haematoma as a result of a fall from a bicycle.

is usually at least 75–100 ml but can be greater (Leestma 1988). Measurement, particularly in unsuspected cases, may be difficult.

There is usually herniation across the midline, uncal herniation and tonsillar herniation. In addition, sectioning of the brain may reveal compression of the posterior cerebral artery with haemorrhagic infarction of the posteroinferior temporal and medial occipital lobes. Secondary brain-stem haemorrhages are commonly seen. If assisted ventilation has taken place, the changes of 'non-perfused' brain may be superadded.

Areas of contusional injury may be seen either in association with a fracture site or beneath an area of impact. In cases where there is additional brain injury SDH with other areas of contusional/lacerating injury may be identified.

Histology

Histological examination of the dura and haematoma is rarely of assistance in cases of EDHs. However, in those cases with a more chronic time course histological assessment may show organizing features relating to the haematoma, e.g. inflammatory cell infiltrate and haemosiderin deposition, as well as granulation tissue with features of organization (Tatagiba *et al.* 1989, Viljoen and Wessels 1990).

Problem areas

MISSED EDH

Although rare, this still occurs. It may be that the individual gives an unclear history in terms of an injury and subsequent loss of consciousness. A skull fracture may not be identified either because radiological examination of the patient was not carried out or the radiograph was misinterpreted. Alcohol, as well as other drugs, may lead to diagnostic problems. Particularly difficult is the EDH in children where fractures are less common and the history may be less clear. In the criminal setting the defence may arise that, had diagnosis and treatment been instituted, the individual need not have died. Expert advice from a neurosurgeon over these issues may well be necessary. However, unless a *novus actus intervenienes* (new intervening act) has been shown to take place, this is unlikely to win the argument although it may play a role in terms of jury 'sympathy'.

TIMING

Again within the medicolegal context timing of injury may be crucial in pinpointing the episode, particularly if only a vague history is available or indeed there is evidence of multiple episodes. Unfortunately, from the pathological appearances alone, one is unlikely to be able to give more than a rough timeframe in terms of hours. Again it may well be that in these cases more detailed clinicopathological assessment is necessary with a neurosurgical opinion rather than the pathologist straying outside his or her area of expertise.

Rarities

A SLOW EVOLUTION/RESOLVING EDH

It is recognized that rarely bleeding may take place very slowly over 10–14 days, with spontaneous resolution over several weeks (Graham *et al.* 2002, and see Chapter 14). Again this clearly has implications in the question of timing of injury.

SPONTANEOUS EXTRADURAL HAEMATOMA

Rarely, EDH may occur spontaneously (Henderson *et al.* 2001) or as a result of an underlying vascular malformation (Benzel 1995, Mustafa and Subramanian 1996, Matsumoto *et al.* 2001). Other rarities include association with craniofacial infection (Griffiths *et al.* 2002), after temporomandibular joint arthrocentesis (Carroll *et al.* 2000) and in association with Langerhans' giant cell histiocytosis (Lee *et al.* 2000). In addition, anticoagulant therapy is known to be associated with development of an EDH including within the spinal canal (Rodriguez *et al.* 1995).

HEAT HAEMATOMA

This is well recognized in the forensic setting in association with fire-related deaths. The head is exposed to intense heat and fracturing of the skull and heat haematomas occur in the extradural space. The mechanism is unknown: the distribution closely follows the pattern of charring of the skull, and blood may possibly arise from the skull itself or the venous sinuses. These are more diffusely spread over the hemispheres, commonly bilateral with a somewhat spongy or honeycomb appearance as a result of the blood boiling (Figure 5.2). Carbon monoxide estimation in difficult cases can be done on the clot to estimate the level and correlate with evidence of the deceased being alive at the time of the fire (Saukko and Knight 2004). If the haematoma occurred during the fire, carbon monoxide correlating with the blood level should be present in the haematoma. However, in grossly damaged bodies, difficulty may still arise in identifying ante-mortem injury from postmortem injury, especially in the presence of heat fractures.

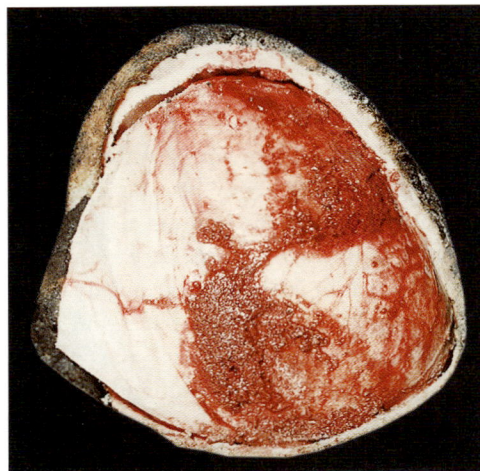

Figure 5.2 Heat haematoma; note more diffuse spread over hemispheres and granular appearance in contrast to the acute extradural haematoma shown in Figure 5.1 and Figure 2.5 (Chapter 2).

SUBDURAL HAEMATOMA

Bleeding occurs between the dura and the surface of the brain. Classification in the past has been into acute, subacute and chronic SDH, but there has been variation in this and unsatisfactory histological attempts at ageing. A more clinical approach has been for classification based on the appearances of the haematoma with acute comprising clotted blood and subacute a mixture of clotted blood and fluid, and chronic being when the haematoma is fluid. Increased use of scanning has led to better definition of the components from the radiological perspective.

Acute subdural haematoma

CIRCUMSTANCES, MECHANISMS AND CLINICAL FEATURES

Subdural haematoma occurs as a result of a different mechanism of injury from that in EDH, which is a focal impact injury. SDH is caused by tearing of the bridging veins between the cortical surface and the dural sinuses. SDH is also seen as a component of a burst lobe where there is associated severe contusional injury (see Chapter 7). Arterial bleeding from superficial cortical arteries may also give rise to acute SDH. Postmortem analysis has indicated that those caused by arterial rupture are more usually temporoparietal, whereas bridging vein rupture is typically in the frontoparietal and parasagittal region (Maxeiner and Wolf 2002). Most acute SDHs occur as a result of a fall or assault (72 per cent), with a lower incidence occurring in vehicular incidents (24 per cent). Experimental studies in non-human primates have shown that acute SDH occurs with injury where there are high rates of angular acceleration or deceleration, as in a fall or assaults, which lead to rupture of the bridging veins. In motor vehicle collisions the deceleration rate is much slower (Gennarelli and Thibault 1982). It has also been suggested that acute SDH is more easily produced with occipital impact. It has been shown experimentally that the largest relative skull–brain motion and strain in the bridging veins occurs when there are anterior–posterior and posterior–anterior angular forces as opposed to lateral, superior–inferior or inferior-posterior forces (Kleiven 2003). SDH can also arise as a result of whiplash injury including the consequences of blast injury (Ommaya and Yarnell 1969, Swalwell 1993, Davis and Robertson 1997). This mechanism has been supported by experimental work (Gennarelli 1993) and has also been reported as a result of repeated episodes of violent shaking during torture (Pounder 1997).

The location of the SDH without consideration of other factors should not be taken as indicating the site of impact. Assessment of other findings, including external wounds, scalp bruising and skull fractures, is necessary. Contusional brain injury and cerebral swelling are commonly associated features with traumatic axonal injury including diffuse axonal injury at the most severe end. The associated parenchymal damage is a major factor in the clinical presentation and prognosis (Wilberger et al. 1991).

Subdural haematomas are 'pure' with no associated brain injury in 13 per cent (Graham et al. 2002). In these cases the clinical presentation is that of an expanding space occupying lesion. There is a more variable time course than that of EDH and clinicopathological correlation with an episode of injury may be difficult.

A number of underlying conditions may predispose to acute SDH. These include underlying brain atrophy, including dementias and alcoholic brain atrophy, as well as underlying coagulopathies including anticoagulant therapy injury. Other conditions may be associated with spontaneous SDH, including arteriovenous malformations, neoplastic lesions and following rupture of an intracranial aneurysm (O'Sullivan et al. 1994, Bromberg et al. 1998, Han et al. 1999).

Ethanol may be sequestered in an SDH and its estimation in a haematoma may aid in interpretation of any intoxicating state (Riggs et al. 1998).

MACROSCOPIC PATHOLOGY

This is shown as fresh blood diffusely spread over one or both hemispheres (Figure 5.3). In approximately 20 per cent of adult cases the SDH is bilateral (Whitwell 2001).

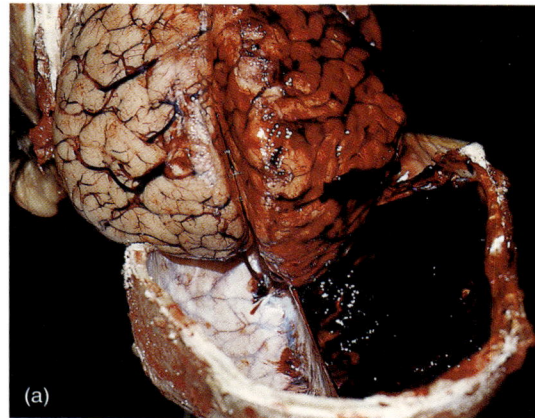

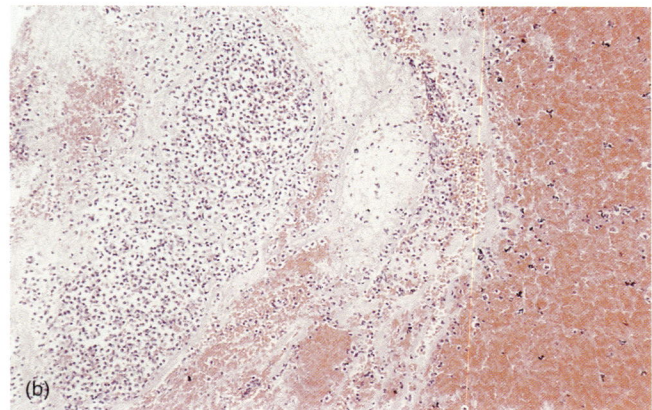

Figure 5.3 (a) Acute subdural haematoma with associated cerebral swelling. (b) Acute subdural haematoma – histology showing collections of red cells, fibrin and leucocytes.

The pure acute SDH causes a mass lesion and death may occur rapidly (DiMaio and DiMaio 2001). Assessment of volume should be done although this can be difficult. Clots of 50 ml or more are likely to cause symptoms and of over 100 ml death (Davis and Robertson 1997). Interhemispheric subdural haemorrhage in adults is suggestive of an underlying coagulopathy (haemophilia, anticoagulation therapy) (Bartels et al. 1995). Examination for intracranial aneurysm should be done before fixation because this makes dissection of the circle of Willis much more difficult. Assessment of cerebral atrophy in the presence of oedema may be difficult in the fresh state, but the brain should be weighed and subsequently reweighed after fixation.

Postmortem studies have been performed in an attempt to identify bridging vein damage – these have demonstrated venous contrast leakage in the subdural space (Maxeiner 1997). The technique, however, requires radiological input and is time-consuming, so it is not currently a readily available postmortem investigation.

Other findings include contusional brain injury, macroscopic markers of traumatic axonal injury and subarachnoid haemorrhage. For histology of acute SDH, see later.

PROBLEM AREAS

Previous injury/disease

Previous head injury/assault and underlying disease that may have significance in the medicolegal situation include alcoholic atrophy, previous contusional injury and epilepsy. Detailed assessment of previous injury history as well as history of seizures may be important in terms of causation. Rare underlying conditions may be identified including vascular malformations and neoplasia (see above). In an individual case it may be difficult to arrive at a firm conclusion on the relative roles of the traumatic and underlying natural or other pathology in the brain.

Mechanism of injury

In the assault situation it may be argued that a fall was the cause of the SDH or other brain pathology rather than directly as a result of a blow. Detailed clinicopathological analysis, as well as details of the alleged assault, are essential. Despite this it may be impossible to differentiate between the various components of the incident.

Infant SDH is covered in Chapter 12, p. 140.

Chronic subdural haematoma

Pathologically, chronic subdural haematoma (CSDH) is characterized by the formation of organizing membranes over the surfaces of the brain, with the potential to increase enlargement of the haematoma over time; histologically these membranes, together with the vascular channels, are highly typical. The capillaries are very thin walled and prone to spontaneous rupture without trauma giving repeated episodes of fresh bleeding. If small, chronic subdural haematoma may be an incidental finding at postmortem examination. With increasing volume, signs and symptoms such as confusion with features similar to a dementia, stroke-like features, as well as the effects of a space-occupying lesion, may occur.

Chronic SDH is increasingly recognized as a common and important treatable neurosurgical condition. It has a higher incidence in the elderly population (Fogelholm and Waltimo 1975, Kudo et al. 1992, Asghar et al. 2002). There appears to be racial variation – a higher incidence is reported in Japanese people. Epidemiological figures have shown a higher incidence in males – probably related to increased trauma in this group.

A significant proportion of CSDHs is seen in association with trauma, although this may be of a minor nature and is reported in around two-thirds of patients (Sambasivan 1997, Missori et al. 2000, Asghar et al. 2002). It is likely that trauma is relatively trivial because in the atrophic brain the bridging veins leaving the cortical surface to enter the venous sinuses are longer and weaker and more vulnerable to tearing. Other predisposing causes include anticoagulation therapy – most studies, however, fail to differentiate acute, subacute and chronic SDH. The risk of SDH increases proportionally to the level of anticoagulation as measured by the prothrombin time ratio (Hylek and Singer 1994).

Intracranial hypotension as a result of shunting or cerebrospinal fluid (CSF) leakage, either iatrogenic or spontaneous, is also recognized to cause SDH (Giamundo et al. 1985, Chen and Levy 2000). In the former increased traction on bridging veins, as well as the large negative intracranial pressures generated when there is large distance between the ventricular catheter and shunt terminus, have been put forward as a contributing factor (Chen and Levy 2000).

Other pre-existing conditions include epilepsy (Fogelholm and Waltimo 1975) and alcohol abuse (Fogelholm and Waltimo 1975, Stroobandt et al. 1995, Sambasivan 1997). Factors such as cortical atrophy and coagulation disorders, as well as alcohol, as risk factors for trauma may all play a role (Soderstrom et al. 1997).

Vascular abnormalities as well as primary and secondary tumours have also been reported in association with CSDH (Popovic et al. 1994, Cinalli et al. 1997, Alimehmeti and Locatelli 2002).

MACROSCOPIC APPEARANCES

This varies with age. The inner arachnoidal membrane is initially thin – it is visible at around 7–10 days. The colour of the haematoma may be uniform or show areas of recent haemorrhage. The membranes are well developed at around 3 weeks or so.

Later the membranes become thicker ('rubber-like') and the contents solid with fluid areas of varying colours (Figure 5.4). The liquefied clot resembles motor oil after several weeks (Figure 5.5).

Other features present include indentation of the brain with or without midline shift. This is often severe in well-developed cases. There may be discoloration of the underlying arachnoid. Associated conditions include atrophy, related to age, dementia or alcohol, as well as other pathologies such as old cerebral infarction or previous head injury.

Bruising of the scalp in older cases may no longer be visible, particularly if there is no definite or a relatively minor traumatic event. However, subscalp bruising can persist for a considerable time, particularly in elderly people

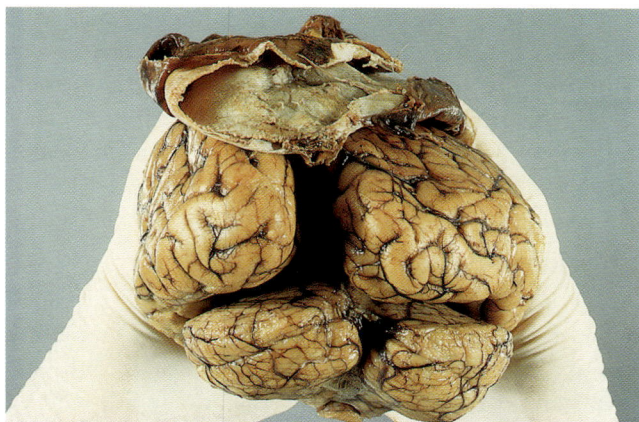

Figure 5.4   Chronic bilateral subdural haematoma.

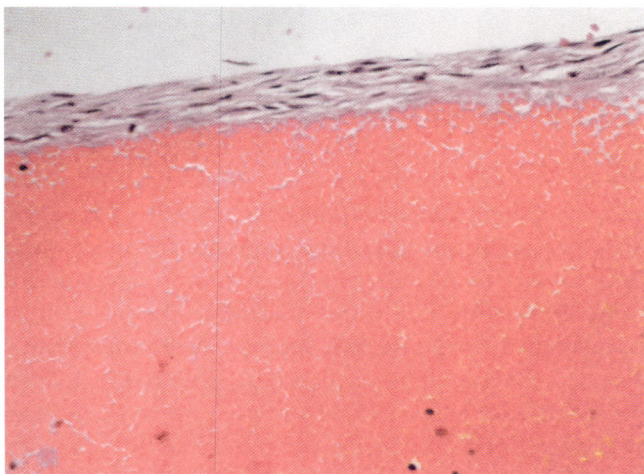

Figure 5.5   Early neomembrane with thin layer of fibroblasts.

(Perper and Wecht 1980) and may lead to haematoma formation within the subscalp tissues. Histology of any bruise should also be undertaken.

HISTOLOGY OF SDH

The literature on ageing SDH is limited. A guide published by Munro in 1936 is still useful, as is the summary outline in *Knight's Forensic Pathology* (Saukko and Knight 2004). A summary of the major features is shown in Table 5.1.

Multiple samples of dura should be taken. In acute SDH the haematoma comprises fresh clot with fresh red cells and fibrin to varying degrees. Within 24–48 h neutrophil polymorphs infiltrate, followed by macrophages within a few days. Recent work has indicated that immunocytochemistry for MHC Class 2 and monocytes/macrophage antigens may be of assistance (Al-Sarraj *et al.* 2004). Erythrocyte lysis may be variable and intact red cells can persist for several days or more. Fibroblasts enter the clot after a few days.

Early CSDHs possess a recognizable neomembrane. This is delicate, often only the thickness of a few cells (Figure 5.5). By 4 weeks the inner and outer membranes are well formed and become gradually hyalinised over 1–3 months (Figure 5.6). Giant capillaries with secondary re-bleeding may be seen (Figure 5.7).

Late appearances are of thickened dura caused by membranes fusing into a fibrous layer. This may calcify or ossify.

These appearances can be taken only as a general guide. It is usually possible to comment that the findings could have developed as a result of a particular episode. There is variation in evolution from individual to individual. Data

Table 5.1   Ageing of subdural haematoma

Up to 24–48 hours	Fresh red blood cells, fibrin.
2–5 days	Some red cell lysis with macrophages. Fibroblasts start to form membrane on the dural aspect.
7 days	Fibroblasts invade the clot. Dural layer of fibroblasts visible macroscopically – approximately 12 cells in thickness. Few cells on the arachnoidal side.
14 days	Progressive red cell lysis with the appearance of giant capillaries. Dural fibroblastic layer approximates half the thickness of the dura. Fibroblastic membrane on the arachnoidal side.
28 days	Clot becomes liquefied with areas of fresh haemorrhage. Fibroblastic layer adjacent to the dura is of similar thickness to the dura with a well-formed arachnoidal layer.
1–3 months	Giant capillaries with secondary haemorrhages. Hyalinisation of the membranes.
Later	Ageing becomes difficult. The membranes become thick and calcification may occur.

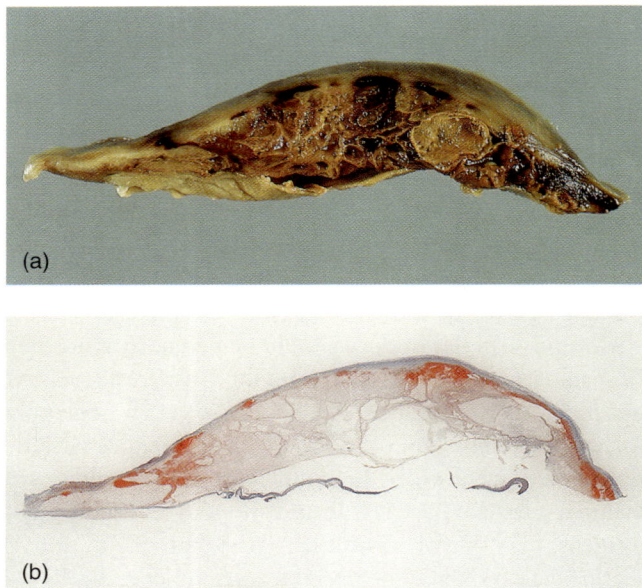

Figure 5.6 Chronic subdural haematoma with neo-membranes and areas of multiloculated haematoma and some bleeding (3 months). (a) Macroscopic appearance. (b) Low power of elastic van Geison stain.

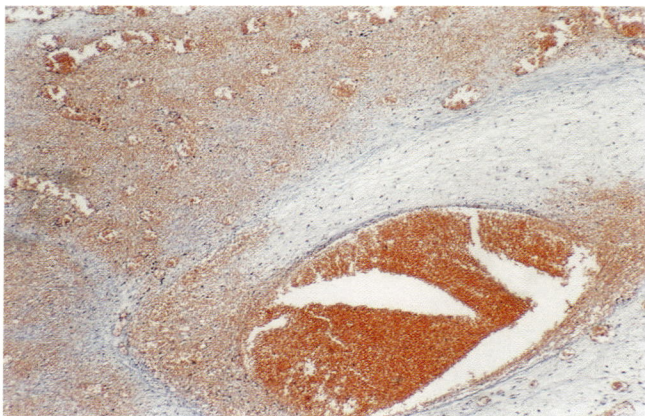

Figure 5.7 Giant capillaries with secondary bleeding.

are lacking in the infant group, although it has been suggested that ageing may occur more rapidly (Leestma 1998).

CHRONIC SDH IN THE INFANT

The most common cause is trauma, either non-accidental or accidental. Distinction between the two may be difficult. Symptoms and signs may be minimal or there may be an acute presentation with apnoea or seizures (Swift and McBride 2000). Underlying conditions include coagulopathies and structural brain abnormalities. Rupture of an arachnoid cyst after minor trauma has been reported

to give rise not only to acute SDH but also to subdural hygroma as well as bleeding into the cyst with varying symptoms and signs (Parsch *et al*. 1997, De *et al*. 2002 Gelabert-González *et al*. 2002). It is important to exclude this as an underlying cause/contributory factor in the context of suspected non-accidental injury.

SUBDURAL HYGROMA

This is a collection of clear, xanthochromic or blood-stained fluid within the subdural space. The mechanism of injury is understood to be leakage of CSF into the subdural space, caused by a tear in the arachnoid, which is valve like in nature (Graham *et al*. 2002). Membranes do not usually form and subdural veins are not present, in contrast to cortical atrophy where they cross a widened subdural space (Kostanian *et al*. 2000).

PROBLEM AREAS IN CSDH

Clinically relating trauma to the development of this type of haematoma may be difficult. It is essential to have a history of the event in question. A classic time course from inception to presentation is 3–4 months in the adult, with gradual expansion of the lesion. In the medicolegal setting this is commonly an assault, such as a mugging in an elderly victim with direct trauma to the head, or an assault that results in a fall, such as when a handbag is snatched. Photographic evidence as well as other documentation may be available from the time of the assault. Evidence of pre- and post-event clinical state is also necessary. This may well demonstrate a decline in cerebral function after the episode. Defence counsel may argue for a different episode being responsible for the CSDH, although in practice a well-documented episode would need to be identified. Pathological examination at most can only be 'consistent with' any episode.

The diagnosis of CSDH may be difficult, particularly in elderly people. If not diagnosed, this may well raise the issue of diagnostic/therapeutic delay in a potentially treatable condition, although in law this is unlikely as such to be of relevance.

REFERENCES

Al-Sarraj, S., Mohamed, S., Kibble, M. *et al*. 2004. Subdural hematoma (SDH): assessment of macrophage reactivity within the dura mater and underlying hematoma. *Clin Neuropathol*, **23**, 62–75.

Alimehmeti, R. and Locatelli, M. 2002. Epidural B cell non-Hodgkin's lymphoma associated with chronic subdural hematoma. *Surg Neurol*, **57**, 179–82.

Asghar, M., Adhiyaman, V., Greenway, M.W. *et al.* 2002. Chronic subdural haematoma in the elderly – a North Wales experience. *J R Soc Med*, 95(6), 290–2.

Bartels, R.H.M.A., Verhagen, W.I.M., Prick, M.J.J. *et al.* 1995. Interhemispheric subdural hematoma in adults; Case reports and a review of the literature. *Neurosurgery*, 36, 1210–14.

Benzel, E.C. 1995. Cervical epidural haematoma secondary to an extradural vascular malformation. *Neurosurgery*, 36, 585–8.

Bricolo, A.P. and Pasut, M.D. 1984. Extradural hematoma: Toward zero mortality. A prospective study. *Neurosurgery*, 14, 8–12.

Bromberg, J.E., Vandertop, W.P. and Jansen, G.H. 1998. Recurrent subdural haematoma as the primary and sole manifestation of chronic lymphocytic leukaemia. *Br J Neurosurg*, 12, 373–6.

Carroll, T.A., Smith, K. and Jakubowski, J. 2000. Extradural haematoma following temporomandibular joint arthrocentesis and lavage. *Br J Neurosurg*, 14, 152–4.

Chen, J.T. and Levy, M.L. 2000. Causes, epidemiology, and risk factors of chronic subdural hematoma. *Neurosurg Clin N Am*, 11, 399–406.

Cinalli, G., Zerah, M. and Carteret, M. 1997. Subdural sarcoma associated with chronic subdural haematoma. Report of two cases and review of the literature. *J Neurosurg*, 86, 553–7.

Davis, R.H. and Robertson, D.M. (eds) 1997. *Textbook of Neuropathology*, 3rd edn. Baltimore: Williams & Wilkins.

De, K., Berry, S. and Denniston, S. 2002. Haemorrhage into an arachnoid cyst: a serious complication of minor head trauma. *Emerg Med J*, 19, 365–6.

DiMaio, V.J. and DiMaio, D. 2001. *Forensic Pathology*, 2nd edn. Boca Raton, FL: CRC Press.

Fogelholm, R. and Waltimo, O. 1975. Epidemiology of chronic subdural haematomas. *Acta Neurochir (Wien)*, 32, 247–50.

Gelabert-González, M., Fernández-Villa, Cutrín-Prieto, J. *et al.* 2002. Arachnoid cyst rupture with subdural hygroma: report of three cases and literature review. *Child Nerv Syst*, 18, 609–13.

Gennarelli, T.A. 1993. Cerebral concussion and diffuse brain injuries. In Cooper, P.R. (ed.), *Head Injury*, 3rd edn. Baltimore: Williams & Wilkins, 137.

Gennarelli, T.A. and Thibault, L.E. 1982. Biomechanics of acute subdural haematoma. *J Trauma*, 22, 680–6.

Giamundo, A., Benvenuti, D. and Lavano, A. 1985. Chronic subdural haematoma after spinal anaesthesia. Case report. *J Neurosurg Sci*, 29, 153–5.

Graham, D.I., Gennarelli, T.A. and McIntosh, T.K. 2002. Trauma. In Graham, D.I. and Lantos, P.L. (eds), *Greenfield's Neuropathology*, 7th edn. London: Arnold, Chapter 14.

Griffiths, S.J., Jatavallabhula, N.S. and Mitchell, R.D. 2002. Spontaneous extradural haematoma associated with craniofacial infections: case report and review of literature. *Br J Neurosurg*, 16, 188–91.

Han, P.P., Theodore, N., Porter, R.W. *et al.* 1999. Subdural haematomas from a type 1 spinal arteriovenous malformation. Case report. *J Neurosurg*, 90, 255–7.

Henderson, R.D., Pittock, S., Piepgras, D.G. and Wijdicks, E.F. 2001. Acute spontaneous spinal epidural hematoma. *Arch Neurol*, 58, 1145–6.

Hylek, E.M. and Singer, D.E. 1994. Risk factors for intracranial hemorrhage in outpatients taking warfarin. *Ann Intern Med*, 120, 897–902.

Kleiven, S. 2003. Influence of impact direction on the human head in prediction of subdural hematoma. *J Neurotrauma*, 20, 365–79.

Koc, R.K., Pasaoglu, A., Menku, A. *et al.* 1998. Extradural hematoma of the posterior cranial fossa. *Neurosurg Rev*, 21, 52–7.

Kostanian, V., Choi, J.C., Liker, M.A. *et al.* 2000. Computed tomographic characteristics of chronic subdural hematomas. *Neurosurg Clin N Am*, 11, 479–89.

Kudo, H., Kuwamura, K. and Izawa, I. 1992. Chronic subdural hematoma in elderly people: Present status on Awaji Island and epidemiological prospect. *Neurol Med Chir (Tokyo)*, 32, 207–9.

Lee, K.W., McLeary, M.S., Zuppan, C.W. and Won, D.J. 2000. Langerhans' cell histiocytosis presenting with an intracranial epidural hematoma. *Pediatr Radiol*, 30, 326–8.

Leestma, J.E. 1988. *Forensic Neuropathology*. New York: Raven Press.

Mandavia, D.P. and Villagomez, J. 2001. The importance of serial neurologic examination and repeat cranial tomography in acute evolving epidural hematoma. *Pediatr Emerg Care*, 17, 193–5.

Matsumoto, K., Akagi, K., Abekura, M. and Tasaki, O. 2001. Vertex epidural hematoma associated with traumatic arteriovenous fistula of the middle meningeal artery: a case report. *Surg Neurol*, 55, 302–4.

Maxeiner, H. 1997. Detection of ruptured cerebral bridging veins at postmortem examination. *Forensic Sci Int*, 89, 103–10.

Maxeiner, H. and Wolf, M. 2002. Pure subdural hematomas: a postmortem analysis of their form and bleeding points. *Neurosurgery*, 50, 503–8 (discussion 508–9).

Mazza, C., Pasqualin, A., Ferriotti, G. *et al.* 1982. Traumatic extradural haematoma in children. *Acta Neurochirurg*, 65, 67–80.

Mealey, J. 1960. Acute extradural haematomas without demonstrable skull fractures. *J Neurosurg*, 17, 27–34.

Missori, P., Maraglino, C., Tarantino, R. *et al.* 2000. Chronic subdural haematomas in patients aged under 50. *Clin Neurol Neurosurg*, 102, 199–202.

Munro, D. and Merritt, H.H. 1936a. Surgical pathology of subdural haematoma. Based on a study of one hundred and five cases. *Arch Neurol Psychiatry*, 35, 64–78.

Mustafa, M. and Subramanian, N. 1996. Spontaneous extra-dural haematoma causing spinal cord compression. *Int Orthopaedics*, 20, 393–4.

Ommaya, A.K. and Yarnell, P. 1969. Subdural haematoma after whiplash injury. *Lancet*, ii, 237–9.

O'Sullivan, M.G., Whyman, M., Steers, J.W. *et al.* 1994. Acute subdural haematoma secondary to ruptured intracranial aneurysm: diagnosis and management. *Br J Neurosurg*, 8, 439–45.

Parsch, C.S., Krass, J., Hoffmann, E. *et al.* 1997. Arachnoid cysts associated with subdural hematomas and hygromas: Analysis of 16 cases, long-term follow-up, and review of literature. *Neurosurgery*, 40, 482–90.

Perper, J.A. and Wecht, C.H. (eds) 1980. *Microscopic Diagnosis in Forensic Pathology*. Springfield, IL: Thomas, 12.

Phonprasert, C., Suwanwela, C., Hongsaprabhas, C. *et al.* 1980. Extradural haematoma: analysis of 138 cases. *J Trauma*, 20, 679–83.

Popovic, E.A., Lyons, M.K. and Scheithauer, B.W. 1994. Mast cell-rich convexity meningioma presenting as a chronic subdural hematoma: Case report and review of the literature. *Surg Neurol*, 42, 8–13.

Pounder, D.J. 1997. Shaken adult syndrome. *Am J Forensic Med Pathol*, 18, 321–4.

Riggs, J.E., Schochet, S.S. Jr and Frost, J.L. 1998. Ethanol level differentiation between post mortem and sub dural haematoma. *Milit Med*, 163, 722–4.

Rodriguez, Y., Baena, R., Gaetani, P. *et al.* 1995. Spinal epidural haematoma during anticoagulant therapy. A case report and review of the literature. *J Neurosurg Sci*, 39, 87–94.

Sambasivan, M. 1997. An overview of chronic subdural hematoma: Experience with 2300 cases. *Surg Neurol*, 47, 418–22.

Saukko, P. and Knight, B. 2004. *Knight's Forensic Pathology*, 3rd edn. London: Arnold.

Servadei, F., Nanni, A., Nasi, M.T. *et al.* 1995a. Evolving brain lesions in the first 12 hours after head injury: analysis of 37 comatose patients. *Neurosurgery* 37, 899–906.

Servadei, F., Vergoni, G., Staffa, G. *et al.* 1995b. Extradural haematomas: How many deaths can be avoided? *Acta Neurochir (Wien)*, 133, 50–5.

Soderstrom, C.A., Smith, G.S., Dischinger, P.C. *et al.* 1997. Psychoactive substance use disorders among seriously injured trauma center patients. *JAMA*, 277, 1769–74.

Stroobantd, G., Fransen, P., Thauvoy, C. and Menard, E. 1995. Pathogenic factors in chronic subdural haematoma and causes of recurrence after drainage. *Acta Neurochir (Wien)*, 137, 6–14.

Swalwell, C. 1993. Letter to editor. Head injuries from short distance falls. *Am J Forensic Med Pathol*, 14, 171.

Swift, D.M. and McBride, L. 2000. Chronic subdural hematoma in children. *Neurosurg Clin N Am*, 11, 439–46.

Tatagiba, M., Sepehrnia, A., El Ax, M. *et al.* 1989. Chronic epidural haematoma – report of eight cases and review of the literature. *Surg Neurol*, 32, 453–8.

Viljoen, J.J. and Wessels, L.S. 1990. Subacute and chronic extradural haematomas. *S Afr J Surg*, 28, 133–7.

Whitwell, H.L. 2001. Head injury. In Rutty, G.N. (ed.), *Essentials of Postmortem examination Practice*, Vol. 1. London: Springer-Verlag.

Wilberger, J.E., Harris, M. and Diamond, D.L. 1991. Acute subdural hematoma: Morbidity, mortality and operative timing. *J Neurosurg*, 74, 212–18.

6

Subarachnoid haemorrhage and cerebro–vascular pathology

Helen L Whitwell and Christopher M Milroy

This chapter covers subarachnoid haemorrhage, both that seen in association with trauma and that resulting from natural disease. Trauma can produce vascular lesions not associated with subarachnoid haemorrhage and this is also covered. Subarachnoid haemorrhage is defined as bleeding within the subarachnoid space, which normally contains cerebrospinal fluid (CSF).

TRAUMATIC SUBARACHNOID HAEMORRHAGE

Subarachnoid haemorrhage in association with traumatic brain injury

Subarachnoid haemorrhage is a frequent finding in many head injuries in conjunction with other pathology including extradural haematoma (EDH), subdural haematoma (SDH) and contusional brain injury. It is typically patchy subarachnoid haemorrhage over the surface of the brain (Figure 6.1). Most commonly this is seen over the hemispheres and cerebellum, and it arises as a result of shearing injury to small blood vessels. There may be single or multiple areas of haemorrhage, which may be more diffuse over the hemispheres in some cases. There is an association between subarachnoid haemorrhage and a poor outcome, which is likely to be related to the initial traumatic damage rather than to the effects of vasospasm and secondary ischaemic brain damage (Servadei *et al.* 2002). The location of subarachnoid bleeding cannot be taken to indicate site of impact and other factors need to be taken into account such as scalp bruising, skull fractures and contusional injury. Aseptic meningitis caused by leakage of blood into the CSF is common which may be clinically apparent.

Histology in the early stage shows areas of haemorrhage in the subarachnoid space with red cells, which later lyse, and lymphocytes and macrophages. Haemosiderin deposition may develop within 36 h or so and last for months or years (Saukko and Knight 2004). There is variable acute and subsequently chronic inflammatory cell infiltrate with eventually residual blood pigments in the meninges and subarachnoid space; these may persist for years. Fibrosis as a result of organization of blood clot may develop, with a potential long-term complication of hydrocephalus (Weller 1995).

True traumatic subarachnoid haemorrhage

True traumatic subarachnoid haemorrhage is a well-recognized forensic entity. The term should be reserved for the specific entity of damage to the vertebrobasilar vasculature in association with trauma, most commonly blows, to the head and neck, giving rise to neck hyperextension and rotation. The recognition that such trauma may result in fatal subarachnoid haemorrhage was first suggested by Ford, and may have been described in the nineteenth century by Wilks, but it was formulated after the work of Simonsen and Contostavlos (Wilks 1859, Ford 1956, Simonsen 1963, 1967, 1984, Contostavlos 1971, Cameron and Mant 1972, Harland *et al.* 1983).

CIRCUMSTANCES

The classic situation is a sudden collapse arising in the context of an assault or a fraças as a result of direct blunt trauma. Postmortem examination reveals massive basal subarachnoid haemorrhage. In terms of injury, the most common site is to the side of the neck (Figure 6.2a). It

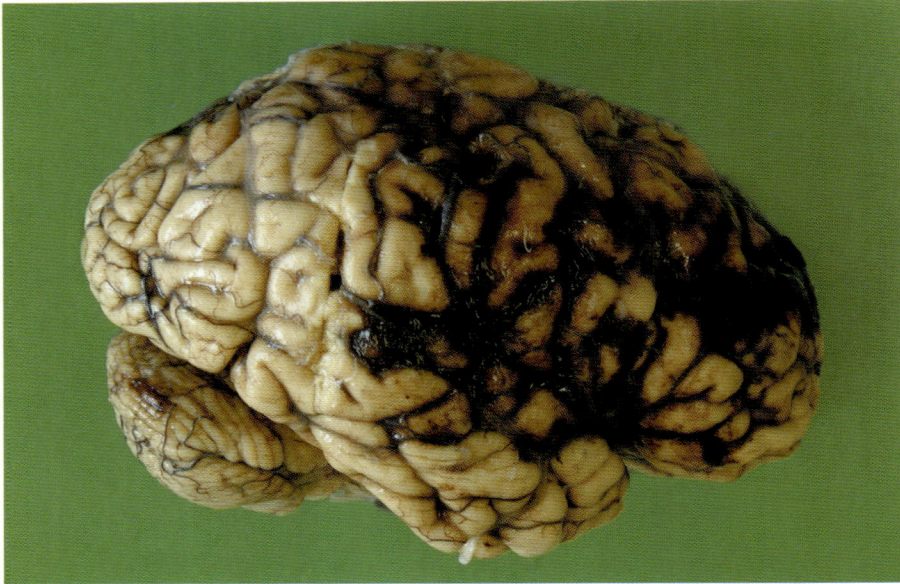

Figure 6.1　Patchy subarachnoid haemorrhage as seen in severe head injury.

should be recognized that about 50 per cent of cases show no external injury. Bruising to the deep muscles, in particular around the sternomastoid, is commonly seen. However, bruising may occur to any point of the head including the chin (Figure 6.2b) and also to the back of the head. Thus, in any particular case, it may be impossible to differentiate whether the haemorrhage was initiated by an action such as a kick or a subsequent fall, or indeed a fall unrelated to the assault.

The degree of trauma is not necessarily severe; this type of haemorrhage is not commonly associated with fractures of the skull unless there is a sustained assault (Contoslavlos 1971, 1995, Deck and Jagadha 1986).

MECHANISM OF INJURY

The early literature reports attributed bleeding to rupture in the extracranial portion of the cranial vessels, as the vertebral artery loops its way around C1 before entering the skull. This concept, however, is not backed up on careful review of the literature, in which in only about a third of the series was there a definite identified point (Leadbeatter 1994). This is coupled with the fact that dissection of the upper neck is notoriously difficult and artefactual tears are easily produced (Figure 6.3). Furthermore, with this concept, it was difficult to understand how blood tracks from the extracranial to the intracranial cavity, in many cases giving rise to massive intraventricular bleeding (Coast and Gee 1984). It is now becoming increasingly recognized that bleeding is seen in the intracranial vessels (Cooper *et al.* 1999, Gray *et al.* 1999) as well as at the point of penetration of the dura (Koszyca *et al.* 2003).

POSTMORTEM EXAMINATION

The demonstration of the site of rupture in traumatic subarachnoid haemorrhage can be one of the most difficult entities at postmortem examination. In some it may be impossible to identify the site of rupture. The frequency of demonstration of the rupture will, as with all postmortem work, depend on the diligence with which the search is carried out, and what methods of demonstration are used, but, even with the most detailed and systematic dissection, no vascular lesion may be identified.

Various techniques have been recommended to find the site of rupture. Bromilow and Burns (1985) developed a method of opening the cervical portion of the vertebral arteries *in situ*, which then allows the whole of the extracranial portion of the artery to be viewed and sampled histologically. This method is of less value in the examination of intracranial ruptures.

A further method described by Leadbeatter (1994) involves removing the brain stem and cerebellum with the base of the skull, including the foramen magnum, in continuity with the cervical vertebral column and the cervical cord with the undissected neck structures. After fixation and decalcification the vertebral arteries can be dissected out. As well as sampling areas of arterial injury, if necessary the whole specimen can be embedded for microscopy. This is time-consuming as well as labour intensive.

A method reported by McCarthy *et al.* (1999) that the authors have found useful, on having opened the skull in a case where a traumatic subarachnoid haemorrhage is suspected, is to ligate the distal basilar artery and then remove the cerebral hemispheres, leaving the brain stem

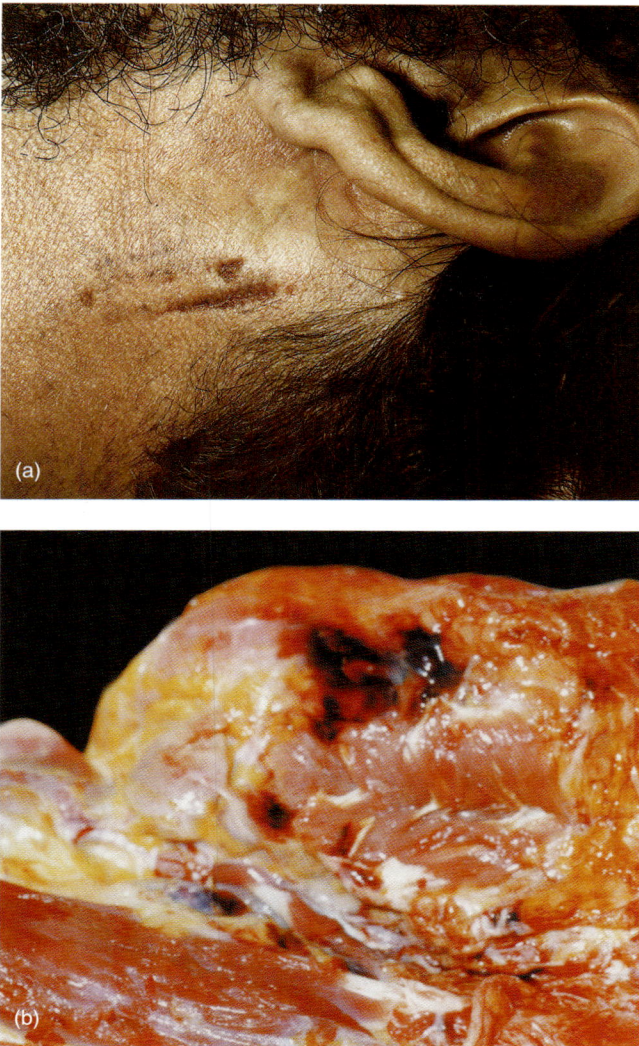

Figure 6.2 (a) Kick to side of neck in traumatic subarachnoid haemorrhage. (b) Internal jaw bruise as a result of a blow causing a traumatic subarachnoid haemorrhage.

Figure 6.3 Artefactual tear of vertebral artery arising during dissection.

and cerebellum *in situ*. The origin of each vertebral artery is identified, the thoracic organs having been left *in situ*. A cannula is passed into one of the vertebral arteries and water injected through into the artery while observing the vasculature at the base of the brain. If there is a rupture in the intracranial portion, water should emerge through the rupture in the artery. Videographic recording is also possible using this method. The site can be removed for histological examination.

Plain radiological examination of the craniocervical region has also been recommended as an adjunct. This has the advantage that congenital bony abnormalities may be identified (Gross 1990).

Postmortem angiography may be of value (Karhunen *et al*. 1990). However, negative angiography does not exclude the possibility of vascular injury. Radiological facilities may be unavailable.

MACROSOPIC NEUROPATHOLOGY

External examination shows predominantly basal subarachnoid haemorrhage (Figure 6.4a), which, depending on the site of rupture, may be seen extending into one or other vertebral canal. With survival adherent clot may be seen (Figure 6.4b). Occasional direct inspection may identify an area of tearing. Further examination before fixation should be undertaken for the presence of an aneurysm or other vascular abnormality. There are rare cases of rupture of an underlying aneurysm occurring with a direct time relationship to head injury (see later).

On section intraventricular haemorrhage is commonly seen. In addition, other features of head injury may be present in cases where there has been significant injury, e.g. contusions.

HISTOLOGY

A number of histological studies have been performed on the vertebral artery. Wilkinson (1972) described the

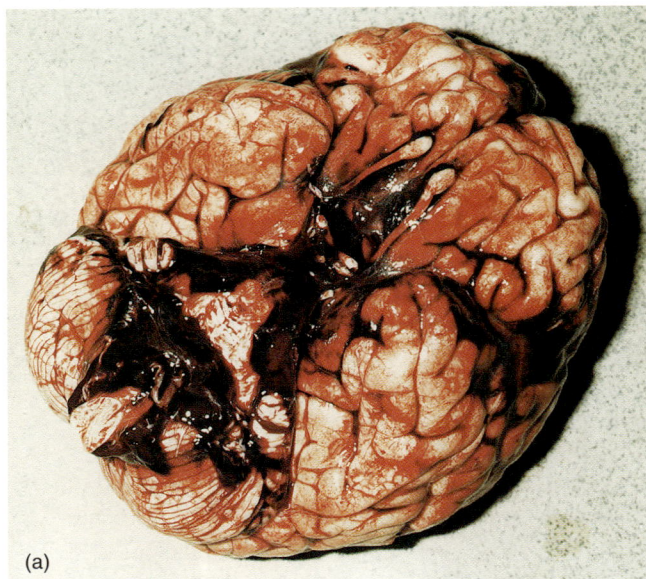

(a)

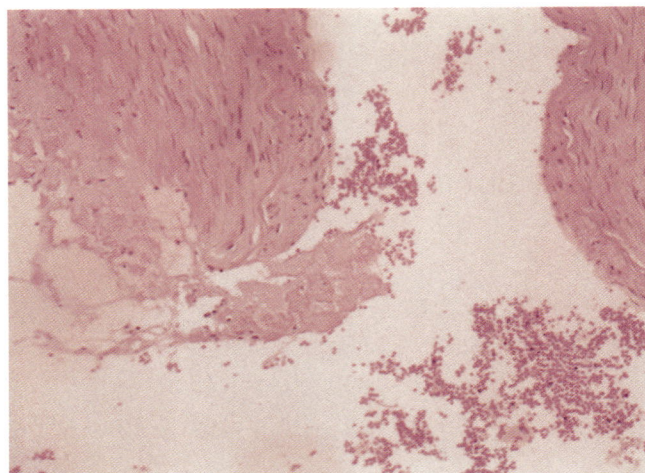

Figure 6.5 Acute tear of basilar artery in traumatic subarachnoid haemorrhage showing disruption of the vessel wall, fibrin and haemorrhage.

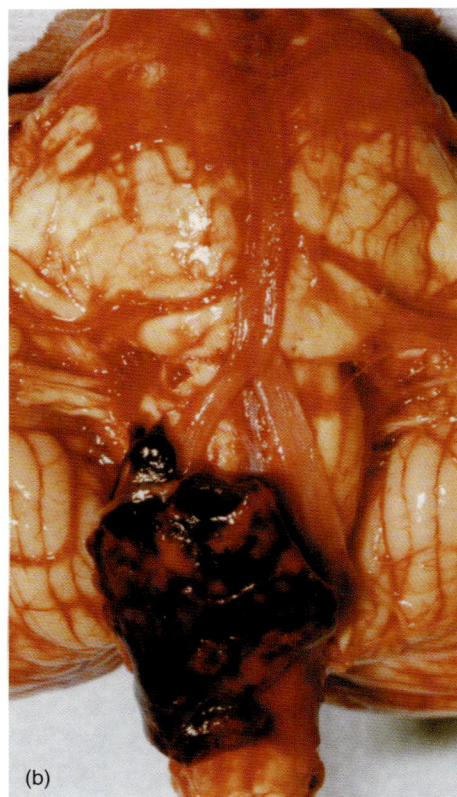

(b)

Figure 6.4 (a) Basal subarachnoid haemorrhage occurring in traumatic subarachnoid haemorrhage. (b) Haemorrhage over right vertebral artery which had ruptured in a case of prolonged survival.

vertebral artery in 20 individuals aged 60–75 years. The extracranial portion is described as having a well-developed adventitia and elastic lamina with a broad media of collagen, smooth muscle and elastic fibres.

About 1 cm before it penetrates the dura, the artery changes in structure, becoming thinner with loss of adventitial collagen and a fragmented or absent external elastic lamina. These findings are confirmed by Coast and Gee (1984). Further studies by Johnson *et al.* (2000, 2001) have demonstrated the development of collagenous scarring with increasing age as well as fragmentation of elastic fibres. Changes are often focal with sparing of the intracranial portions. Atheromatous degeneration also increased with age.

Most published cases of traumatic subarachnoid haemorrhage do not include histology. However, it should be mandatory because there may be medicolegal implications as to the role or otherwise of underlying natural disease. The vascular lesions seen are mainly tears (Figure 6.5), which may be incomplete and often associated with intramural dissection. Perivascular haemorrhage is frequently seen. Vessel wall abnormalities have been described in some cases including cystic medial necrosis (Dowling and Curry 1988, Opeskin and Burke 1998). Injury to the tunica media in fatal vertebral artery rupture has been described, suggesting distortional injury of the artery (Pollanen *et al.* 1996).

Neuropathological examination of the brain in typical cases, where the traumatic subarachnoid haemorrhage is the sole pathology, commonly show no other significant features. β-Amyloid precursor protein (APP) is usually negative in the acute setting. Microhaemorrhages are frequently seen in the deeper structures and there may be periventricular haemorrhage as well as choroid plexus haemorrhage.

MEDICOLEGAL ISSUES

Alcohol and traumatic subarachnoid haemorrhage

The association with alcohol is an interesting one. Various series show that almost all individuals who die of this condition have variable levels of blood alcohol (Simonsen 1963, Hillbom and Kaste 1981, Gray *et al.* 1999). However, it is also well recognized in individuals who may be unconscious from another cause, e.g. as a result of a head injury. The laxity of the neck muscles is thought to play a major role in that the neck is more easily jerked in an unconscious or intoxicated individual.

It is recognized that ethanol has direct effects on blood vessels, decreasing vasospasm after injury and causing vasodilatation (Barry and Scott 1979). In addition there is delayed muscle reaction and poor coordination. However, alcohol is a feature of many homicides and other violent deaths (Whittington 1980).

Degree of trauma

As mentioned earlier, the degree of trauma required to produce this lesion is said to be only of moderate force. Injury can be caused by a single blow or impact (Leadbeatter 1994, Gray *et al.* 1999). The authors have personal experience of relatively minor trauma to the face/neck producing the classic picture. It is also recognized to occur with sudden movement of the neck without the occurrence of impact (personal experience of the authors', Gee 1982). It may be difficult in an individual case to attribute any particular injury to the fatal haemorrhage.

Time of survival

Most of these cases are associated with sudden collapse and death. The mechanism of death is unclear but may relate to direct brain-stem disruption (Lindenberg and Freytag 1970, Leadbeatter 1994). However, there are cases in the literature of more prolonged survival before collapse and this is the authors' experience. Clearly if an artery is only partially damaged, the onset of haemorrhage may be delayed (Marek 1981, Deck and Jagadha 1986, Leadbeatter 1994).

SUBARACHNOID HAEMORRHAGE: NATURAL

Bleeding from a saccular or berry aneurysm accounts for 70–90 per cent of naturally occurring subarachnoid haemorrhages (Kalimo *et al.* 2002). Other causes include ruptured vascular malformations, mycotic aneurysm rupture and subarachnoid haemorrhage in conjunction with an intracerebral haemorrhage. Aneurysms may also rarely be atherosclerotic, infective, and occur in association with drug abuse (Chen *et al.* 2003). It should be recognized, however, that, in around 20 per cent of cases of subarachnoid haemorrhage with no definite history of trauma, no lesion is identified – so-called aneurysm-negative cases (Rinkel *et al.* 1993). Associated conditions include fibromuscular dysplasia (Cloft *et al.* 1998), various connective tissue disorders (Stehbens 1989), polycystic kidneys (Rinkel *et al.* 1993) and vascular abnormalities including arteriovenous malformations (Weller 1995). There also is a familial tendency (Alberts 1999) in some early onset cases. There is a recognized association with alcohol and cigarette smoking for subarachnoid haemorrhage (Juvela *et al.* 1993). It is less clear whether these factors can be definitely implicated in the formation of aneurysms as such. Increased alcohol consumption has been shown to impair outcome with increased rebleeding and delayed ischaemia (Juvela 1992). The role of chronic hypertension is somewhat unclear. Many studies have not shown a definitive correlation with aneurysmal formation (Kalimo *et al.* 2002). Seasonal variation in aneurysmal subarachnoid haemorrhage has been noted (Weller 1995), with rupture in males peaking in autumn and in females in spring (Chyatte *et al.* 1994). Rupture is also more common during the day and in association with physical stress.

Pathology

Cerebral aneurysms are saccular in nature, occurring particularly at the branching points of the intracranial cerebral arteries: 85–90 per cent occur in the territory of the terminal internal carotid arteries or the anterior part of the circle of Willis and around 5–10 per cent occur on the posterior circulation (Figure 6.6). The size varies from a millimetre or so to much larger with 85 per cent presenting as rupture of less than 10 mm (Forget *et al.* 2001). Giant aneurysms are classified as such with a diameter of greater than 25 mm (Figure 6.7).

Post mortem, identification of these aneurysms is optimal in the unfixed brain, with careful dissection along the paths of the various intracranial arteries having washed away any large amounts of clot. They may, however, be difficult to identify because, in particular if thin walled, they can be almost destroyed as they bleed. Macroscopically varying degrees of degenerative features may be seen in the wall including calcification, arteriolosclerosis and evidence of previous thrombus with organization. There may be evidence of previous leakage from the aneurysm with brown subarachnoid staining. The site of rupture may be indicated by the location of haemorrhage basal cisterns, (circle of Willis aneurysm), haematoma

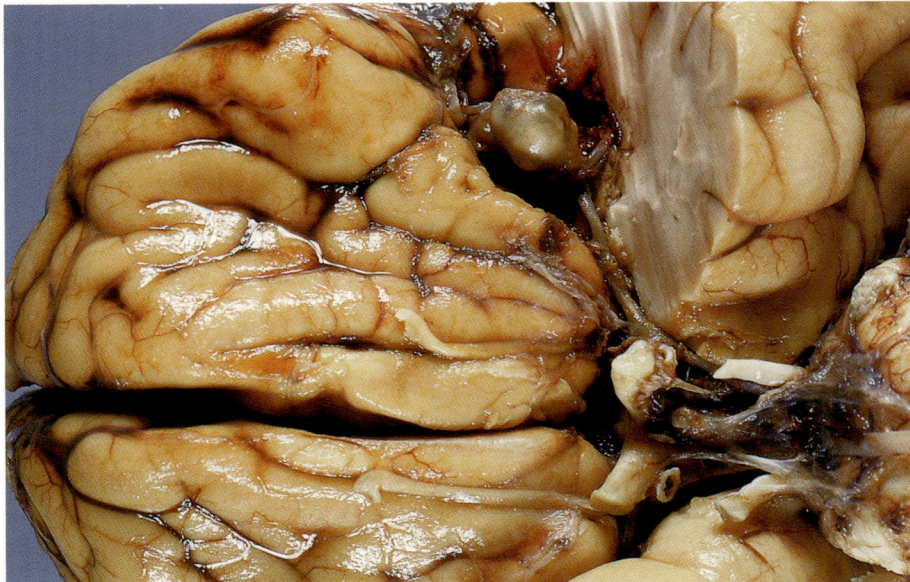

Figure 6.6 Middle cerebral artery aneurysm.

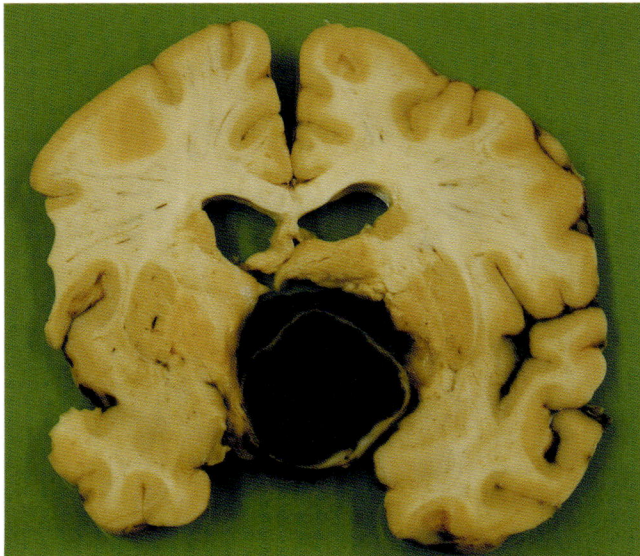

Figure 6.7 Giant aneurysm arising from the internal carotid/middle cerebral artery.

Figure 6.8 Histological appearances of a 'berry' aneurysm.

of survivability. The wall comprises fibrous tissue of varying thickness with atheromatous degeneration commonly present (Figure 6.8). Evidence of previous rupture and haemorrhage may also be seen.

Problem areas and medicolegal issues

RELATIONSHIP OF RUPTURE OF A SACCULAR ANEURYSM AND TRAUMA

This can be a difficult area and requires detailed clinico-pathological assessment in relation to the alleged assault. This situation is uncommon but recognized as evidenced by the detailed series of Bowen (1984) and Gonsoulin *et al.* (2002). Knight (1979) reports a case that illustrates the legal issues of 'beyond reasonable doubt' in English law where the victim collapses within moments of trauma to the head and neck and dies as a result of rupture of a

between the frontal lobes (anterior cerebral or anterior communicating artery aneurysm) and sylvian fissure (middle cerebral artery aneurysm). Rupture into brain parenchyma can occur which may extend into the ventricular system.

Histological features

The site of rupture may be identifiable with variable acute inflammation and other features depending on the period 'berry' aneurysm on the posterior communicating artery. The authors have seen rare isolated cases where the time relationship between assault and collapse was instantaneous. It may be postulated that the transient hypertension of such a situation is the underlying mechanism. It may also be impossible to exclude the possibility that direct trauma such as a heavy blow to the head or neck could not cause rupture of a thin walled aneurysm.

TRAUMATIC VASCULAR INJURY

Other types of traumatic vascular injury are recognized other than the acute traumatic subarachnoid haemorrhage, including traumatic dissection, thrombosis, arteriovenous fistula and traumatic aneurysms.

Arterial dissection

Acute arterial dissection with massive subarachnoid haemorrhage is discussed earlier in this chapter. Arterial dissection giving rise to superadded thrombosis is a cause of ischaemic strokes in young adults. In the older population the picture is often complicated by underlying arteriosclerosis (Figure 6.9). Dissection may be difficult to identify. It is likely that this entity is underdiagnosed as a result of closed head injury and may be responsible for clinical signs rather than as a direct result of trauma (Rommel et al. 1999). The extracranial cerebrovascular system is the major location including the carotid system and vertebral arteries although traumatic dissection of intracranial arteries is also recognized as a cause of infarction (Rutherfood et al. 1996). Incomplete tears of the basilar artery have also been described following trauma not necessarily associated with massive basal subarachnoid haemorrhage (Djokic et al. 2003). Carotid dissection usually begins above the carotid bifurcation. It is thought to be susceptible to dissection in this location because it can be stretched by hyperextension and rotation. Presentation may be delayed (Mokri et al. 1988, Nunnink 2002). Direct or penetrating trauma may also cause dissection (de Recondo et al. 1995).

Trauma to the vertebral artery causing dissection is well described and commonly involves the third segment at the level of the axis and atlas, extending intracranially (proximally). This may be complicated by superadded thrombosis or aneurysm formation (Martin and Enevoldson 1998). Penetrating injury or spinal fracture may also cause dissection. Situations known to be associated with cervical arterial dissection include chiropractic neck manipulation, head injury, including motor vehicle collisions, spinal fracture, direct blunt neck trauma, hanging and strangulation, as well as a number of sporting activities (Vanezis 1986, Hinse et al. 1991, Lee et al. 1995, Prabhu et al. 1996, Norris et al. 2000, Rothwell et al. 2001). Pathological examination in those cases where there is a traumatic history do not usually show underlying disease. There is an association with fibromuscular dysplasia and other heritable disorders of collagen and elastin. There also appears to be increased risk in cases of hypertension, atherosclerosis, smoking, oral contraception and migraine, although the literature is somewhat unclear (Haldeman et al. 1999,

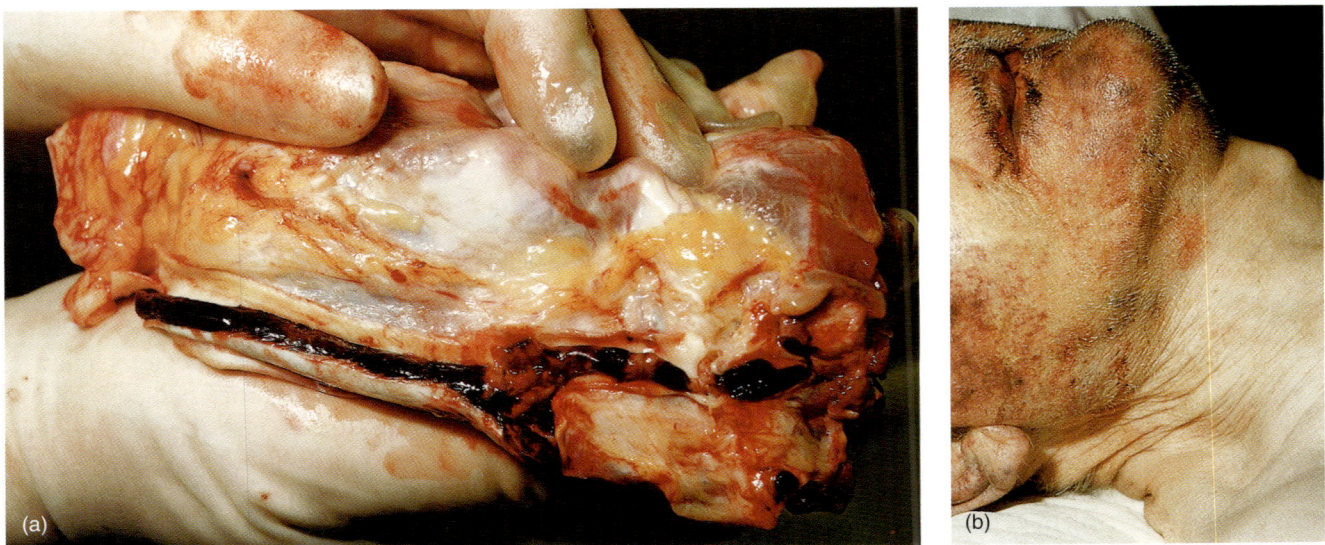

Figure 6.9   (a) Carotid artery thrombosis occurring following trauma to the neck. (b) Bruising to face/neck after assault with subsequent internal carotid artery thrombosis.

Mitchell 2002). In medicolegal terms each case will need careful assessment of all factors including a history and detailed pathology. The latter may necessitate serial sectioning of the entire vessel (Leadbeatter 1994).

Traumatic aneurysm

This type of aneurysm forms less than 1 per cent of all aneurysms, and are more common in children and males, the latter probably reflecting the increased incidence of trauma (Larson et al. 2000).

AETIOLOGY AND PATHOGENESIS
Traumatic aneurysms occur after trauma, either blunt or penetrating in nature. Causes of blunt trauma include road traffic accidents, falls and assaults. Penetrating injuries include penetrating stab wounds and missile injury (Holmes and Harbaugh 1993, Cogbill and Sullivan 1995, Saito et al. 1995, Aarabi 1988, Amirjamshadi et al. 1996, Kieck and de Villiers 1994). There are also reports of aneurysms occurring as a complication of surgery (Ventureyra and Higgins 1994, Bavinski et al. 1997). The mechanism of injury involves either direct vascular injury or stretching of the vessel by adjacent forces, and relates to the anatomical location of the involved artery. They are not infrequently related to skull fractures. Anatomically classification is:

- Proximal to the circle of Willis: these include infra- and supraclinoid carotid artery aneurysms as well as vertebrobasilar aneurysms.
- Distal to the circle of Willis, which includes cortical and subcortical aneurysms.

The clinical presentation of such aneurysms is variable, including haemorrhage, focal neurological deficits, seizures and acute decrease in consciousness level. The time from trauma to presentation in acute intracranial haemorrhage is variable and may be from months to years. Presentation may also be the result of local effects of a mass lesion such as cranial nerve palsies, visual disturbance, hydrocephalus and growing skull fracture (Endo et al. 1980).

Aneurysms may be true or false although some authors also include a mixed group (Larson et al. 2000). True aneurysms are as a result of localized weakening of the wall with local disruption of the intima, and variable involvement of the intimal elastic layer and media but leaving the adventitia intact. False aneurysms (or pseudoaneurysms)

arise as a result of full-thickness laceration to the arterial wall with haematoma in continuity with the lumen. The wall is formed by organization of blood to fibrous tissue. False aneurysms are the most common histological type.

Differentiation from natural 'berry' aneurysms may be difficult. The following factors may aid: peripheral location, site away from a branching point, irregular outline of the sac, absence of a neck, delayed filling and emptying, and location adjacent to the falx edge (Opeskin 1995).

Histological examination can be identical to saccular aneurysms, particularly in the case of true aneurysms. A false aneurysm is most usually traumatic in origin. Features suggestive of trauma include dissection of the arterial wall and foci of medial and intimal fibrosis away from the site of the aneurysm (Paul et al. 1980). A history of trauma, which, as commented above, may be years previously, should be sought. Aneurysms in childhood, particularly if peripheral, are commonly traumatic or inflammatory in origin (Paul et al. 1980, Buckingham et al. 1988).

Arteriovenous fistula: traumatic

These are rare; most commonly they are carotid–cavernous fistulae where the internal carotid artery passes adjacent to the venous cavernous sinuses. These most commonly occur in young men and may be delayed some time after the trauma (Kalimo et al. 2002, Tsutsumi et al. 2002). A method for exposing the intraosseous portion of the carotid arteries has recently been described by Langlois and Little (2003) as an aid to postmortem examination. Other sites for traumatic arteriovenous fistula include the middle meningeal artery with a meningeal vein and the extradural vertebral artery.

External pressure to the neck and neuropathological aspects

External pressure to the neck such as occurs with strangulation may lead to neuropathological complications in survivors. These are most commonly related to either the effects of global hypoxia/ischaemia or due to arterial pathology with thrombosis and arterial territory infarction. Milligan and Anderson (1980) reported two cases of stroke within days of attempted strangulation with thrombotic occlusion of the internal carotid artery. Global hypoxia/ischaemia has been reported giving rise to bilateral basal ganglia pathology (Simpson et al. 1987, Kirton 2001, Lumb 2001).

The time relationship to trauma as well as any underlying disease need to be evaluated (Figure 6.10).

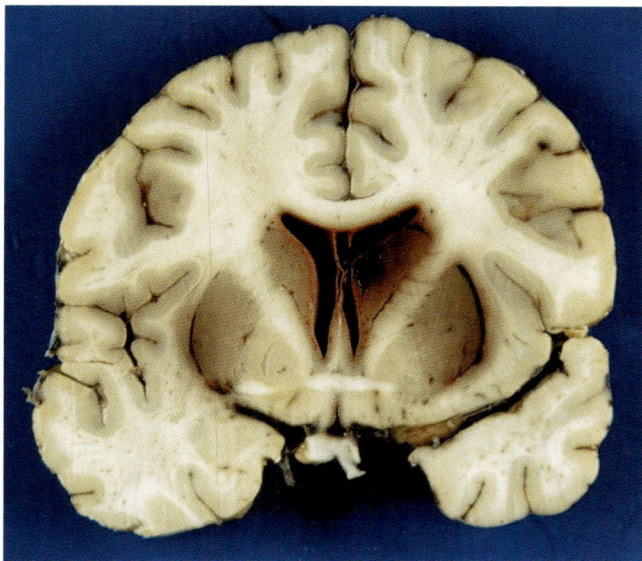

Figure 6.10 Haemorrhagic softening of the right caudate nucleus with similar changes bilaterally in the globus pallidum following survival of 24–48 hours after attempted strangulation.

REFERENCES

Aarabi, B. 1988. Traumatic aneurysms of brain due to high velocity missile head wounds. *Neurosurgery*, **22**(6 Pt 1), 1056–63.

Alberts, M.J. 1999. Subarachnoid hemorrhage and intracranial aneurysms. In Alberts, M.J. (ed.), *Genetics of Cerebrovascular Disease*. New York: Futura, 237–59.

Amirjamshidi, A., Rahmat, H. and Abbassioun, K. 1996. Traumatic aneurysms and arteriovenous fistulas of intracranial vessels associated with penetrating head injuries occurring during war: principles and pitfalls in diagnosis and management. A survey of 31 cases and review of the literature. *J Neurosurg*, **84**, 769–80.

Barry, K.J. and Scott, R.M. 1979. Effect of intravenous ethanol on cerebral vasospasm produced by subarachnoid blood. *Stroke*, **10**, 535–7.

Bavinski, G., Viller, M., Knoop, E. *et al*. 1997. False aneurysms of the intracavernous carotid artery – report of 7 cases. *Acta Neurochir*, **139**, 37–43.

Bowen, D.A.L. 1984. Ruptured berry aneurysms: a clinical, pathological and forensic review. *Forensic Sci Int*, **26**, 227–34.

Bromilow, A. and Burns, J. 1985. Technique for removal of the vertebral arteries. *J Clin Pathol* **38**, 1400–2.

Buckingham, M.J., Crone, K.R., Ball, W.S. *et al*. 1988. Traumatic intracranial aneurysms in childhood: two cases and a review of the literature. *Neurosurgery*, **22**, 398–408.

Cameron, J.M. and Mant, A.K. 1972. Fatal subarachnoid haemorrhage associated with cervical trauma. *Med Sci Law*, **12**, 66–70.

Chen, H-J., Liang, C-L., Lu, K. and Lui, C-C. 2003. Rapidly growing internal carotid artery aneurysm after amphetamine abuse. Case report. *Am J Forensic Med Pathol*, **24**, 32–4.

Chyatte, D., Chen, D.L., Bronstein, K. and Brass, L.M. 1994. Seasonal fluctuations in the incidence of intracranial aneurysm rupture and its relationship to changing climatic conditions. *J Neurosurg*, **81**, 525–30.

Cloft, H.J., Kallmes, D.F., Kallmes M.H. *et al*. 1998. Prevalence of cerebral aneurysms in patients with fibromuscular dysplasia: a reassessment. *J Neurosurg*, **88**, 436–40.

Coast, G.C. and Gee, D.J. 1984. Traumatic subarachnoid haemorrhage; an alternative source. *J Clin Pathol*, **37**, 1245–8.

Cogbill, T.H. and Sullivan, H.G. 1995. Carotid artery pseudoaneurysm and pellet embolism to the middle cerebral artery following a shotgun wound of the neck. *J Trauma*, **39**, 763–7.

Contostavlos, D.L. 1971. Massive subarachnoid haemorrhage due to laceration of the vertebral artery associated with fracture of the transverse process of the atlas. *J Forensic Sci*, **16**, 40–56.

Contostavlos, D.L. 1995. Isolated basilar traumatic subarachnoid haemorrhage: an observer's 25 year re-evaluation of the pathogenetic possibilities. *Forensic Sci Int*, **73**, 61–74.

Cooper, P.N., Sunter J.P. and McCarthy J.H. 1999. Traumatic subarachnoid haemorrhage: a 10 year case study. *J Pathol*, **187**(S), 7A.

de Recondo, A., Woimant, F., Ille, O. *et al*. 1995. Post-traumatic common carotid artery dissection. *Stroke*, **26**, 705–6.

Deck, J.H. and Jagadha, V. 1986. Fatal subarachnoid hemorrhage due to traumatic rupture of the vertebral artery. *Arch Pathol Lab Med*, **110**, 489–93.

Djokic, V., Savic, S. and Atanasijevic, T. 2003. Medicolegal diagnostic value and clinical significance of traumatic incomplete tears of the basilar artery. *Am J Forensic Med Pathol*, **24**, 208–13.

Dowling, G. and Curry, B. 1988. Traumatic basal subarachnoid haemorrhage: report of six cases and a review of the literature. *Am J Forensic Med Pathol*, **9**, 23–31.

Endo, S., Takaku, A., Aihara, H. *et al*. 1980. Traumatic cerebral aneurysm associated with underlying skull fracture. Report of two infancy cases. *Child's Brain*, **6**, 131–9.

Ford, R. 1956. Basal subarachnoid haemorrhage and trauma. *J Forensic Sci*, **1**, 117–26.

Forget, T.R., Benitez, R., Veznedaroglu, E. *et al*. 2001. A review of size and location of ruptured intracranial aneurysms. *Neurosurgery*, **49**, 1322–5.

Gee, D.J. 1982. Traumatic subarachnoid haemorrhage. Proceedings of the 12th Congress of the International Academy of Forensic and Social Medicine. Vienna: Egermann, 495–8.

Gonsoulin, M., Barnard, J.J. and Prahlow, J.A. 2002. Death resulting from ruptured cerebral artery aneurysm: 219 cases. *Am J Forensic Med Pathol*, **23**, 5–14.

Gray, J.T., Peutz, S.M., Jackson, S.L. and Green, M.A. 1999. Traumatic subarachnoid haemorrhage: a 10 year case study and review. *Forensic Sci Int*, **105**, 13–23.

Gross, A. 1990. Traumatic basal subarachnoid haemorrhages: postmortem examination material analysis. *Forensic Sci Int*, **45**, 53–61.

Haldeman, S., Kohlbeck, F.J. and McGregor, M. 1999. Risk factors and precipitating neck movements causing vertebrobasilar artery dissection after cervical trauma and spinal manipulation. *Spine*, **24**, 785–94.

Harland, W.A., Pitts, J.F., Watson, A.A. 1983. Subarachnoid haemorrhage due to upper cervical trauma. *J Clin Pathol*, **36**, 1335–41.

Hillbom, M. and Kaste, M. 1981. Does alcohol intoxication precipitate aneurysmal subarachnoid haemorrhage? *J Neurol Neurosurg Psychiatry*, **44**, 523–6.

Hinse, P., Thie, A. and Lachenmayer, L. 1991. Dissection of the extracranial vertebral artery: report of four cases and review of the literature. *J Neurol Neurosurg Psychiatry*, **54**, 863–9.

Holmes, B. and Harbaugh, R.E. 1993. Traumatic intracranial aneurysms: a contemporary review. *J Trauma-Injury Infect Crit Care*, **35**, 855–60.

Johnson, C.P., How, T., Scraggs, M. *et al.* 2000. J. A biochemical study of the human artery with implications for fatal arterial injury. *Forensic Sci Int*, **109**, 169–82.

Johnson, C.P., Baugh, R., Wilson, C.A. and Burns, J. 2001. Age related changes in the tunica media of the vertebral artery: implications for the assessment of vessels injured by trauma. *J Clin Pathol*, **54**, 139–45.

Juvela, S. 1992. Alcohol consumption as a risk factor for poor outcome after aneurysmal subarachnoid haemorrhage. *BMJ*, **304**, 1663–7.

Juvela, S., Hillbom, M., Numminen, H. and Koskinen, P. 1993. Cigarette smoking and alcohol consumption as risk factors for aneurysmal subarachnoid hemorrhage. *Stroke*, **24**, 639–46.

Kalimo, H., Kaste, M. and Haltia, M. 2002. In Graham, D.I. and Lantos, P.L. (eds), *Greenfield's Neuropathology*, 7th edn. London: Arnold, Chapter 6.

Karhunen, P.J., Kauppila, A., Penttila, A. and Erkinjuntti, T. 1990. Vertebral artery rupture in traumatic subarachnoid haemorrhage detected by postmortem angiography. *Forensic Sci Int*, **44**, 107–15.

Kieck, C.F. and de Villiers, J.C. 1984. Vascular lesions due to transcranial stab wounds. *J Neurosurg*, **60**, 42–46.

Kirton, C.A. and Riopelle, R.J. 2001. Meige syndrome secondary to basal ganglia injury: a potential cause of acute respiratory distress. *Canadian J Neurol Sciences*, **28**, 167–73.

Knight, B. 1979. Trauma and ruptured cerebral artery aneurysm. *BMJ*, i, 1430–1.

Koszyca, B., Gilbert, J.D. and Blumbergs, P. 2003. Traumatic subarachnoid hemorrhage and extracranial vertebral artery injury. A case report and review of the literature. *Am J Forensic Med Pathol*, **24**, 114–18.

Langlois, N.E. and Little, D. 2003. A method for exposing the intraosseous portion of the carotid arteries and its application to forensic case work. *Am J Forensic Med Pathol*, **24**, 35–40.

Larson, P.S., Reisner, A, Morassutti, D.J. *et al.* 2000. Traumatic intracranial aneurysms. *Neurosurg Focus*, **8**, Article 4.

Leadbeatter, S. 1994. Extracranial vertebral artery injury – evolution of a pathological illusion. *Forensic Sci Int*, **67**, 35–40.

Lee, K.P., Carlini, W.G., McCormick, G.F. and Albers, G.W. 1995. Neurologic complications following chiropractic manipulation: A survey of California neurologists. *Neurology*, **45**, 1213–15.

Lindenberg, R. and Freytag, E. 1970. Brainstem lesions characteristic of traumatic hyperextension of the head. *Arch Pathol*, **90**, 509–15.

Lumb, P.D., Milroy, C.M. and Whitwell, H.L. 2001. Neuropathological changes in delayed death after strangulation. In: Graham, D.I. Proceedings of the Centenary Meeting of the British Neuropathological Society held at Goldsmiths' Hall, London, January 2001. *Neuropath and Applied Neurobiol*, **27**, 151.

McCarthy, J.H., Sunter, J.P. and Cooper, P.N. 1999. A method for demonstrating the source if bleeding in cases of traumatic subarachnoid haemorrhage. *J Pathol*, **187**, 30A.

Marek, Z. 1981. Isolated sub-arachnoid hemorrhage as a medic-legal problem. *Am J Forensic Med Pathol*, **2**, 19–22.

Martin, P.J. and Enevoldson, T.P. 1998. Vertebral aneurysm due to isolated spontaneous dissection of the intracranial vertebral artery. *J Neurol Neurosurg Psychiatry*, **65**, 946.

Milligan, N. and Anderson, M. 1980. Conjugal disharmony: A hitherto unrecognized cause of strokes. *BMJ*, **281**, 421–422.

Mitchell, J. 2002. Vertebral artery atherosclerosis: a risk factor in the use of manipulative therapy? *Physiother Res Int*, **7**, 122–35.

Mokri, B., Piepgras, D.G. and Houser, O.W. 1988. Traumatic dissections of the extracranial internal carotid artery. *J Neurosurg*, **68**, 189–97.

Norris, J.W., Beletsky, V. and Nadaereishvili, Z.G. 2000. Sudden neck movement and cervical artery dissection. *CMAJ*, **163**(1), 38–40.

Nunnink, L. 2002. Blunt carotid artery injury. *Emerg Med (Fremantle)*, **14**, 412–21.

Opeskin, K. 1995. Traumatic pericallosal artery aneurysm. *Am J Forensic Med Pathol*, **16**, 11–16.

Opeskin, K and Burke, M.P. 1998. Vertebral artery trauma. *Am J Forensic Med Pathol*, **19**, 206–17.

Paul, G.A., Shaw, C-M. and Wray, L.M. 1980. True traumatic aneurysm of the vertebral artery: case report. *J Neurosurg*, **53**, 101–5.

Pollanen, M.S., Deck, J.H.N. and Blenkinsop, B. 1996. Injury of the tunica media in fatal rupture of the vertebral artery. *Am J Forensic Med Pathol*, **17**, 197–201.

Prabhu, V., Kizer, J., Patil, A. *et al.* 1996. Vertebrobasilar thrombosis associated with nonpenetrating cervical spine trauma. *Trauma*, **40**(1), 130–7.

Rinkel, G.J.E., Gijn, J.V. and Wijdicks, E.F.M. 1993. Subarachnoid haemorrhage without detectable aneurysm. A review of causes. *Stroke*, **24**, 1403–9.

Rommel, O., Niedeggen, A., Tegenthoff, M. *et al.* 1999. Carotid and vertebral artery injury following severe head or cervical spine trauma. *Cerebrovasc Dis*, **9**, 2002–9.

Rothwell, D.M., Bondy, S.J. and Williams, J.I. 2001. Chiropractic manipulation and stroke: a population-based case-control study. *Stroke*, **32**, 1054–60.

Rutherfood, S.G., Dada, M.A. and Nel, J.P. 1996. Cerebral infarction and intracranial arterial dissection in closed head injury. *Am J Forensic Med Pathol*, **17**(1), 53–7.

Saito, K., Baskaya, M.K., Shibuya, M. *et al*. 1995. False traumatic aneurysm of the dorsal wall of the supraclinoid internal carotid artery. *Neurol Med Chir (Tokyo)*, **35**, 886–91.

Saukko, P. and Knight, B. 2004. *Knight's Forensic Pathology*, 3rd edn. London: Arnold.

Servadei, F., Murray, G.D., Teasdale, G.M. *et al*. 2002. Traumatic subarachnoid hemorrhage: demographic and clinical study of 750 patients from the European brain injury consortium survey of head injuries. *Neurosurgery*, **50**, 261–7; discussion 267–9.

Simpson, R.K. Jr, Goodman, J.C., Rouah, E. *et al*. 1987. Late neuropathological consequences of strangulation. *Resuscitation*, **15**, 171–85.

Simonsen, J. 1963. Traumatic subarachnoid haemorrhage in alcohol intoxication. *J Forensic Sci*, **8**, 97–116.

Simonsen, J. 1967. Fatal subarachnoid haemorrhage in relation to minor head injuries. *J Forensic Med*, **14**, 146–55.

Simonsen, J. 1984. Fatal subarachnoid haemorrhage in relation to minor injuries in Denmark from 1967 to 1981. *Forensic Sci Int*, **24**, 57–63.

Stehbens, W.E. 1989. Etiology of intracranial berry aneurysms. *J Neurosurg*, **70**, 823–31.

Tatsuno, Y. and Lindenberg, R. 1974. Basal subarachnoid hematomas as sole intracranial traumatic lesions. *Arch Pathol*, **97**, 211–15.

Tsutsumi, M., Kazekawa, K., Tanaka, A. *et al*. 2002. Traumatic middle meningeal artery pseudoaneurysm and subsequent fistula formation with the cavernous sinus: case report. *Surg Neurol*, **58**, 325–8.

Vanezis, P. 1986. Vertebral artery injuries in road traffic accidents: a post-mortem study. *J Forensic Sci*, **26**, 281–91.

Ventureyra, E.C. and Higgins, M.J. 1994. Traumatic intracranial aneurysms in childhood and adolescence. Case reports and review of the literature. *Child's Nerv Syst*, **10**, 361–79.

Weller, R.O. 1995. Subarachnoid haemorrhage and myths about saccular aneurysms. *J Clin Pathol*, **48**, 1078–81.

Whittington, R.M. 1980. Alcohol related deaths. Birmingham Coroner's records. *BMJ*, **284**, 1162.

Wilkinson, I.M.S. 1972. The vertebral artery. Extracranial and intracranial structure. *Arch Neurol*, **27**, 392–6.

Wilks, S. 1859. Sanguineous meningeal effusion. *Guy's Hosp Rep*, **5**, 119.

7

Contusional brain injury and intracerebral haemorrhage: traumatic and non-traumatic

Helen L Whitwell

Contusional and lacerating brain injury is characteristic of blunt head trauma. The pia arachnoid is not breached in contusional injury; however, with lacerating injury both the pia arachnoid and the underlying brain substance are torn. Contusional injury may produce varying sizes of intracerebral haematoma.

Intracerebral haemorrhage (ICH) is most commonly natural rather than traumatic. This chapter covers both natural and traumatic intracerebral haemorrhages concentrating, in particular, on diagnostic issues at neuropathological examination.

CONTUSIONAL BRAIN INJURY

Macroscopic appearances

Acute contusions typically occur over the crests of the gyri and have a conical appearance with the widest part in the superficial position and the apex deeper (Figure 7.1). They have a red or red–purple appearance when fresh. Larger contusions extend into white matter and can give rise to significant intracerebral haemorrhage. There is usually associated subarachnoid haemorrhage. Complications such as burst lobe and delayed intracerebral haemorrhage may occur. Old contusions appear gold–brown (Figure 7.2). They are not infrequently seen as an incidental finding (Graham *et al.* 2002).

Methods for quantification of the amount of contusional injury have been developed, initially by Adams *et al.* (1985). Their method is based on the extent and depth of the contusional injury in various regions of the brain. A contusion index can be given for the whole brain as well as for various anatomical sites. A further quantitative approach

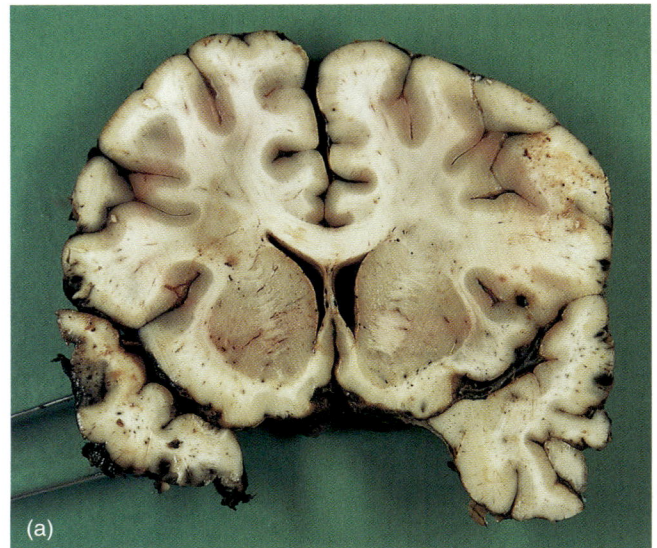

(a)

(b)

Figure 7.1 (a) Bilateral temporal lobe contusions. (b) Close-up of an area of temporal lobe contusion.

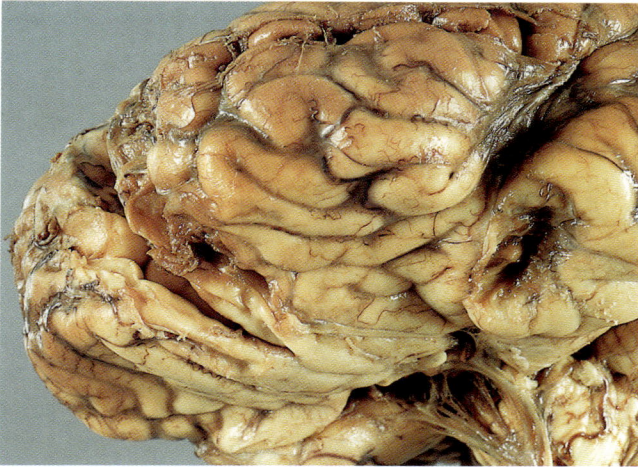

Figure 7.2 Old contusions of the frontal and temporal regions.

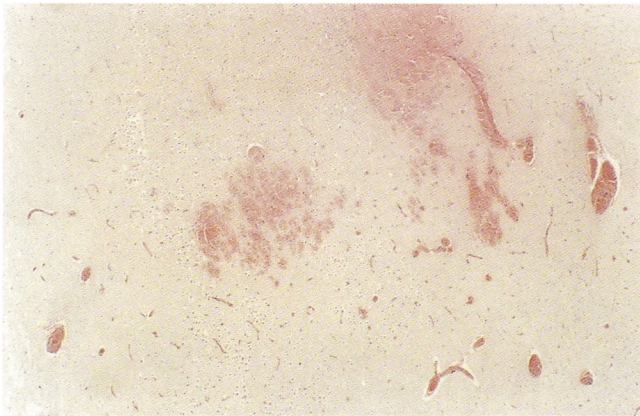

(a)

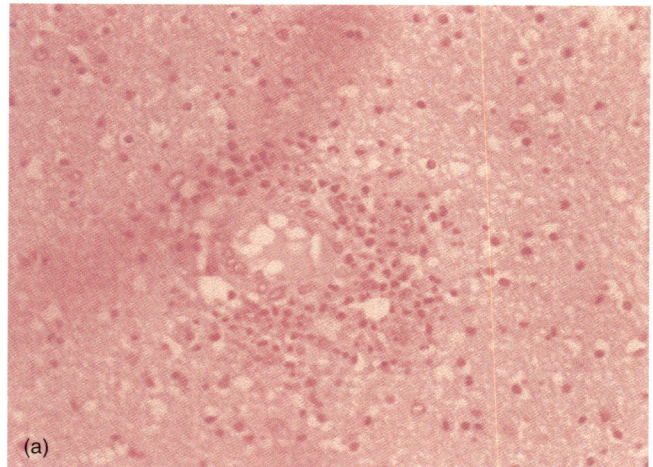

(b)

Figure 7.4 (a) Neutrophil polymorphs in contusional injury 2 days in age. (b) CD68 – macrophages in an area of contusion after several days.

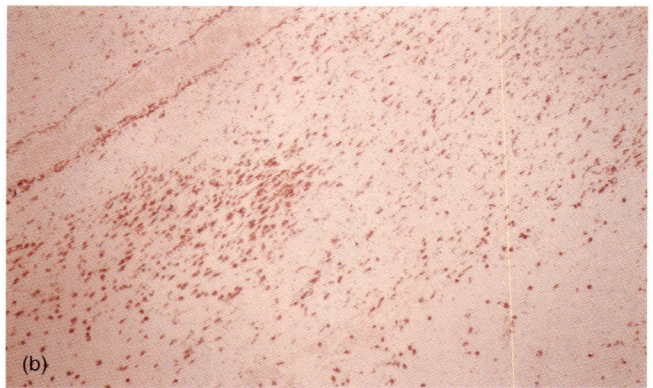

Figure 7.3 Perivascular haemorrhages – early contusional injury.

by Ryan *et al.* (1994) uses a sector scoring method. Both of these studies have provided considerable data for individual cases as well as series of cases and have been extremely useful in biomechanical correlation. However, for most medicolegal cases, description with diagrams and photographs will suffice. Digital technology has made photographic recording relatively simple.

Microscopic appearances

Perivascular haemorrhages with larger areas of haemorrhage are seen in the early stages (Figure 7.3). Infiltration by inflammatory cells/leukocyte polymorphs is the first stage to be seen, initially with intravascular accumulation, but subsequently in the perivascular location (Holmin *et al.* 1998, Oehmichen *et al.* 2003, Figure 7.4a).

The time course is variable – some authors report intra-parenchymal leukocyte polymorphs within a few hours (Anderson and Opeskin 1998). Within days the intra-parencymal infiltrate comprises a mixture of T lymphocytes, reactive microglia and monocytes/macrophages, as well as the neutrophil polymorphs (Holmin *et al.* 1998, Engel *et al.* 2000). Generally these appear from around 72 h. Haemosiderin may be seen as early as 48 h or so, although detailed pathological studies are lacking. Capillary proliferation, astrocytosis and accumulation of haemosiderin macrophages are identifiable with increasing survival, after about 4–5 days (Figure 7.4b).

Neuronal changes, as are seen in hypoxic–ischaemic injury, can be observed in adjacent neurons.

It should be said that the time frames can only be approximate and factors such as the age of the individual may play a role. Old contusions histologically resemble areas of previous infarction with meningeal fibrosis and gliosis, as well as haemosiderin macrophages (Figure 7.5). The underlying white matter shows variable demyelination.

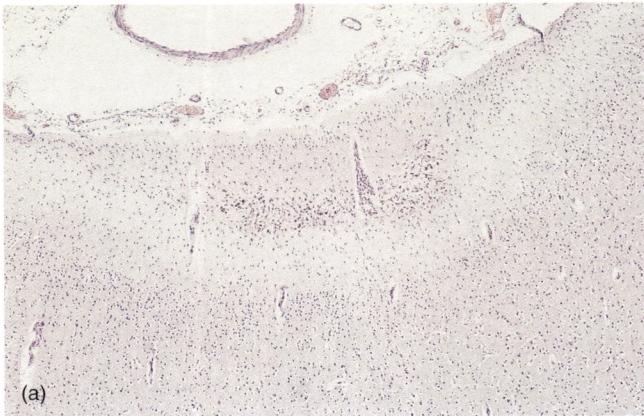

(a)

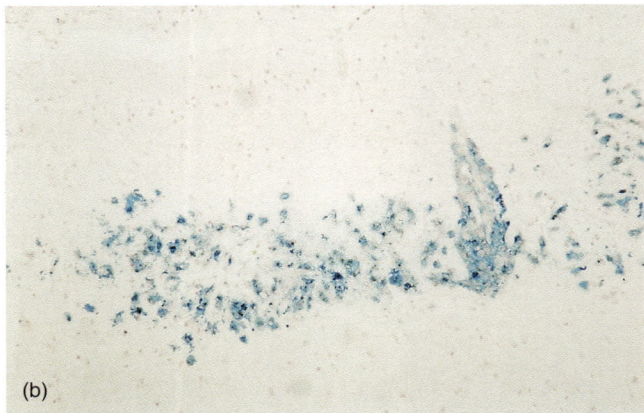

(b)

Figure 7.5 (a) Small old contusion. (b) Perls' stain showing haemosiderin macrophages.

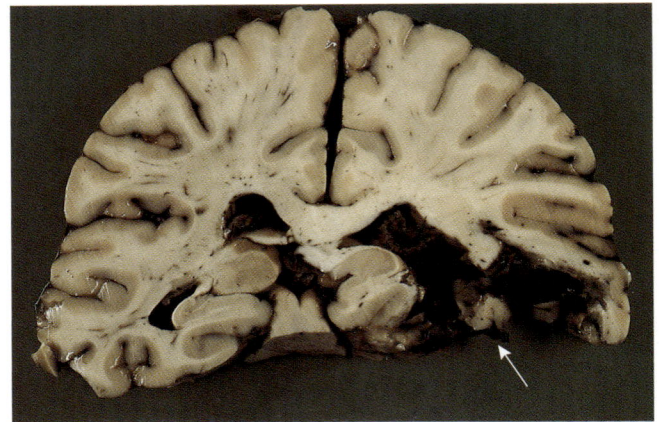

Figure 7.6 'Burst' lobe haematoma in continuity with cerebral surface.

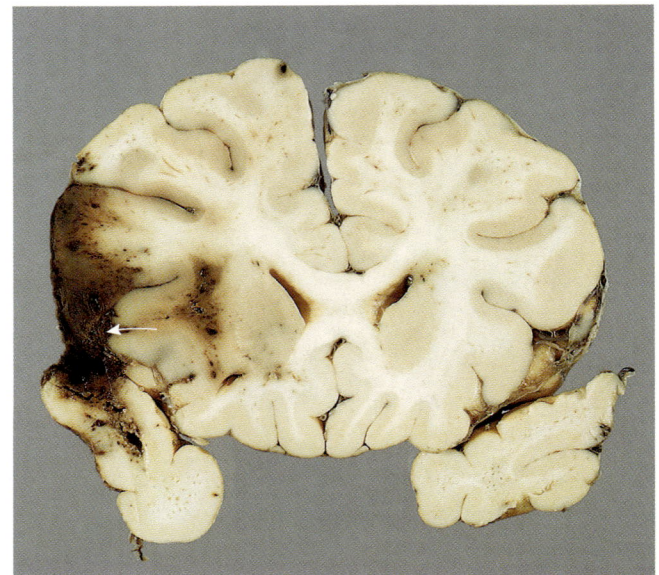

Figure 7.7 Coup contusions beneath area of multiple hammer blows.

It is well recognized that intracerebral haematomas in association with contusional injury may be delayed. Presentation is usually within 48 h of diagnosis, but occasionally longer (Baratham and Dennyson 1972, Nannassis *et al.* 1989, Graham *et al.* 2002). This may mean that a patient with relatively minor contusional injury presents with a rapidly deteriorating picture including features of a space-occupying lesion. CT or MRI will reveal the development of such a haematoma. The relationship between development of such a haematoma and pre-existing disease, e.g. hypertension, is unclear. Other complicating issues include the role of alcohol, drugs such as cocaine and amphetamines, as well as decreased blood coagulation as a result of either anticoagulant therapy or disease such as chronic liver disease (Mackenzie 1996).

Burst lobe

This occurs when a haematoma, most commonly in the temporal region, but also seen in the frontal region, is in continuity with the cerebral surface and adjacent acute subdural haematoma (Figure 7.6). This occurs in the setting of contusional injury and represents a severe end of the spectrum that may be seen in both coup and contre-coup injuries.

Coup contusions

These are contusions that occur beneath the site of impact. They are less frequent than contre-coup contusions (Dawson *et al.* 1980). Experimental work has indicated that tissue damage is related to the mechanical loading (Shreiber *et al.* 1999). The common sites are over the hemispheres or cerebellum where the brain is exposed to the relevant object (Figure 7.7). This contrasts with contre-coup injury where the base of the brain and temporal/frontal poles are the typical sites. This type of injury is most commonly seen where a relatively small object strikes the skull, causing the skull to 'bend' (transient deformation) into the brain and resulting in 'suction' forces as well as direct compressive injury (DiMaio and DiMaio 2001). A classic example would be as the result of a blow from an object such as a hammer. In general terms there is correlation between the degree of injury sustained and the mass/velocity relationship of the object (Leestma 1988). The external appearance of the wound may indicate the implement (e.g. hammer, baton), but it should be recognized that the features of any wound may not always be specific. Thus, a hammer blow may simply leave a bruise depending on the force used. Further aid to the identification of a particular weapon may be the characteristic of any skull fracture and external wounds (see Chapter 4).

Contre-coup contusions

These are generally said to occur at sites opposite to the site of impact. The characteristic locations are the frontal poles, the orbital surfaces of the frontal lobes and the temporal lobes including the temporal poles (Figure 7.8). However, it is now recognized that frontal and temporal contusions may be found when the impact site is not directly opposite, and it has been suggested that a more correct definition of a contre-coup contusion is one that is not immediately beneath the impact site, irrespective of its exact location (Graham *et al.* 2002). The situation in which this distribution of contusional injury is classically seen is when the moving head strikes the ground. This may produce a combination of an occipital laceration with surrounding abrasion, scalp bruising, underlying skull fracture, subdural haematoma and contusional brain injury. In forensic practice this may follow a punch or blow to the face.

The mechanism of contre-coup contusional injury has been the subject of considerable work. It is recognized that the brain rebounds in the skull after ground impact, with the anterior parts coming into contact with the rough floor of the anterior fossa, wing of sphenoid and petrous temporal bone (Saukko and Knight 2004, DiMaio and

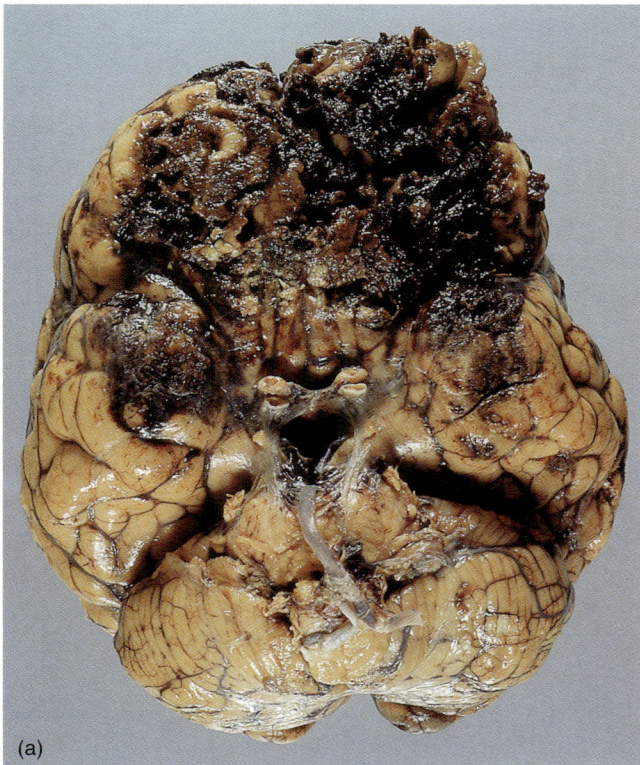

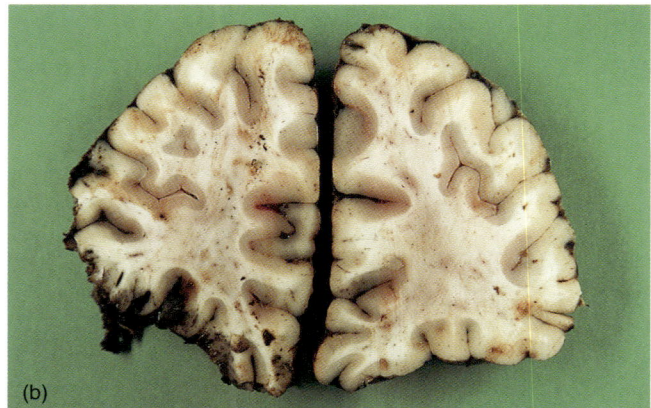

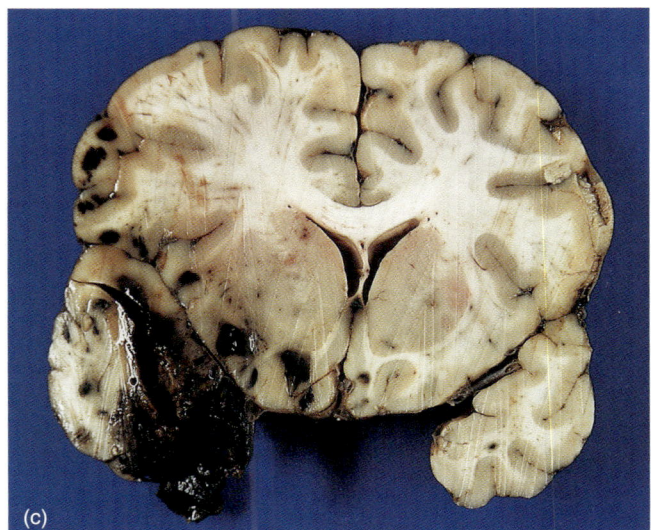

Figure 7.8 (a) Inferior aspect of brain with contre-coup contusional/lacerating injury. (b) Frontal contre-coup contusions. (c) Extensive temporal contre-coup contusions.

DiMaio 2001). Other work has also identified that a 'vacuum' effect occurs at the site of a contre-coup lesion, with pressure momentarily increased at the site of impact and decreased diametrically opposite (Yanagida *et al.* 1989). It is also recognized that movement of the brain towards the impact site causes opposite tensile strains. If these forces are larger than vascular tolerance, contusional injury will occur (Graham *et al.* 2002). Other factors include the anatomy of the skull with rough irregular bony areas in the orbital roofs and temporal fossae. Experimentally it has been shown that frontal pole contusions occur before temporal contusions with increasing mechanical input (Graham *et al.* 2002). Where there is a side impact to the head contusions are seen in the opposite side of the brain. Contre-coup contusional injury is rare in the occipital or parietal region. Contusional injury is more severe in the presence of a skull fracture.

Larger objects such as baseball bats, tree branches and other heavy objects may involve much greater kinetic energy and produce contre-coup as well as coup injuries (Leestma 1988, Morrison *et al.* 1998). This is particularly so when the fixed head is struck, i.e. when the victim is on the ground or against an unyielding surface. It would, however, be exceptional in this context not to see some coup injury. In the assault situation, which not infrequently involves a number of participants, the contribution of an individual injury to the final picture may be difficult to establish from the pathology alone.

Fracture contusions

These occur directly beneath the site of a skull fracture as a direct result of such (Figure 7.9). If the fracture is linear, the contusion tends to follow a similar pattern. Fracture

contusions may not relate to the point of impact because the fracture may originate some distance away. The frontal region tends to be an atypical site for coup contusional injury, the majority of cases are contre-coup. Posterior fossa contusional injury is almost entirely related to skull fractures. These may expand to produce haematomas (Vrankovic *et al.* 2000). Contre-coup posterior fossa contusions are extremely rare because the cerebellum and occipital lobes appear to be well protected possibly as a result of the relatively smooth surface of the skull, as well as the ability of the frontal structures to absorb energy from an impact.

Herniation contusions

These are either damage to the parahippocampal gyrus impacting against the tentorium cerebelli or the cerebellar tonsils impacting against the foramen magnum at the time of injury. They are usually associated with significant injury such as impact with a vehicle. They are also seen in gunshot wounds.

LACERATING INJURY

Lacerating injury may be seen in either coup or contre-coup contusional injuries. However, lacerating injury is well recognized as occurring in the absence of contusional injury, particularly where falls from significant heights are involved with open skull fractures. The location of these lacerations does not follow the pattern of coup or contre-coup injuries as described above and may involve any part of the brain (Figure 7.10).

OTHER TRAUMATIC INTRACEREBRAL HAEMATOMAS

These include the types listed below. The pathogenesis of the first three listed is outlined in Chapter 8 (Primary traumatic brain injury), because these are characteristic markers for traumatic axonal injury. Brainstem haemorrhages are seen in association with effects of raised pressure within the cranial cavity as well as in primary traumatic brain injury (Chapters 8 and 9). Not infrequently in an individual case a variety of findings may be present. The relationship between drugs, including alcohol and ICH, is covered in Chapter 14.

1. Petechial white matter haemorrhages (diffuse vascular injury)

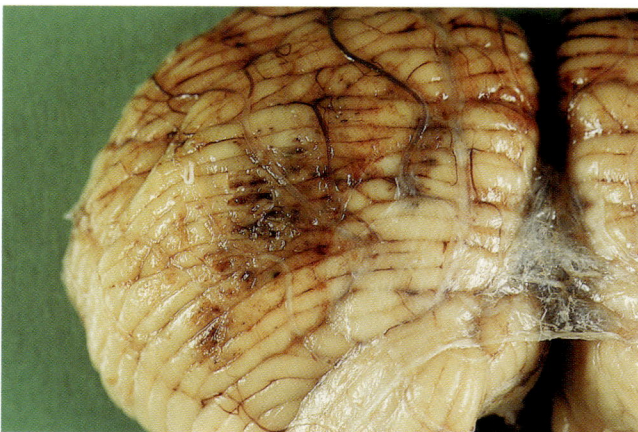

Figure 7.9 Inferior cerebellar hemisphere showing contusion in association with occipital fracture.

2. Tissue tear haemorrhages
3. Basal ganglia haematomas
4. Traumatic intraventricular haemorrhage
5. Brainstem haemorrhages
6. Cerebellar haemorrhages.

Petechial white matter haemorrhages (diffuse vascular injury)

These occur as a result of very severe head injury with short survival. The pattern is seen macroscopically and more readily microscopically with small haemorrhages occurring around blood vessels (see Chapter 8, p. 97). These are particularly prominent in the frontal and temporal lobes, adjacent to midline structures, and in the stem adjacent to the aqueduct and floor of the fourth ventricle. They are often associated with 'tissue tear' haemorrhages (see

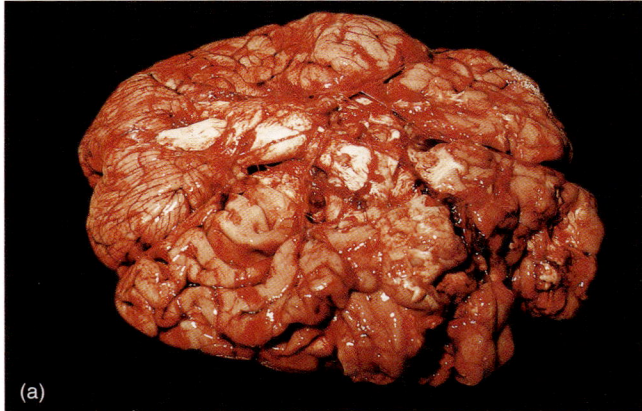

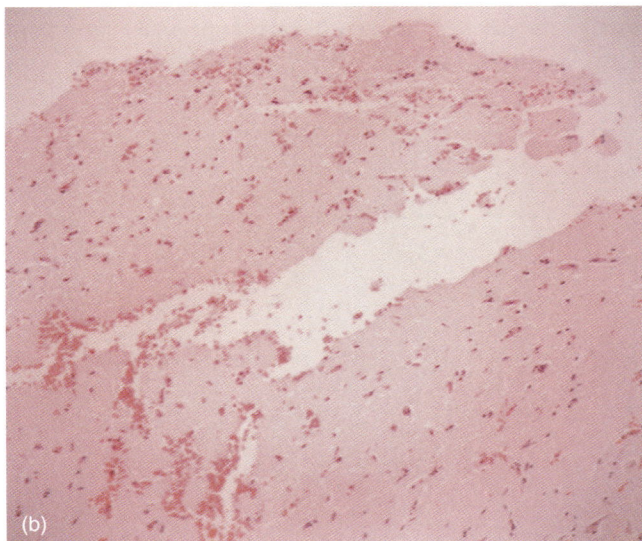

below). The precise aetiology is unclear but is probably related to vascular damage at the time of injury in association with primary axotomy. The majority of cases occur in road traffic deaths. Macroscopic differential diagnosis includes rare conditions such as diffuse demyelinating encephalomyelopathy (see Chapter 13, p. 162).

Tissue tear haemorrhages

These are associated with traumatic axonal injury and are macroscopic indicators of that diagnosis. They are areas of ICH of varying size caused by injury to vascular structures and axons as a result of shearing injury. They represent maximal areas of acceleration-induced brain injury and are discussed in detail in Chapter 8. Larger linear or flame-shaped haemorrhages are known as gliding contusions and occur at the severe end of injury (see Chapter 8, p. 94). Locations are predominantly in the parasagittal white matter, corpus callosum, centrum semi-ovale, and deep grey matter as well as the rostral brain stem and pons. Not infrequently in lesser degrees of traumatic axonal injury foci may only be seen in the corpus callosum (Figure 7.11). Histological examination shows variable haemorrhage, often in a perivascular location, and associated axonal pathology with changes in APP staining as covered in Chapter 8. Axonal pathology is almost invariably widespread elsewhere, further supporting the diagnosis.

Basal ganglia haematomas

The occurrence of these has been better documented with the advent of computed tomography (CT) and magnetic resonance imaging (MRI). They occur deep within the

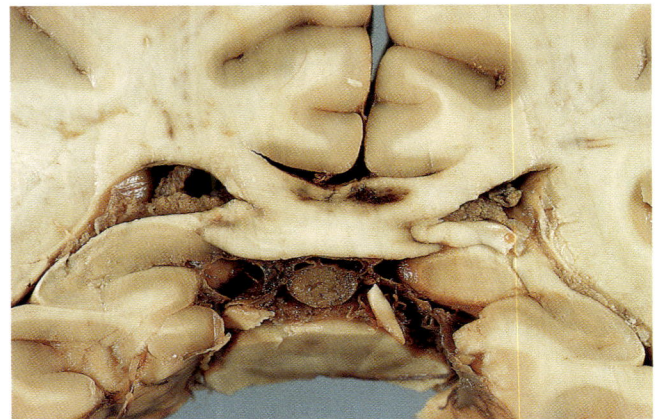

(a)

(b)

Figure 7.10 (a) Lacerating injury in association with extensive skull fracture as a result of a fall from a height. (b) Histology of acute lacerating injury.

Figure 7.11 Haemorrhages in the corpus callosum in traumatic axonal injury as a result of severe blunt head injury during an assault.

brain tissue in the basal ganglia region, varying in size from being relatively small (>1 cm) to considerably larger. This type of haemorrhage is frequently associated with diffuse axonal injury and thus a poor prognosis (Adams *et al.* 1986). They are most commonly identifiable shortly after injury with a period of lucidity being rare. This contrasts with haematomas in association with contusional injury, in which there may be variable periods of lucidity before presentation (see below). There is some evidence to indicate that they enlarge over a period of time after injury (Boto *et al.* 2001). The importance of these, in the context of a road traffic death, is that they may resemble primary hypertensive ICHs and bring into question the cause of the incident. Distinguishing features include the following: evidence of other traumatic brain injury such as corpus callosum, brain-stem and white matter haemorrhages in the locations associated with traumatic axonal injury (Figure 7.12). Impact injury may also be seen with skull fractures as well as contusional brain damage. Spontaneous ICH tends to occur in an older age group (Siddique *et al.* 2002).

It is difficult to substantiate a diagnosis of a primary hypertensive haemorrhage where there is no documented evidence of hypertension or postmortem findings to support a diagnosis, in particular left ventricular hypertrophy (Hangartner *et al.* 1985) or microscopic vascular changes to indicate hypertension (Mackenzie 1996). Full toxicological analysis should be undertaken because it is increasingly recognized that alcohol as well as other drugs may cause/

contribute to a primary ICH, as well as coagulation disorders including anticoagulant therapy (Yajima *et al.* 2001; see Chapter 14, p. 172). Examination of available medical records as well as correlation of the circumstances of the incident may also aid in interpretation.

Traumatic intraventricular haemorrhage

This is seen in a variety of situations:

- In association with the entity of traumatic subarachnoid haemorrhage caused by rupture or tear to the vertebrobasilar system (see Chapter 5, p. 71).
- In association with severe head injury – the precise source cannot usually be identified macroscopically but histology may show haemorrhages in association with vessels in the interventricular septum or choroid plexus. Rupture of ependymal vessels has also been identified (Makino *et al.* 2001). These tend to be in the posterior aspects of the ventricular system (Figure 7.13).
- Extension of haemorrhage from within the brain parenchyma secondary to extension of ICH.

Brain-stem haemorrhages

These may occur as part of diffuse axonal injury and the location of these differs from those associated with raised intracranial pressure and primary hypertensive haemorrhage. However, cerebral swelling with diffuse axonal injury occurs and precise identification may be difficult. Haemorrhages secondary to raised pressure are commonly centrally placed within the medulla and pons.

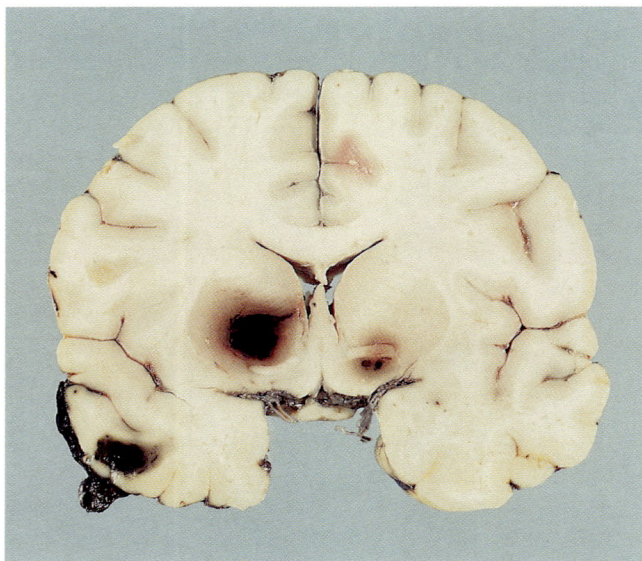

Figure 7.12 Basal ganglia haematoma caused by diffuse axonal injury. Note evidence of other areas of traumatic haemorrhage, which help to differentiate from a primary hypertensive haemorrhage.

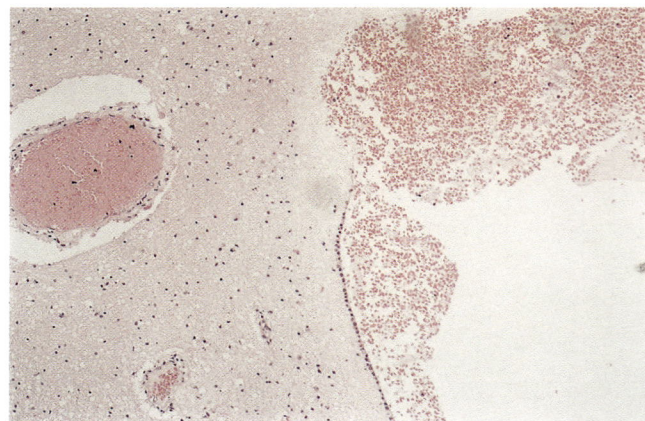

Figure 7.13 Intraventricular haemorrhage with disruption of the ventricular lining in a case of rapid death due to head injury.

Primary traumatic haemorrhages classically involve the dorsolateral quadrant of the rostral brain stem (see Chapter 8, p. 94).

Cerebellar haemorrhages

These most commonly are either traumatic in origin, usually in association with cerebellar contusional injury, particularly in the presence of a fracture contusion (Vrankovic *et al.* 2000), or primarily hypertensive. This latter condition affects the central cerebellum including the dentate nucleus. Other rarer causes include malformations (see later).

NATURAL INTRACEREBRAL HAEMORRHAGE

Spontaneous ICH together with subarachnoid haemorrhage accounts for around 15 per cent of all strokes (Mendelow 1991). The mortality is high and related to site, size and presenting Glasgow Coma Scale (GCS) score (Mackenzie 1996). Brain-stem haematomas are almost uniformly fatal whereas there is a lower mortality for basal ganglia or lobar haematomas. Hypertension is still the major cause of natural ICH, although the incidence has declined with the advent of therapeutic drugs.

HYPERTENSIVE INTRACEREBRAL HAEMORRHAGE

This is the most common cause of ICH and accounts for about 50 per cent of all causes of ICH (Kalimo *et al.* 2002). The majority occur in the deep cerebral structures, with cortical and subcortical areas less commonly involved. In terms of the aetiology, this was widely attributed to the rupture of microaneurysms described in particular detail by Cole and Yates (1967). However, these are rarely identified in postmortem tissue (Takebayashi and Kanekao 1983), although other workers have identified microaneurysms at operation and histologically (Wakai and Nagai 1989). Further studies suggest that these may be complex vascular coils (Challa *et al.* 1992). These vascular coils appear to represent vascular tortuosities rather than true microaneurysms. These are most common at the grey–white interface. The cause of the haemorrhage in hypertension is as yet unclear, although one explanation is fibrinoid necrosis of the small arteries and arterioles, which appears to correspond to areas of focal dilatation experimentally (Fredriksson *et al.* 1988). More

recently arterial dissections of penetrating lenticulostriate arteries in hypertension-induced cerebral haemorrhage have been recorded in surgical specimens (Mizutan *et al.* 2000).

Macroscopic findings

Hypertensive ICH occurs most commonly in the putamen and thalamus (60 per cent), with a lesser proportion in the hemispheres (lobar 20 per cent), cerebellum (13 per cent) and pons (7 per cent) (Kalimo *et al.* 2002) (Figure 7.14a). The common vessels to rupture in the hemispheres are small vessels such as a lenticulostriate vessel or pial perforating artery. Bleeding is usually a one-off event in contrast to the rebleeding that occurs with aneurysms, vascular malformations and amyloid angiopathy.

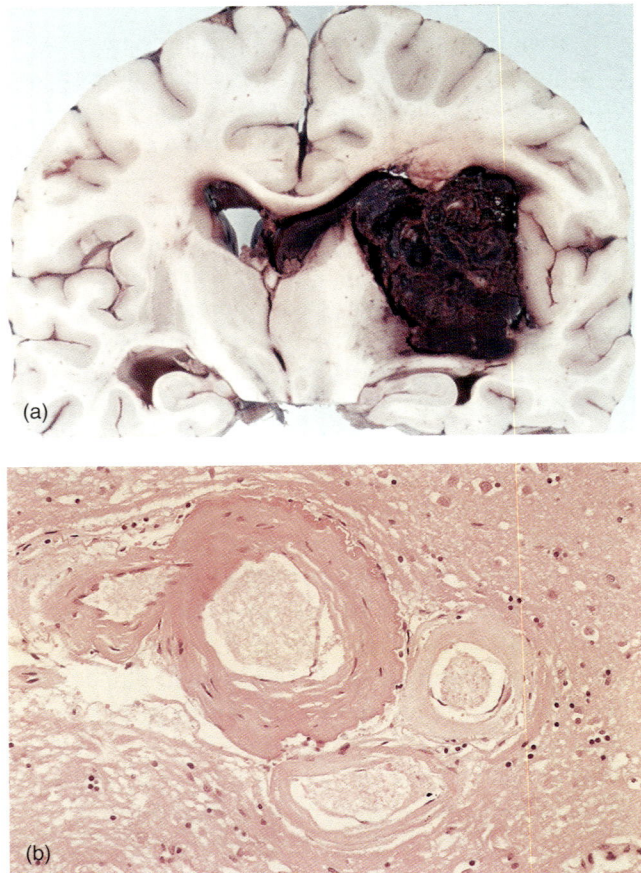

Figure 7.14   (a) Hypertensive haemorrhage with intraventricular extension. (Courtesy of Dr Colin Smith.) (b) Histology of microvascular pathology in hypertension showing arteriosclerotic change and hyaline vessel wall thickening.

Histology

As discussed above microaneurysms are almost never identifiable. Examination of the small vessels shows thickened walls in long-standing hypertension with eosinophilic homogenization which may be caused by either true fibrinoid change or collagenous fibrosis (hyalinosis) (Figure 7.14b). The latter probably results from initial fibrinoid necrosis. Differentiation may be made using stains for fibrin and collagen. Perivascular microinfarcts are commonly associated (Leestma 1988).

Examination of the haematoma shows initially fresh red cells with adjacent neuronal hypoxic–ischaemic change developing after several hours, with variable acute inflammatory infiltrate over 24–48 h. Later changes include macrophage accumulation with haemosiderin deposition. Intact red cells may persist for days to weeks (Perper and Wecht 1980).

Problem areas

- Any attempt to time the age of a haemorrhage is difficult and for the most part may not be necessary in the context of a medicolegal case.
- The issue of aetiology of, in particular, a deep basal haematoma has been discussed above.
- The role of 'stress' is complex. One hypothesis is that acute rise in blood pressure may cause rupture of perforating cerebral vessels (Caplan 1988). This is reported in hypertensive as well as non-hypertensive individuals in the context of emotional or physical stress. This may be mediated in part by catecholamine release. Lammie et al. (2000) reported a case of ICH in an elderly man with known hypertension after severe emotional upset. Small-vessel fibrinoid necrosis of recent origin was identified adjacent to red cell extravasation and remote from the haematoma. It appears likely that altered cerebral vasomotor reactivity or autoregulation that occurs as part of the ageing process and chronic hypertension may predispose a vessel to rupture as well as to increased blood vessel fragility. In any individual case detailed clinicopathological correlation is essential (Black and Graham 2002). However, these cases can be extremely problematic in the context of a criminal case.

ICH IN ASSOCIATION WITH CEREBRAL AMYLOID ANGIOPATHY

Cerebral amyloid angiopathy (CAA) is characterized by extracellular deposition of amyloid in the walls of blood vessels in the meninges and brain. The amyloid deposited most commonly is β-amyloid (Aβ) which is the same as that deposited in the cortical tissues in Alzheimer's disease and Down's syndrome. ICH in association with amyloid accounts for around 12 per cent of all causes of ICH. The incidence of amyloid angiopathy increases with age. The blood vessels affected become rigid and friable which is why re-bleeding is not infrequent. There is a complex association with both APOEε4 and -ε2 alleles (Greenberg et al. 1995, McCarron et al. 1999a, 1999b). APOEε4 is associated with asymptomatic CAA whereas APOEε2 is over-represented in patients who have CAA haemorrhage, often in association with anticoagulant therapy. The condition is often associated with Alzheimer-type pathology but may be seen in isolation. CAA is associated with hypertension in 30 per cent of cases, which may exacerbate the tendency to haemorrhage (Vinters 1987).

Macroscopic findings

The location of this type of haemorrhage is commonly subcortical in a lobar pattern involving the parietal or occipital regions, which may be a guide, at brain examination, that this is not a primarily hypertensive haemorrhage or a traumatic ICH (Figure 7.15a). The haemorrhage may extend into the subarachnoid space and be seen as the dura is reflected. Haemorrhages may be multiple and in some cases petechial in nature.

Histology

The blood vessels most frequently affected are the small-to-medium size arterioles in the cerebral cortex and leptomeninges. It is usually most severe in the occipital and parietal cortex as well as meningeal vessels. There is usually sparing of the hippocampus, basal ganglia and white matter. These are easily identifiable on haematoxylin and eosin (H&E) examination, with the typical pink hyaline appearance of amyloid involving the tunica media and adventitia. Congo red can be used to confirm this (Figure 7.15b) although immunohistochemical staining for Aβ protein is the most sensitive.

Problem areas

- The relationship of haemorrhage to 'stress' (either physical or emotional). This may predispose to ICH, possibly along a similar mechanism to that outlined above with regard to hypertensive haemorrhage.

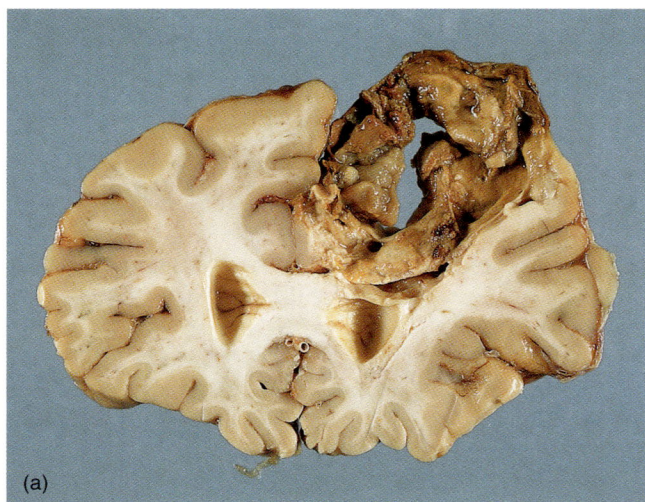

(a)

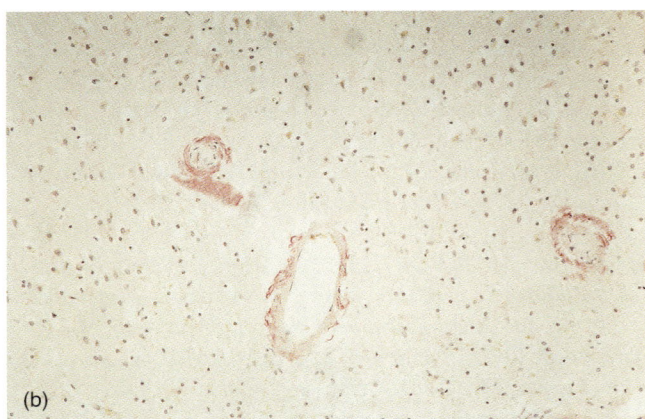

(b)

Figure 7.15 (a) Old haematoma occurring in the background of cerebral amyloid angiopathy. (b) Congo red showing amyloid angiopathy.

In the case of physical trauma this may not necessarily be severe in the background of vascular fragility (Opeskin 1996).

- Association with warfarin treatment: CAA is an important contributor to ICH in patients on warfarin therapy. This has been demonstrated by the histological diagnosis as well as the elevated frequency of *APOEε2* allele in patients with CAA haemorrhage (Rosand *et al.* 2000).

OTHER CAUSES OF INTRACEREBRAL HAEMATOMA

Vascular malformations

There are various types of vascular malformations that may cause intracerebral haemorrhage and, to a varying

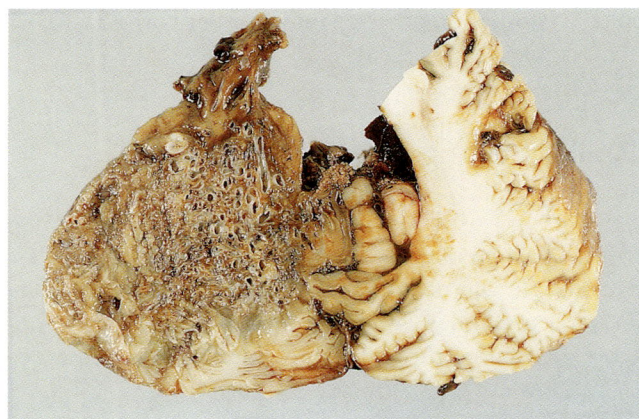

Figure 7.16 Cerebellar vascular malformation – mixed arteriovenous with cavernous areas.

degree, subarachnoid haemorrhage. Precise classification into a specific group may be difficult at times.

Arteriovenous malformations

These generally present in patients aged 20–40 years, affecting 0.01–0.5 per cent of the population (Fleetwood 2002). These mainly arise on the surface of the cerebral hemispheres, particularly in the middle cerebral territories, and consist of abnormal arterial and venous structures usually in association with rather gliotic, hyalinized brain (Figure 7.16). They also occur in deep locations such as the thalamus or basal ganglia where there is a higher incidence of ICH (Fleetwood 2003). They are an unpredictable cause of haemorrhage with a significant morbidity and mortality rate of between 1 and 4 per cent (Mackenzie 1996). They may be associated with aneurysms.

Cavernous angiomas

These occur most often in children. Identification may be difficult angiographically. However, on MRI a cortical mass of abnormal vessels surrounded by cerebral parenchyma with haemosiderin is identifiable. Histologically, these consist of vessels with collagenous walls with no muscle or elastic tissue. Presentation as an intracerebral haemorrhage is less common than that of seizures, headache or other neurological deficits.

Venous angiomas

These consist of veins surrounded by normal brain tissue, which drains either into the leptomeninges or into the central venous system. They may present with seizures or haemorrhage.

Capillary telangiectasias

These are normally incidental findings at postmortem examination presenting as areas of haemorrhagic blush frequently in the pons (Challa *et al.* 1995) and bleeding rarely occurs. They comprise dilated small vessels randomly distributed throughout the brain parenchyma.

CONCLUSION

Intracerebral haemorrhages have a variety of causes. Particularly problematic is the relationship between natural and traumatic disease, as well as drugs and alcohol.

REFERENCES

Adams, J.H., Doyle, D., Graham, D.I. *et al.* 1985. The contusion index: a re-appraisal in man and experimental non-missile head injury. *Neuropathol Appl Neurobiol*, **11**, 299–308.

Adams, J.H., Doyle, D. and Graham, D.I. 1986. Deep intracerebral (basal ganglia) haematomas in fatal non-missile head injury in man. *J Neurol Neurosurg Psychiatry*, **49**, 1039–43.

Anderson, R.McD. and Opeskin, K. 1998. Timing of early changes in brain trauma. *Am J Forensic Med Pathol*, **19**, 1–9.

Baratham, G. and Dennyson, W.G. 1972. Delayed traumatic intracerebral haemorrhage. *J Neurol Neurosurg Psychiatry*, **35**, 698–706.

Black, M. and Graham, D.I. 2002. Sudden unexplained death in adults caused by intracranial pathology. *J Clin Pathol*, **55**, 44–50.

Boto, G.R., Lobato, R.D., Rivas, J.J., Gomez, P.A., de la Lama, A. and Lagares, A. 2001. Basal ganglia hematomas in severely head injured patients: clinicoradiological analysis of 37 cases. *J Neurosurg*, **94**, 224–32.

Caplan, L.R. 1988. Intracerebral haemorrhage revisited. *Neurology*, **38**, 624–7.

Challa, V.R., Moody, D.M. and Bell, M.A. 1992. The Charcot–Bouchard aneurysm controversy: impact of a new histological technique. *J Neuropathol Exp Neurol*, **51**, 264–71.

Challa, V.R., Moody, D.M. and Brown, W.R. 1995. Vascular malformations and the central nervous system. *J Neuropathol Exp Neurol*, **54**, 609–21.

Cole, F.M. and Yates, P.O. 1967. The occurrence and significance of intracerebral microaneurysms. *J Pathol Bacteriol*, **93**, 393–411.

Dawson, S.L., Hirsch, C.S., Lucas, F.V. and Sebek, B.A. 1980. The contre-coup phenomenon. Reappraisal of a classic problem. *Human Pathol*, **11**, 155–66.

DiMaio, V.J. and DiMaio, D. 2001. Trauma to the skull and brain: craniocerebral injuries. In *Forensic Pathology*, 2nd edn. Boca Raton, FL: CRC Press.

Engel, S., Schluesener, H., Mittelbronn, M. *et al.* 2000. Dynamics of microglial activation after human traumatic brain injury are revealed by delayed expression of macrophage-related proteins MRP8 and MRP14. *Acta Neuropathol*, **100**, 313–22.

Fleetwood, I.G., Marcellus, M.L., Levy, R.P., Marks, M.P. and Steinberg, G.K. 2003. Deep arteriovenous malformations of the basal ganglia and thalamus: natural history. *J Neurosurg*, **98**, 747–50.

Fleetwood, I.G. and Steinberg, G.K. 2002. Arteriovenous malformations. *Lancet*, **359**, 863–73.

Fredericksson, K., Nordberg, C., Kalimo, H. *et al.* 1988. Cerebral microangiopathy in stroke-prone spontaneously hypotensive rats. An immunohistochemical and ultrastructural study. *Acta Neuropathol*, **75**, 241–52.

Graham, D.I., Gennarelli, T.A. and McIntosh, T.K. 2002. Trauma. In Graham, D.I. and Lantos, P.L. (eds), *Greenfield's Neuropathology*, 7th edn. London: Arnold, Chapter 14, 823–98.

Greenberg, S.M., Rebeck, G.W., Vonsattel, J.P.G. *et al.* 1995. Apolipoprotein E 4 and cerebral haemorrhage associated with amyloid angiopathy. *Ann Neurol*, **38**, 254–9.

Hangartner, J.R.W., Marley, N.J., Whitehead, A., Thomas, A.C. and Davies, M.J. 1985. The assessment of cardiac hypertrophy at postmortem examination. *Histopathology*, **9**, 1295–306.

Holmin, S., Söderland, J., Biberfeld, P. and Mathiesen, T. 1998. Intracerebral inflammation after human brain contusion. *Neurosurgery*, **42**, 291–8.

Kalimo, H., Kaste, M. and Haltia, M. 2002. In Graham, D.I. and Lantos, P.L. (eds), *Greenfield's Neuropathology*, 7th edn. London: Arnold, Chapter 6.

Lammie, G.A., Lindley, R., Weir, S. and Wiggam, I.M. 2000. Stress-related primary intracerebral haemorrhage: postmortem examination clues to underlying mechanism. *Stroke*, **31**, 1426–8.

Leestma, J.E. 1988. *Forensic Neuropathology*. New York: Raven Press.

McCarron, M.O., Nicholl, J.A.R., Ironside, J.W. *et al.* 1999a. Cerebral amyloid angiopathy-related haemorrhage. Interaction of APOEe2 with putative risk factors. *Stroke*, **30**, 1643–6.

McCarron, M.O., Hoffmann, K.L., DeLong, D.M., Gray, L., Saunders, A.M. and Alberts, M.J. 1999b. Intracerebral hemorrhage outcome: Apolipoprotein E genotype, hematoma, and edema volumes. *Am Acad Neurol*, **53**, 2176–9.

Mackenzie, J.M. 1996. Intracerebral haemorrhage. *J Clin Pathol*, **49**, 360–4.

Makino, Y., Sannohe, S., Kita, T. and Kuroda, N. 2001. Morphological changes of cerebral ventricular wall in head injuries observed with large histological specimens – Can it be the evidence of the impact to the brain? 7th Indo-Pacific Congress on Legal Medicine and Forensic Sciences, Melbourne.

Mendelow, A.D. 1991. Spontaneous intracerebral haemorrhage. *J Neurol Neurosurg Psychiatry*, **54**, 193–5.

Mizutani, T., Kojima, H. and Miki, Y. 2000. Arterial dissections of penetrating cerebral arteries causing hypertension-induced cerebral haemorrhage. *J Neurosurg*, **93**, 859–62.

Morrison, A.L., King, T.M., Korell, M.A. *et al.* 1998. Acceleration–deceleration injuries to the brain in blunt force trauma. *Am J Forensic Med Pathol*, **19**, 109–12.

Nanassis, K., Frowein, R.A. and Karimi, A. 1989. Delayed post-traumatic intracerebral bleeding. *Neurosurg Rev*, **12**(suppl 1), 243–51.

Oehmichen, M., Walter, T., Meissner, C. and Friedrich, H.J. 2003. Time course of cortical hemorrhages after closed traumatic brain injury: statistical analysis of post-traumatic histomorphological alterations. *J Neurotrauma*, **20**(1), 87–103.

Opeskin, K. 1996. Cerebral amyloid angiopathy: A review. *Am J Forensic Med Pathol*, **17**, 248–54.

Perper, J.A. and Wecht, C.H. (eds) 1980. *Microscopic Diagnosis in Forensic Pathology*. Springfield, IL: Thomas, 12.

Rosand, J., Hylek, E.M., O'Donnell, H.C. and Greenberg, S.M. 2000. Warfarin-associated hemorrhage and cerebral amyloid angiopathy: A genetic and pathologic study. *Am Acad Neurol*, **55**, 947–51.

Ryan, G.A., McLean, A.J. and Vilenius, A.T.S. 1994. Brain injury patterns in fatally injured pedestrians. *J Trauma*, **36**, 469–76.

Saukko, P. and Knight, B. 2004. *Knight's Forensic Pathology*, 3rd edn. London: Arnold.

Shreiber, D.I., Bain, A.C., Ross, D.T. *et al.* 1999. Experimental investigation of cerebral contusion: histopathological and immunochemical evaluation of dynamic cortical deformation. *J Neuropathol Exp Neurol*, **58**, 153–64.

Siddique, M.S., Gregson, B.A., Fernandes, H.M. *et al.* 2002. Comparative study of traumatic and spontaneous intracerebral hemorrhage. *J Neurosurg*, **96**, 86–9.

Takebayashi, S. and Kanekao, M. 1983. Electron microscopic studies of ruptured arteries in hypertensive intracerebral haemorrhage. *Stroke*, **14**, 28–36.

Vintners, H.V. 1987. Cerebral amyloid angiopathy. A critical review. *Stroke*, **18**, 311–24.

Vrankovic, D., Splavski, B., Hecimovic, I. *et al.* 2000. Anatomical cerebellar protection of contrecoup hematoma development. Analysis of the mechanism of 30 posterior fossa coup hematomas. *Neurosurg Rev*, **23**, 156–60.

Wakai, S. and Nagai, M. 1989. Histological verification of microaneurysm as a cause of cerebral haemorrhage in surgical specimens. *J Neurol Neurosurg Psychiatry*, **52**, 595–9.

Yajima, Y., Hayakawa, S., Mimasaka, M. *et al.* 2001. Intracerebral haematoma: traumatic or non-traumatic. *J Clin Forensic Med*, **8**, 163–5.

Yanagida, Y., Fjiwara, S. and Mizoi, Y. 1989. Differences in the intracranial pressure caused by a 'blow' and/or a 'fall' – an experimental study using physical models of the head and neck. *Forensic Sci Int*, **41**, 135–45.

8

Primary traumatic brain injury

Jennian F Geddes

With current technology, the only types of primary traumatic brain injury that are detectable are damage to blood vessels and damage to axons.

The principal topic of this chapter is damage to axons covering the areas of diffuse axonal injury and the less severe degrees of traumatic axonal injury. However, damage to blood vessels commonly occurs with axonal damage in the more severe head injuries, so that small haemorrhages in characteristic sites act as useful macroscopic indicators – to clinicians, on a scan, or to a pathologist, examining a brain – of associated axonal pathology. The actual damage depends on the forces involved. A range of scattered haemorrhages occur: the most severe are the widespread petechial white matter haemorrhages ('diffuse vascular injury') seen in people who have died almost immediately after very severe head trauma (Figure 8.1a) – the most minor, isolated, tiny tissue tear haemorrhages in the corpus callosum. In general terms, a collection of small haemorrhages in characteristic sites, such as the parasagittal white matter, corpus callosum, deep grey matter, cerebellar peduncles and the upper brain stem, is a more sinister finding than contusions, because of the association with severe axonal injury. It is important to look carefully, because many such haemorrhages are macroscopically trivial lesions, which are easily missed when the brain is sliced, particularly if it is cut fresh at the time of the postmortem examination.

In the parasagittal region of the frontal and parietal lobes, severe acceleration produces typical linear or flame-shaped traumatic haemorrhages in the subcortical white matter, which are (rather confusingly) known as 'gliding contusions'. The term is confusing because they differ radically from the sort of contusions described in Chapter 7, in that they are not superficial, do not involve the cortex and are

caused by movement, rather than impact. Gliding contusions, which are often bilateral, are illustrated in the Figure 8.2b.

Excessive movement also damages the corpus callosum and this is another typical site for traumatic 'tissue tear' haemorrhages. These may be extremely small and insignificant-looking, as may the small deep haemorrhages in the basal ganglia that are frequently also associated with axonal damage (Figure 8.2b) (Adams *et al.* 1986, Geddes and Graham 2004), but they should not be ignored. If they occur in conjunction with small haemorrhages in the rostral midbrain and pons, usually its dorsolateral sector (Figure 8.3), they are pathognomonic of diffuse axonal injury or DAI (Adams *et al.* 1989a, Graham *et al.* 2002). Note that these traumatic brain-stem lesions are different in extent and topography from the secondary 'Duret' haemorrhages seen as a result of raised intracranial pressure (Figure 8.4) (see Chapter 9, p. 111).

DAI AND LESSER DEGREES OF TRAUMATIC AXONAL INJURY

The fact that the axons might be damaged in a head injury was first demonstrated nearly 50 years ago by Strich (1956, 1961), who performed a neuropathological study of the brains of patients who had been vegetative after a head injury. She described widespread microscopic damage or 'diffuse degeneration of the cerebral white matter' and suggested that it resulted from shearing of axons at the time of injury. It was not until the 1980s, however, that the conditions necessary to cause such brain damage were known: working on primates, Gennarelli *et al.* (1982)

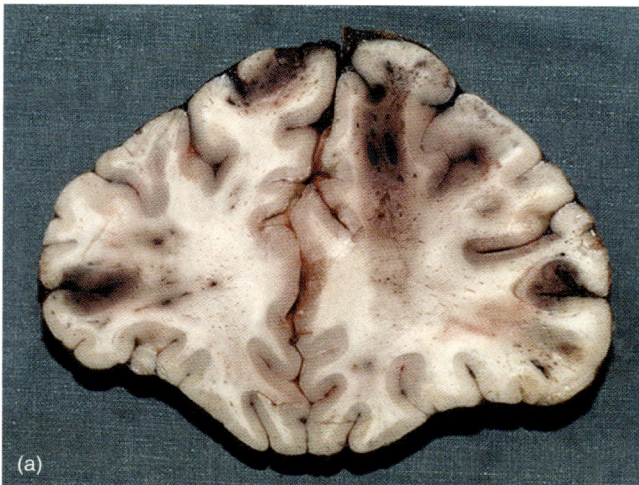

Figure 8.1   Diffuse vascular injury in the brain of a road traffic collision victim who died of a head injury almost immediately after the incident. (a) A gliding contusion is present in the frontal lobe of the hemispheric slice, adjacent to the midline, and there were numerous petechial haemorrhages scattered through the central white matter. (b) The lesions in the brain stem are in a distribution typical of diffuse axonal injury (DAI), i.e. in the dorsal part of the brain stem. Damage as widespread as this is assumed to be incompatible with life. (Reproduced from Geddes and Graham 2004, by kind permission of Butterworth-Heinemann.)

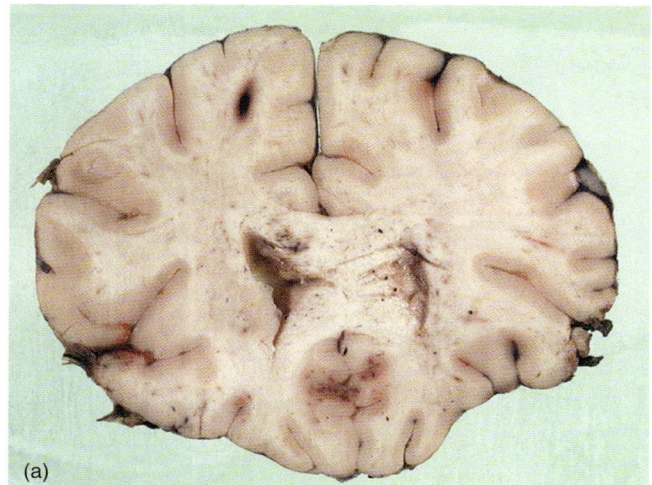

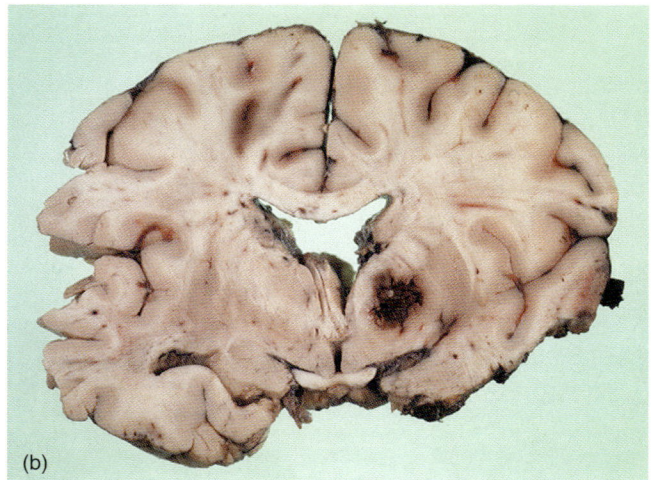

Figure 8.2   (a) and (b) Macroscopic markers of diffuse axonal injury (DAI) at brain cut: gliding contusions in the parasagittal white matter of both slices; there is a lesion in the corpus callosum at both levels and also a typical small deep grey matter haemorrhage, not exerting mass effect. Such a patient would be likely to have brain stem lesions as well. (Reproduced from Geddes and Graham 2004 by kind permission of Butterworth Heinemann.)

were able to show that DAI was the result of angular or rotational acceleration of high magnitude, i.e. that it was primarily a non-impact phenomenon. This explained the typical situations in which DAI occurred in human head injury – high-speed road traffic deaths and falls from a height – where excessive movement of the brain was likely to occur (Adams *et al.* 1991). Meticulous neuropathological studies by the Glasgow group of Adams and Graham (Adams *et al.* 1984, 1989a, 1989b, Graham *et al.* 1983,

Figure 8.3 Diffuse axonal injury (DAI): dorsolateral quadrant haemorrhages in the brain stem.

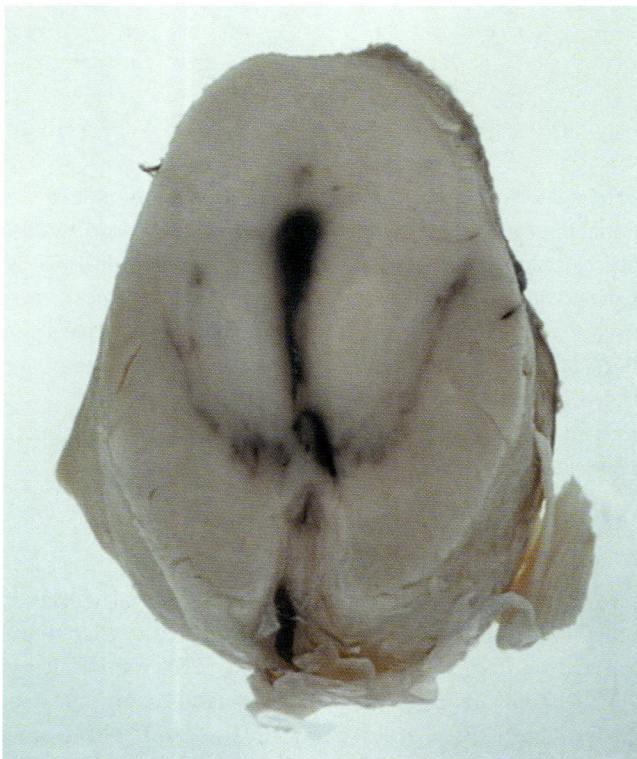

Figure 8.4 A secondary 'Duret'-type haemorrhage in the brain stem, resulting from mass effect and shift. Contrast with Figure 8.3.

1989, 1992, 1993, 1995) helped to provide a clear picture of this very severe type of pathology. DAI was defined primarily as a clinicopathological entity resulting from trauma, where the victim was unconscious from the time of injury – typically, without structural changes on the scan to account for his or her loss of consciousness. The widespread nature of the damage caused prolonged coma, with survivors remaining either vegetative or with severe neurological disability.

Once the conditions necessary for DAI were established, it was possible to develop cheap and reproducible non-primate models, to enable axonal damage to be studied at a cellular level. A number of such models have been used and have helped greatly in the understanding of what happens to axons after injury (Gennarelli 1996). This understanding has in turn led to the development of immunocytochemical markers of axonal damage, of which the most used in diagnostic neuropathology is an anti-serum to amyloid precursor protein (APP, formerly termed β-APP) (Sherriff et al. 1994). Systematic neuropathological studies employing this antibody have shown that a degree of damage to axons probably occurs in the vast majority of head injuries (Gentleman et al. 1995) and that the subsequent clinical picture is dictated by the amount and distribution of that damage. It is now understood that traumatic axonal injury (TAI) comprises a spectrum (Geddes et al. 2000). At the mild end of the spectrum, there is good experimental evidence to suggest that some axonal damage may be reversible (Tomei et al. 1990, Maxwell et al. 1994) and, indeed, that mild reversible axonal injury may be the neuropathological substrate of concussion (Gennarelli et al. 1998).

At the other end of the spectrum lies DAI, where strains on axons are so severe that they lead to irreversible damage in a widespread ('diffuse') distribution, in the brain stem as well as in the cerebral hemispheres, leading to the clinical features described above. In between these two extremes of TAI lies a range of damage that falls short of DAI in severity, for which we do not know the clinical correlates. However, rare opportunities to study systematically the brains of individuals who suffered a recent documented mild head injury, but died of an unrelated cause, suggest that scattered traumatic axonal damage, usually hemispherical, may be a common finding after mild head injury in patients who may have had few neurological or psychological symptoms (Blumbergs et al. 1994, 1995; unpublished personal observations). Biomechanical work supports the experience of neuropathologists by showing that proportionally greater force is required to produce the axonal damage in the brain stem which is the cardinal feature of DAI (Smith et al. 2000).

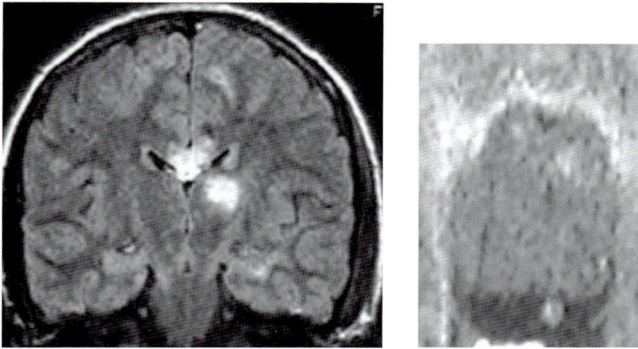

Figure 8.5 Images such as these, obtained by magnetic resonance imaging, confirm a clinical diagnosis of severe diffuse traumatic axonal injury. There are lesions in the corpus callosum, deep grey matter/posterior limb of the internal capsule and in the rostral midbrain. Compare the brain-stem image with Figure 8.3. (Reproduced from Graham *et al.* 2002 by kind permission of Arnold.)

There are sometimes clues in the history that DAI was present. Someone who was unconscious from the moment of injury and who is in a coma that is not explained either by a lesion such as a haematoma on scan or by severe brain swelling must have diffuse brain damage, which is most likely to be DAI. This likelihood is increased by the presence of gliding contusions and the small haemorrhages described above (Figure 8.5). In such situations the only real alternative diagnosis is that the diffuse damage responsible for coma is the result of widespread hypoxic damage. Histology with widespread sampling of appropriate areas is necessary to be sure.

Experimental work shows that extreme forces are necessary for axons to rupture immediately and that in virtually all head injuries axotomy is likely to be a secondary event, occurring some hours after injury as a result of damage to the axonal cytoskeleton. In people dying rapidly after a very severe head injury, whose brains show widespread petechial haemorrhages, it is generally assumed that primary axotomy has occurred all over the brain and that this is incompatible with life. This is an assumption made because there is evidence of widespread vascular damage; it cannot of course be confirmed because survival of at least 2 h is necessary to be able to do so by immunohistochemistry. Note that apart from such circumstances DAI is rarely a cause of death on its own. It is, however, a cause of prolonged coma and an important cause of vegetative state (Graham *et al.* 1983, Kinney and Samuels 1994, Adams *et al.* 1999) or severe post-traumatic neurological disability. TAI localized to the brain stem *may* be a cause of respiratory failure – indeed, this is probably an important mechanism of death in infant head injury

(Geddes *et al.* 2001) and in those who have high cervical injuries.

SAMPLING FOR WIDESPREAD PARENCHYMAL DAMAGE

How relevant is all this to forensic practice? Sometimes not at all. If there is a skull fracture and a large extradural haematoma, or an acute subdural with mass effect and shift, the cause of death will be obvious and whether or not TAI is present will almost certainly be irrelevant. However, in medicolegal work it is seldom just a question of documenting pathology, and histology may be essential in order to provide answers to questions posed by others, which may arise some considerable time later in the course of legal proceedings.

As TAI may be very widespread, it follows that any attempt to diagnose axonal damage must be based on very extensive systematic sampling of the brain. As with most other forms of generalized cerebral pathology, whether global hypoxia, encephalitis, hypoglycaemia or neurodegenerative disease, areas of the brain differ in their susceptibility and one or two samples can give little indication of what is going on in the brain as a whole (Graham *et al.* 2004). In trauma, as has already been said, it is the midline structures that are particularly vulnerable to microscopic damage, which means that parasagittal white matter, corpus callosum, posterior limb of the internal capsule and upper brain stem must all be sampled, as a minimum. It is only by examining sections from a number of blocks from anatomically defined areas that one can estimate the extent of traumatic pathology in a given brain. Only then can one say for certain whether:

- there is no axonal injury
- DAI is present
- there is traumatic axonal damage, but not distributed widely enough to be called DAI
- there is axonal pathology which may not be traumatic.

To assess the extent of hypoxic damage (and this is not infrequently an issue), different areas of the brain need to be examined. Recommended blocks to cover all types of diffuse brain injury in a case of head injury are given in the appendices to Chapter 2. If the brain has to be returned for burial and not retained for any length of time, it is advisable to take all the samples listed so that any questions that arise subsequently can be properly addressed.

AXONAL DAMAGE LOCALIZED TO THE CRANIOCERVICAL JUNCTION

Injuries that cause hypertension/hyperflexion of the neck cause direct damage to the brain stem, which at its most extreme involves transection of the pontomedullary junction (Keane 1978, Gennarelli *et al.* 1981, Simpson *et al.* 1989, Ezzat *et al.* 1995). Full pontomedullary rents are rare (and, it should be said, easily created when inappropriate force is used to remove the brain *post mortem*); partial transection of the lower brain stem is even rarer. However, although complete transection is immediately fatal, survival after partial damage has been described (Pilz *et al.* 1982, Pilz 1983). If there has been any survival, the damage can be confirmed as genuine by both the presence of associated bleeding, caused by tearing of blood vessels at the time of injury and, histologically, immunohistochemistry and routine stains – provided that about 2 h have elapsed since injury.

Lesser degrees of stress or damage centred on the craniocervical junction may lead to non-disruptive injury to the neuraxis in that region, only detectable histologically, i.e. *localized* TAI. This is a separate phenomenon from the more widespread axonal damage caused by acceleration of the head, described above and it may occur in the absence of axonal injury elsewhere in the brain. The mechanism of damage is likely to be stretch to the lower brain stem, where long tracts are particularly vulnerable, principally the corticospinal tracts in the lower pons and medulla. Damage to long tracts has been reported in cases from all age groups – in infants presumed to have suffered inflicted head injury, and in adults with high cervical dislocation or hyperextension injuries to the neck (Lindenberg and Freytag 1970, Hardman 1979, Riggs and Schochet 1995, Geddes *et al.* 2000, 2001). Blocks of several levels of lower brain stem may be needed to detect it, because the pathology varies greatly in severity from case to case (Figure 8.6).

EVOLUTION OF TRAUMATIC AXONAL DAMAGE

Neuropathological and animal studies have shown clearly that axonal damage starts at the time of injury and continues to progress in the hours, days, weeks and months that follow (Povlishock 1993, Geddes *et al.* 1997, 2000, Gennarelli 1997, Maxwell *et al.* 1997, Gennarelli *et al.* 1998, Graham *et al.* 2002). They have also established that few if any axons are transected at the time of head injury. The primary event is damage to the axonal membrane

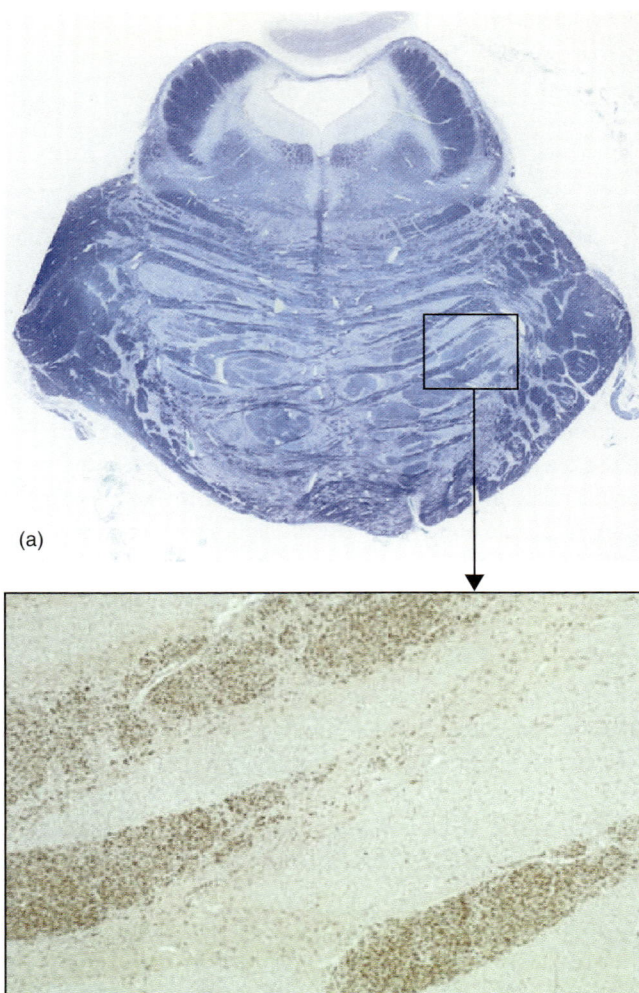

(a)

Figure 8.6 Localized TAI at the craniocervical junction, demonstrated by immunohistochemistry for APP. The low power views show the anatomical restriction of the axonal damage. (a) Damage in the corticospinal tracts in a 32-year-old man, with fracture-dislocation of C1–C2.

and the underlying cytoskeleton. This varies according to the type of strains to which the axon is subjected, but in most cases results in rapid interruption of axonal transport. Immunohistochemistry for axonal damage capitalizes on this, by using antibodies to proteins that are normally transported down the axon in very small quantities, which then accumulate at any point at which transport is impeded. APP is one such molecule. Its use has been systematically studied by a number of different workers and it is known to be detectable at approximately 2 h after injury. Cases in which survival has been very short provide the greatest diagnostic challenges (discussed below). At 2 h (Figure 8.7) axonal immunoreactivity for APP can be seen in scattered axons, some of which may appear enlarged – even resembling small bulbs. At this early stage

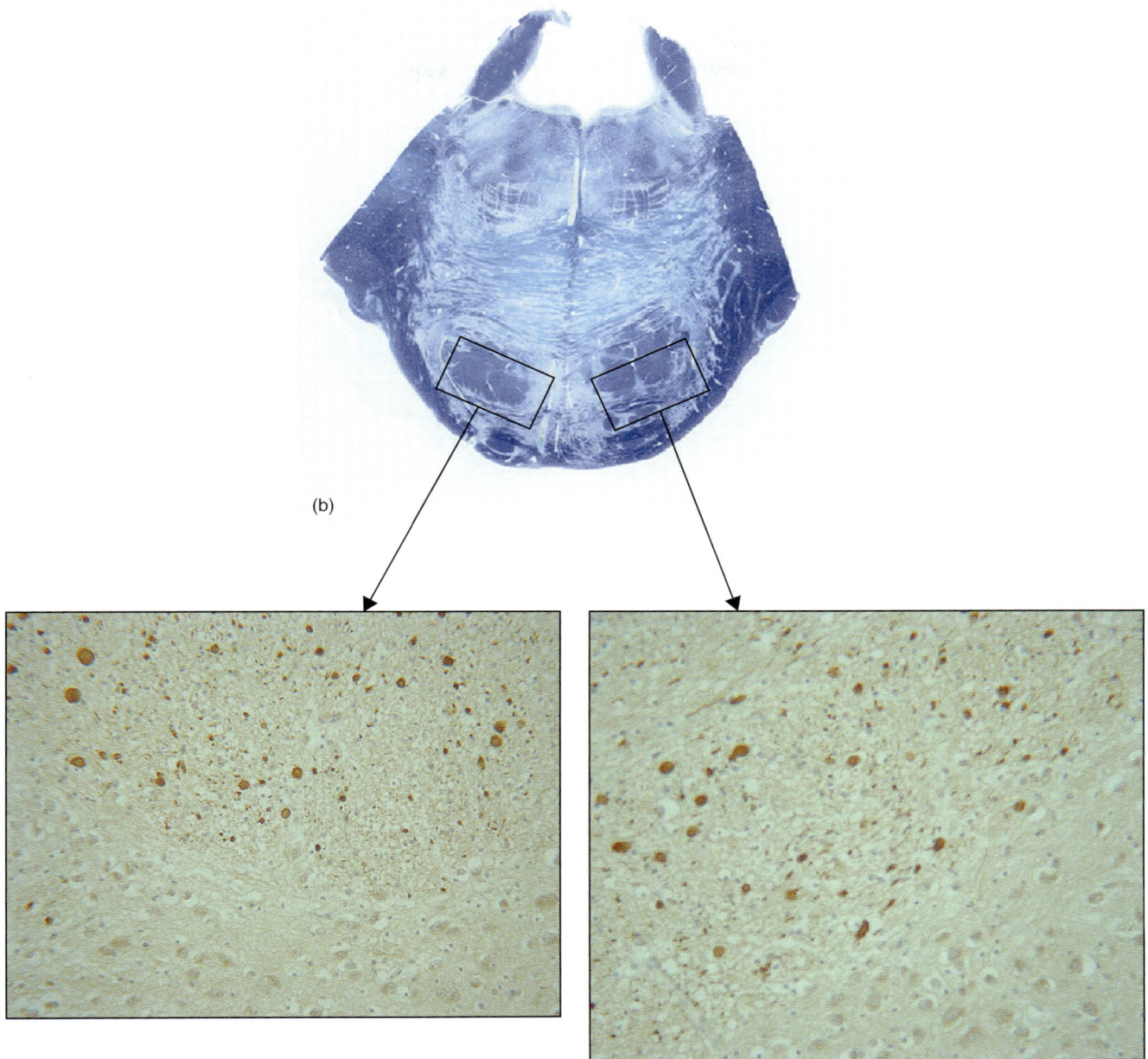

Figure 8.6 Localized TAI at the craniocervical junction. (b) Immunoreactive axonal swellings in the same tracts in the pons of a 2-month-old boy, who died of an inflicted injury. Though fewer than in (a) the swellings are still strikingly restricted to the descending long tracts.

pathology cannot be detected on haematoxylin and eosin (H&E) or traditional silver stains.

A single study of fatal paediatric head injury in road traffic deaths has documented finding damaged (APP-positive) axons at exceptionally short survival times in the internal capsule, thalamus and brain stem in an infant of 3 months and in the internal capsule of a 2-year-old boy. Times of death were well documented for both cases, survival being only 35 min for the infant and 45 min for the older child (Gorrie *et al.* 2002). Such findings may reflect factors such as immaturity of axons in young children or increased vulnerability of certain anatomical brain locations. They indicate that caution should be exercised in interpreting APP expression in children's brains, especially infants, as there are no studies addressing the temporal evolution of axonal damage in the paediatric population.

(c)

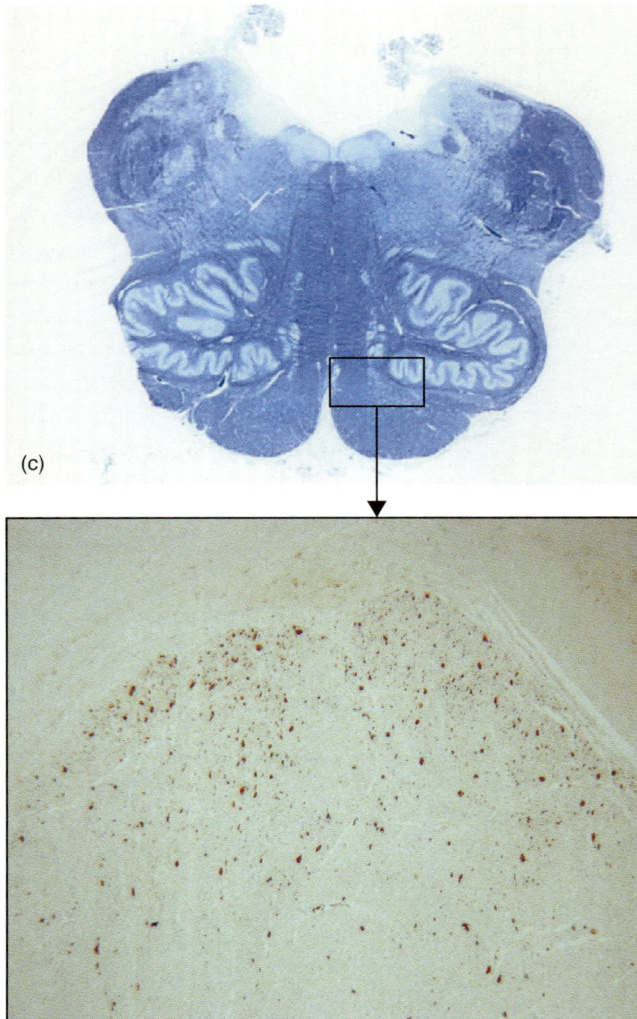

Figure 8.6 Localized TAI at the craniocervical junction. (c) A further example of localized TAI. Damaged axons in the left pyramid of a 31-year-old man who died after falling 25 feet. There was no axonal damage elsewhere in the brain of any of the three cases shown in Figure 8.6, and in each the damage was bilateral. The severity of damage seen in (a) is unusual.

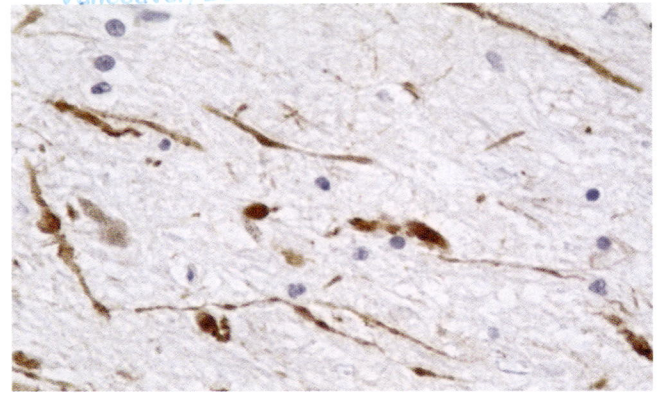

Figure 8.7 With around 2 h survival after injury, damaged axons can be detected with immunohistochemistry (here, with APP). The axonal swellings are not fully formed and would be very difficult to find on staining with haematoxylin and eosin (H&E). (Reproduced from Geddes and Graham 2004 by kind permission of Butterworth Heinemann.)

Whatever the age group, however, care should be taken when interpreting findings in cases of very short survival (see below).

The temporal evolution of axonal damage in terms of morphology and cellular reaction has been well described in adults (Geddes *et al.* 1997, Graham *et al.* 2002), although it must be emphasized that the time course given in such publications is only an outline and takes no account of factors such as age or axonal size and for that reason should not be used to give more than a rough indication of the timing of the pathology. The situation is further confounded by the fact that there is some indication that axons of different sizes may react at different rates and that secondary axotomy may continue to occur well beyond the first 24 h after injury (Maxwell *et al.* 2003, Stone *et al.* 2004). For this reason alone, the suggestion that measuring an axonal bulb can give an indication of the timing of the injury (Wilkinson *et al.* 1999) is unlikely to prove useful – in fact, the idea has recently been rebutted by a careful study (Leclercq *et al.* 2002).

By about 12–18 h after injury, the damaged axons demonstrated by anti-APP will have swollen greatly, as a result of accumulation of substances that have continued to be transported down from the cell body. At this stage the axonal swellings or bulbs are sufficiently large to detect by routine methods such as H&E or silver preparation. By 18–24 h many of these swollen axons will have become severed from their distal portion, i.e. have undergone secondary axotomy – and the damage is irreversible. Traditionally, bulbs are demonstrated by silver preparations, but axons often appear to have varicosities on silver, which may be difficult to interpret. To indicate the extent of the damage most neuropathologists now find immunohistochemistry for APP to be a more helpful adjunct to H&E staining.

Once the bulbs are a few days old, there is no difficulty distinguishing traumatic from ischaemic axonal damage, because the usual tissue reaction will be present if there is early infarction. With about a week's survival after injury, APP staining of bulbs becomes much less uniform and begins to fade, until eventually they are no longer immunoreactive, although some axons adjacent to the bulbs remain strongly positive for APP (Figure 8.8) (Geddes *et al.* 1997).

In a situation in which there are many bulbs *not* staining with anti-APP while nearby axons are still positive, it is probably safe to say that injury occurred at least a week before death.

The other feature that is typical of evolving axonal damage is the arrival of microglia. These cells, which can be well demonstrated by antibodies to CD68 (the author finds the PG-M1 epitope demonstrates microglia better

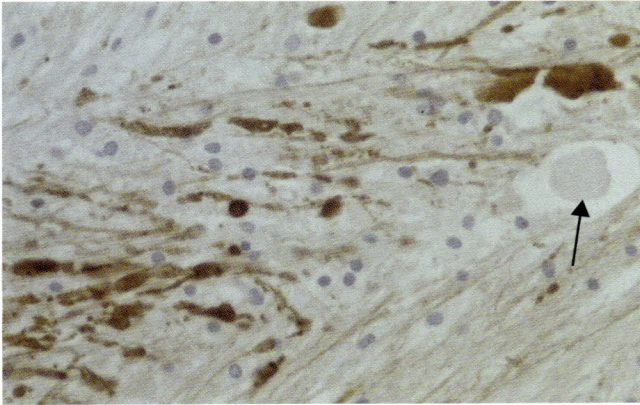

Figure 8.8 Immunoreactivity for APP tends to start fading from the axonal swellings or 'bulbs' from about 5 days after injury, although strong expression continues to be seen in axons and varicosities. This high-power field from the internal capsule of a man who had a 10-day survival shows a large bulb on the right, which is completely negative for APP (arrowhead), whereas adjacent damaged axons still stain.

than KP1), are an important feature of the brain's reaction to injury. For this reason routine staining for CD68 is helpful in cases where there is survival of more than a few days. The initial increase in microglial numbers after injury is subtle and not constant: non-specific upregulation in the white matter after head injury can be detected from about 48 h (although to be certain of this a number of large blocks is needed as well as familiarity with the numbers of microglia normally present in the non-traumatized brain). However, at about 10 days, microglia start collecting round damaged axons in clusters or nodules, which are well shown by CD68 immunohistochemistry. Microglial clusters remain long after the axonal swellings have disappeared and can be seen scattered through the white matter (Figure 8.9) several months after the original injury, as tombstones, marking the site of axonal damage. Note, however, if microglial nodules are found scattered through the white matter in someone who has died of a very recent head injury (or, indeed, of any other cause), it cannot be assumed that the nodules are the result of previous traumatic damage. If there was a definite history of trauma, they may indeed be that, but microglial aggregates occur in a number of conditions, of which previous episodes of severe global hypoxia (Figure 8.10) and viral encephalitis are the two main examples. Isolated microglial clusters, encountered on their own, are commonly seen in routine neuropathological practice; they should be ignored. However, in the absence of a history of a devastating previous condition such as global

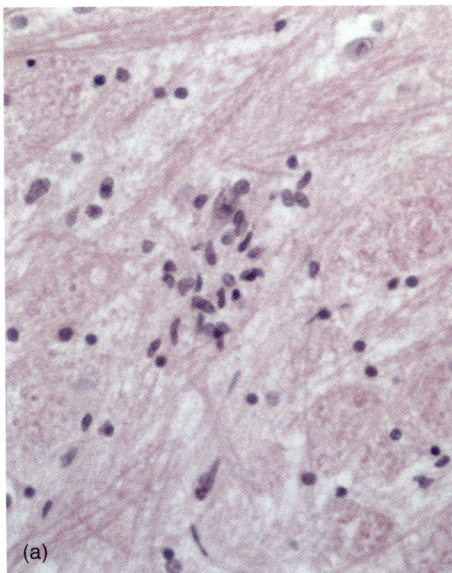

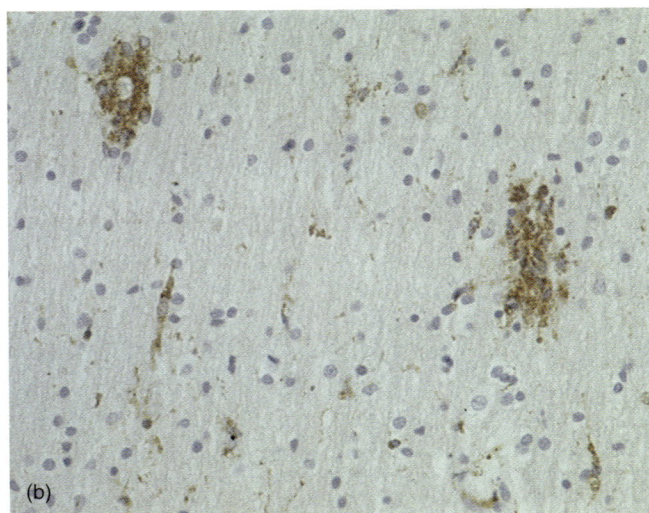

Figure 8.9 Microglial clusters in a patient with diffuse axonal injury (DAI) who survived 5 months. (a) Aggregates of microglia in the medulla can be detected on H&E staining, but are more easily demonstrated by (b) immunohistochemistry (anti-CD68/PG-M1, in the hemispheric white matter of the same patient).

hypoxia or fat embolism, microglial nodules scattered throughout the white matter would be compatible with a previous head injury.

As the central white matter degenerates, secondary long tract degeneration starts. This takes place in the months following injury and is seen in descending, particularly

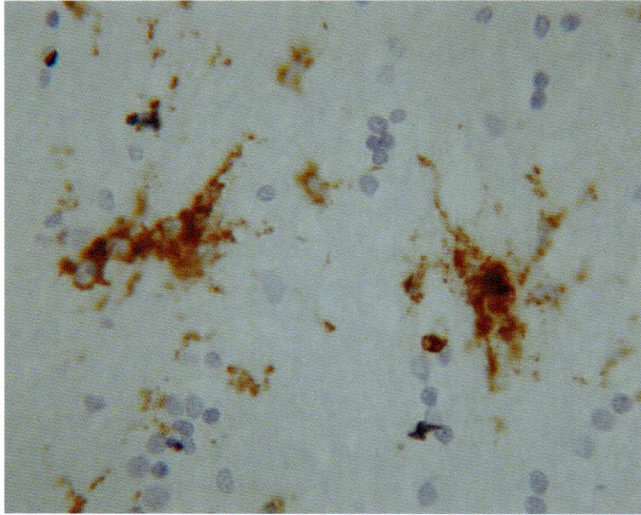

Figure 8.10 Microglial clusters scattered through the brain are not specific for trauma. The image comes from the hemispheric white matter of a patient with no previous history of head injury, but who was severely disabled after an episode of global hypoxia some months earlier.

corticospinal, tracts in the brain stem. The traditional neuropathological method of showing long tract degeneration in such cases is to demonstrate the tell-tale breakdown products of myelin in long tracts by the Marchi technique, although immunohistochemistry for CD68 (PG-M1) is quicker and serves just as well, because it will show the large numbers of foamy macrophages involved in the degeneration of the axons, anatomically confined to the descending brain-stem tracts (Figure 8.11b).

PITFALLS IN THE INTERPRETATION OF AXONAL DAMAGE

The arrival of immunohistochemistry for the detection of axonal injury has made things easier for a pathologist looking at a case of head injury, but also more difficult (Reichard *et al.* 2005). There are two reasons why immunohistochemistry may cause problems: first, with antibodies to proteins such as APP it is possible to detect damaged axons very rapidly after injury, well before any other pathology can be detected by light microscopy. The second is that APP immunoreactivity is not diagnostic of trauma, but is seen in axonal damage of any aetiology. APP expression indicating axonal damage may be seen in severe hypoglycaemic damage and carbon monoxide poisoning, as well as multiple sclerosis plaques, HIV encephalitis, cerebral malaria and Binswanger-type white matter lesions.

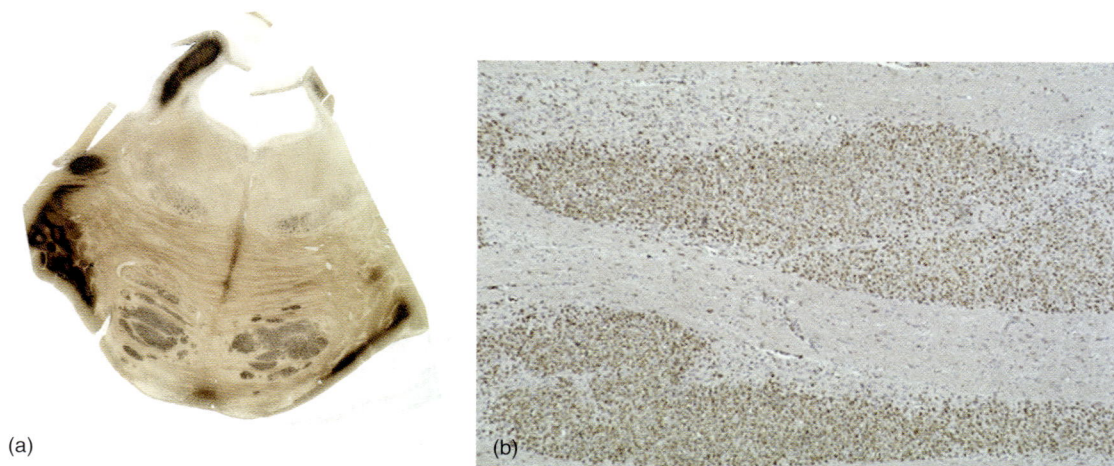

(a)   (b)

Figure 8.11 Long tract degeneration in a patient with diffuse axonal injury (DAI) who survived 5 months after his head injury. (a) The breakdown of the corticospinal tracts and, to a lesser extent, the medial lemniscus, is well demonstrated by the Marchi impregnation technique (note that the darker staining round the edges and in the superior cerebellar peduncle, is an artefact often seen in such preparations.) (b) Immunohistochemistry for CD68 (PG-M1) has been performed on a section from a block taken slightly more rostrally in the brain stem. Once again, the corticospinal tracts are outlined by the immunoreactivity. (Reproduced from Geddes and Graham 2004 by kind permission of Butterworth Heinemann.)

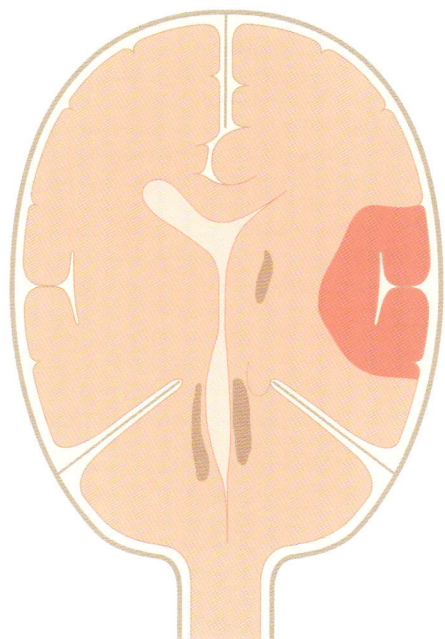

Figure 8.12 Schematic representation of the effects of swelling and shift on axons. A sylvian fissure haematoma (in red) has resulted in pronounced lateral and transtentorial shift of the affected hemisphere. The darker areas indicate sites at which it is common to see incipient ischaemic damage to white matter on APP immunohistochemistry. Note that the sites are the same as those in which traumatic axonal damage may be seen. (Reproduced from Geddes *et al.* 2000 by kind permission of Blackwell Publishing Ltd.)

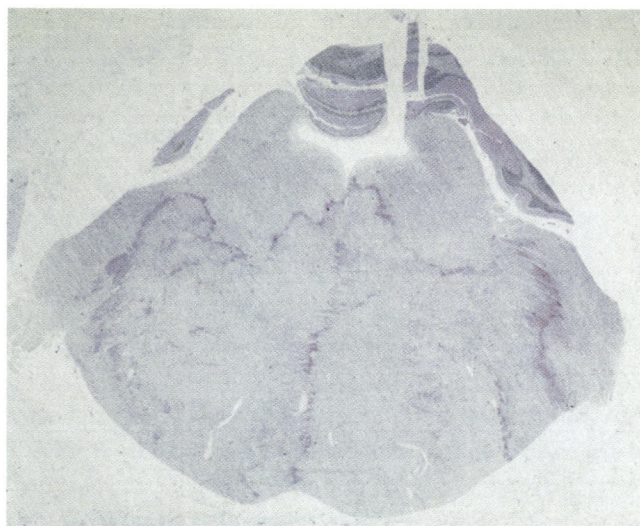

Figure 8.13 A low-power view of a section of pons, from the brain of a patient who died of raised intracranial pressure after a severe head injury. The stain is an immunohistochemical one for APP and, although faint, clearly demarcates a large area of the pontine basis. Such staining is not the result of traumatic axonal damage, but of early ischaemic damage secondary to brain swelling and shift. Similar patterns of immunoreactivity, often involving the midline of the brain stem, can be seen in patients who have not had a head injury.

Axonal damage is increasingly recognized to occur in many generalized diseases of the central nervous system (CNS), particularly those characteristic of severe metabolic derangement (Graham *et al.* 2004). In the context of trauma, however, the principal differential diagnosis is hypoxic–ischaemic axonal damage resulting from raised intracranial pressure. (Global hypoxia does not usually damage axons unless there is also brain swelling. (Dolinak *et al.* 2000)) Once intracranial pressure is raised, penetrating blood vessels are liable to become compressed, leading to ischaemia and infarction in the areas that they supply. Foci of vascular compromise caused by brain swelling are common in head injury and axons around such areas of incipient infarction will be damaged and so are positive with anti-APP. Most brains in fatal head injury are swollen by the time of death, so both traumatic and ischaemic axonal damage are often present in the same brain. The challenge is to work out which is which.

It does not help that roughly the same anatomical areas are vulnerable (Figure 8.12) to axonal injury and to vascular damage from brain swelling. As a general rule, geographical expression of APP tends to be vascular; this is particularly true in the base of the pons, where linear APP immunoreactivity, particularly but not always involving the midline, tends to outline areas of early ischaemia (Figure 8.13). Not infrequently neurons inside such areas show strong cytoplasmic eosinophilia or APP expression, both of which reinforce the impression of vascular damage – although, once again, no systematic studies of somatic APP positivity have been performed and one can only assume that it indicates acute neuronal stress of some sort, probably hypoxic rather than traumatic. Often with several days' survival after injury, it is possible to detect axonal damage of different 'ages', e.g. traumatic damage in typical sites, visible on H&E as large bulbs, as well as histologically more recent damage in a typical vascular pattern, which is the result of terminal brain swelling and resultant foci of ischaemia. That said, the distinction is often difficult to detect, despite suggestions that vascular and traumatic axonal injury differ in their staining characteristics on immunohistochemistry (Lambri *et al.* 2001). The author's experience is that such staining characteristics ('granular pattern with a dirty background' and 'beaded and thick filaments with clear background') are not always easy to identify with certainty, or entirely reliable and that a combination of morphology and topography is more useful. Even so, with short survival it is not always possible to establish the cause of the axonal damage shown by APP immunoreactivity and the most honest thing to do is to admit uncertainty. Fortunately interpretation of axonal damage is often irrelevant to the outcome.

There is little or no information available about the temporal evolution of axonal pathology caused by ischaemia or extreme vascular congestion. No studies have been performed and it may well be, possibly because the subcellular pathology is different, that in such circumstances axonal immunoreactivity for APP appears considerably earlier than the 2 h believed to be necessary in trauma cases. Anecdotal reports, not published, suggest that in exceptional cases axonal damage of a vascular nature may be detected in the brains of people who have survived much less than this time, without a history of head injury. Until a formal neuropathological study has been performed, it is impossible to say for certain whether any significance can be placed on such a finding. Similarly, a recent immunohistochemical study of paediatric optic nerves using anti-APP (Reichard *et al.* 2004) has suggested that immunoreactivity is a specific indicator for trauma, on the basis of the findings in a small sample. No systematic study of APP staining the first cranial nerve has been done, and until we have an idea of the frequency of its expression in different conditions – in particular in the presence of raised intracranial pressure, both episodic and sustained – it would be unwise to conclude anything from APP expression, even in cases where there is subdural and retinal bleeding.

The literature contains many examples of reports of DAI produced on the basis of an inadequate neuropathological examination (Vowles *et al.* 1987, Imajo and Kazee 1992, Pounder 1997, Raisanen *et al.* 1999, Niess *et al.* 2002) – in some cases, DAI was diagnosed on as few as one or two blocks (Geddes and Whitwell 2003). In view of what has been said above, it is clear that authors of such studies have misunderstood the concept of DAI – which denotes widespread pathology, in both supra- and infratentorial compartments – and their papers only confuse the issue further. Some have failed to appreciate that other causes of axonal damage, particularly ischaemia secondary to brain swelling and raised intracranial pressure, may confound the picture (Gleckman *et al.* 1999) and must be properly controlled for. Indeed, on occasion the term 'DAI' has been applied to axonal damage of other aetiologies, resulting in statements such as 'trauma is not the only and may not even be the main cause for DAI' (Niess *et al.* 2002) – something that is nonsense, given that DAI is by definition caused by trauma. Although the term 'diffuse axonal injury' should perhaps be changed to the more accurate 'diffuse traumatic axonal injury', there is no doubt that both clinically and neuropathologically the concept of DAI is a valuable one. For that reason, it is important that its use is restricted to trauma and that axonal injury from other causes is referred to as 'widespread (or "diffuse") hypoglycaemic axonal injury', 'widespread (or "diffuse") vascular injury', etc.

Finally, small pink granular globules mimicking axonal swellings are sometimes found around cerebral blood vessels – a characteristic site for foci of axonal damage. These round structures are often periodic acid – Schiff (PAS) positive, but negative with antisera to APP and neurofilament and are not argyrophilic. Such staining characteristics suggest that they are not axonal in origin. In fact, they are sometimes rather obviously situated in the perivascular space and must be assumed to be derived from plasma.

CONCLUSION: THE USE OF MICROSCOPY FOR PARENCHYMAL DAMAGE

So, in summary, what information relevant to diffuse brain trauma can be gained from doing detailed histology of the brain?

- In cases in which there is doubt about whether a head injury has occurred, or in cases in which the macroscopic findings are insufficient to explain the clinical state, the finding of TAI may be helpful.
- The morphology and staining characteristics of any traumatic axonal damage may help in establishing the timing of a head injury.
- If the TAI detected is sufficient in extent to be classified as diffuse traumatic axonal injury, or DAI (i.e. present in a widespread distribution above and below the tentorium), it may well account for a victim's prolonged unconsciousness.
- If survival has been longer than at least 2 h and there is no axonal injury in multiple samples, one can be confident that there is no significant parenchymal traumatic damage.
- As in other situations, histology may highlight underlying brain disease that may be relevant to the case.

Finally, conclusions such as these can be arrived at only if the examination of the brain has been detailed and the findings interpreted in the light of full clinical information. An inadequate neuropathological examination may be extremely misleading and as a rule it is better not to attempt histology if a proper examination cannot be guaranteed.

REFERENCES

Adams, J.H., Doyle, D., Graham, D.I., Lawrence, A.E. and McLellan, D.R. 1984. Diffuse axonal injury in head injuries caused by a fall. *Lancet*, **ii**, 1420–2.

Adams, J.H., Doyle, D., Graham, D.I., Lawrence, A.E. and McLellan, D.R. 1986. Deep intracerebral (basal ganglia) haematomas in fatal non-missile head injury in man. *J Neurol Neurosurg Psychiatry*, **49**, 1039–43.

Adams, J.H., Doyle, D., Ford, I., Gennarelli, T.A., Graham, D.I. and McLellan, D.R. 1989a. Diffuse axonal injury in head injury: definition, diagnosis and grading. *Histopathology*, **15**, 49–59.

Adams, J.H., Doyle, D., Ford, I., Graham, D.I., McGee, M. and McLellan, D.R. 1989b. Brain damage in fatal non-missile head injury in relation to age and type of injury. *Scot Med J*, **34**, 399–401.

Adams, J.H., Graham, D.I., Gennarelli, T.A. and Maxwell, W.L. 1991. Diffuse axonal injury in non-missile head injury. *J Neurol Neurosurg Psychiatry*, **54**, 481–3.

Adams, J.H., Jennett, B., McLellan, D.R., Murray, L.S. and Graham, D.I. 1999. The neuropathology of the vegetative state after head injury. *J Clin Pathol*, **52**, 804–6.

Blumbergs, P.C., Scott, G., Manavis, J., Wainwright, H., Simpson, D.A. and McLean, A.J. 1994. Staining of amyloid precursor protein to study axonal damage in mild head injury. *Lancet*, **344**, 1055–6.

Blumbergs, P.C., Scott, G., Manavis, J., Wainwright, H., Simpson, D.A. and McLean, A.J. 1995. Topography of axonal injury as defined by amyloid precursor protein and the sector scoring method in mild and severe closed head injury. *J Neurotrauma*, **12**, 565–72.

Dolinak, D., Smith, C. and Graham, D.I. 2000. Global hypoxia *per se* is an unusual cause of axonal injury. *Acta Neuropathol (Berl)*, **100**, 553–60.

Ezzat, W., Ang, L.C. and Nyssen J. 1995. Pontomedullary rent. A specific type of primary brainstem traumatic injury. *Am J Forensic Med Pathol*, **16**, 336–9.

Geddes, J.F. and Graham, D.I. 2004. Trauma to the central nervous system. In Gray, F., Di Girolami, U. and Poirier, R. (eds), *Escourolle and Poirier's Manual of Basic Neuropathology*, 4th edn. Philadelphia: Butterworth-Heinemann (Elsevier).

Geddes, J.F. and Whitwell, H.L. 2003. Shaken adult syndrome revisited. *Am J Forensic Med Pathol*, **24**, 310–11.

Geddes, J.F., Vowles, G.H., Beer, T.W. and Ellison, D.W. 1997. The diagnosis of diffuse axonal injury: implications for forensic practice. *Neuropathol Appl Neurobiol*, **23**, 339–47.

Geddes, J.F., Whitwell, H.L. and Graham, D.I. 2000. Traumatic axonal injury: practical issues for diagnosis in medicolegal cases. *Neuropathol Appl Neurobiol*, **26**, 105–16.

Geddes, J.F., Vowles, G.H., Hackshaw, A.K., Nickols, C.D., Scott, I.S. and Whitwell, H.L. 2001. Neuropathology of inflicted head injury in children. II. Microscopic brain injury in infants. *Brain*, **124**, 1299–306.

Gennarelli, T.A. 1996. The spectrum of traumatic axonal injury. *Neuropathol Appl Neurobiol*, **22**, 509–13.

Gennarelli, T.A. 1997. The pathobiology of traumatic brain injury. *Neuroscientist*, **3**, 73–81.

Gennarelli, T.A., Adams, J.H. and Graham, D.I. 1981. Acceleration induced head injury in the monkey. I. The model, its mechanical and physiological correlates. *Acta Neuropathol (Berlin)*, Suppl 7, 23–5.

Gennarelli, T.A., Thibault, L.E., Adams, J.H., Graham, D.I., Thompson, C.J. and Marcincin, R.P. 1982. Diffuse axonal injury and traumatic coma in the primate. *Ann Neurol*, **12**, 564–74.

Gennarelli, T.A., Thibault, L.E. and Graham, D.I. 1998. Diffuse axonal injury: an important form of traumatic brain damage. *Neuroscientist*, **4**, 202–15.

Gentleman, S.M., Roberts, G.W., Gennarelli, T.A. *et al.* 1995. Axonal injury: a universal consequence of fatal closed head injury? *Acta Neuropathol*, **89**, 537–43.

Gleckman, A.M., Bell, M.D., Evans, R.J. and Smith, T.W. 1999. Diffuse axonal injury in infants with nonaccidental craniocerebral trauma: enhanced detection by beta-amyloid precursor protein immunohistochemical staining. *Arch Pathol Lab Med*, **123**, 146–51.

Gorrie, C., Oakes, S., Duflou, J., Blumbergs, P. and Waite, P.M. 2002. Axonal injury in children after motor vehicle crashes: extent, distribution and size of axonal swellings using b-APP immunohistochemistry. *J Neurotrauma*, **19**, 1171–82.

Graham, D.I., McLellan, D., Adams, J.H., Doyle, D., Kerr, A. and Murray, L.S. 1983. The neuropathology of the vegetative state and severe disability after non-missile head injury. *Acta Neurochir Suppl*, **32**, 65–7.

Graham, D.I., Lawrence, A.E., Adams, J.H., Doyle, D., McLellan, D. and Gennarelli, T.A. 1989. Pathology of mild head injury. In Hoff, J.T., Anderson, T. and Cole, T. (eds), *Mild to Moderate Head Injury*. Boston: Blackwell Scientific, 63–75.

Graham, D.I., Clark, J.C., Adams, J.H. and Gennarelli, T.A. 1992. Diffuse axonal injury caused by assault. *J Clin Pathol*, **45**, 840–1.

Graham, D.I., Adams, J.H. and Gennarelli, T.A. 1993. Pathology of brain damage in head injury. In Cooper, P.R. (ed.), *Head Injury*. Baltimore: Williams & Wilkins, 91–113.

Graham, D.I., Adams, J.H., Nicoll, J.A.R., Maxwell, W.L. and Gennarelli, T.A. 1995. The nature, distribution and causes of traumatic brain injury. *Brain Pathol*, **5**, 397–406.

Graham, D.I., Gennarelli, T.A. and McIntosh, T.K. 2002. Trauma. In Graham, D.I. and Lantos, P.L. (eds), *Greenfield's Neuropathology*, 7th edn. London: Arnold, 767–1052.

Graham, D.I., Smith, C., Reichard, R., Leclercq, P.D. and Gentleman, S.M. 2004. Trials and tribulations of using beta-amyloid precursor protein immunohistochemistry to evaluate traumatic brain injury in adults. *Forensic Sci Int*, **146**, 89–96.

Hardman, J.M. 1979. The pathology of traumatic brain injuries. In Thompson, R.A. and Green, J.R. (eds), *Complications of Nervous System Trauma*. New York: Raven Press, 15–50.

Imajo, T. and Kazee, A.M. 1992. Diffuse axonal injury by simple fall. *Am J Forensic Med Pathol*, **13**, 169–72.

Keane, J.R. 1978. Intermittent see-saw eye movements. Report of a patient in coma after hyperextension head injury. *Arch Neurol*, **35**, 173–4.

Kinney, H.C. and Samuels, M.A. 1994. Neuropathology of the persistent vegetative state. A review. *J Neuropathol Exp Neurol*, **53**, 548–58.

Lambri, M., Djurovic, V., Kibble, M., Cairns, N. and Al-Sarraj, S. 2001. Specificity and sensitivity of betaAPP in head injury. *Clin Neuropathol*, **20**, 263–71.

Leclercq, P.D., Stephenson, M.A., Murray, L.S., McIntosh, T.K., Graham, D.I. and Gentleman, S.MS. 2002. Simple morphometry of axonal swellings cannot be used in isolation for dating lesions after traumatic brain injury. *J Neurotrauma*, **19**, 1183–92.

Lindenberg, R. and Freytag, E. 1970. Brainstem lesions characteristic of traumatic hyperextension of the head. *Arch Pathol*, **90**, 509–15.

Maxwell, W.L., Islam, M.N., Graham, D.I. and Gennarelli, T.A. 1994. A qualitative and quantitative analysis of the response of the retinal ganglion cell soma after stretch injury to the adult guinea-pig optic nerve. *J Neurocytol*, **23**, 379–92.

Maxwell, W.L., Povlishock, J.T. and Graham, D.I. 1997. A mechanistic analysis of nondisruptive axonal injury: a review. *J Neurotrauma*, **14**, 419–40.

Maxwell, W.L., Domleo, A., McColl, G. and Graham, D.I. 2003. Post-acute alterations in the axonal cytoskeleton after traumatic axonal injury. *J Neurotrauma*, **20**, 151–68.

Niess, C., Grauel, U., Toennes, S.W. and Bratzke, H. 2002. Incidence of axonal injury in human brain tissue. *Acta Neuropathol (Berl)*, **104**, 79–84.

Pilz, P. 1983. Survival after ponto-medullary junction trauma. *Acta Neurochir Suppl*, **32**, 75–8.

Pilz, P., Strohecker, J. and Grobovschek, M. 1982. Survival after traumatic ponto-medullary tear. *J Neurol Neurosurg Psychiatry*, **45**, 422–7.

Pounder, D.J. 1997. Shaken adult syndrome. *Am J Forensic Med Pathol*, **18**, 321–4.

Povlishock, J.T. 1993. Pathobiology of traumatically induced axonal injury in animals and man. *Ann Emerg Med*, **22**, 980–6.

Raisanen, J., Ghougassian, D.F., Moskvitch, M. and Lawrence, C. 1999. Diffuse axonal injury in a rugby player. *Am J Forensic Med Pathol*, **20**, 70–2.

Reichard, R.R., White, C.L., Hogan, R.N., Hladik, C.L., Dolinak, D. 2004. Beta-amyloid precursor protein immunohistochemistry in the evaluation of pediatric traumatic optic nerve injury. *Ophthalmology*, **111**, 822–7.

Reichard, R.R., Smith, C. and Graham, D.I. 2005. The significance of beta-APP immunoreactivity in forensic practice. *Neuropathol Appl Neurobiol*, **31**, 304–13.

Riggs, J.E. and Schochet, S.S. 1995. Spastic quadriparesis, dysarthria and dysphagia following cervical hyperextension: a traumatic pontomedullary syndrome. *Milit Med*, **160**, 94–5.

Sherriff, F.E., Bridges, L.R., Gentleman, S.M., Sivaloganathan, S. and Wilson, S. 1994. Markers of axonal injury in post mortem human brain. *Acta Neuropathol*, **88**, 433–9.

Simpson, D.A., Blumbergs, P.C., Cooter, R.D., Kilminster, M., McLean, A.J. and Scott, G. 1989. Pontomedullary tears and other gross brainstem injuries after vehicular accidents. *J Trauma*, **29**, 1519–25.

Smith, D.H., Nonaka, M., Miller, R. *et al.* 2000. Immediate coma following inertial brain injury dependent on axonal damage in the brainstem. *J Neurosurg*, **93**, 315–22.

Stone, J.R., Okonkwo, D.O., Dialo, A.O., Rubin, D.G., Mutlu, L.K., Povlishock, J.T. and Helm, G.A. 2004. Impaired axonal transport and altered axolemmal permeability occur in distinct populations of damaged axons following traumatic brain injury. *Exp Neurol*, **190**, 59–69.

Strich, S.J. 1956. Diffuse degeneration of the cerebral white matter in severe dementia following head injury. *J Neurol Neurosurg Psychiatry*, **19**, 163–85.

Strich, S.J. 1961. Shearing of nerve fibres as a cause of brain damage due to head injury. *Lancet*, **2**, 443–8.

Tomei, G., Spagnoli, D., Ducati, A. *et al.* 1990. Morphology and neurophysiology of focal axonal injury experimentally induced in the guinea pig optic nerve. *Acta Neuropathol*, **80**, 506–13.

Vowles, G.H., Scholtz, C.L. and Cameron, J.M. 1987. Diffuse axonal injury in early infancy. *J Clin Pathol*, **40**, 185–9.

Wilkinson, A.E., Bridges, L.R. and Sivaloganathan, S. 1999. Correlation of survival time with size of axonal swellings in diffuse axonal injury. *Acta Neuropathol*, **98**, 197–202.

9

Brain swelling and oedema, raised intracranial pressure, and the non-perfused brain

Helen L Whitwell

Brain swelling and cerebral oedema are the common reactions of the brain to injury of whatever aetiology. A consequence of this is the formation of various intracranial herniations.

If these are severe the non-perfused (respirator) brain, where there is cessation of cerebral perfusion, may be the final outcome.

BRAIN SWELLING

Simple swelling of the brain is a relatively common finding in deaths encountered in medicolegal practice. The pathophysiology is not clearly understood but in simple terms there is an increase in the volume of the brain, usually related to an increase in intravascular blood. Conditions in which this is recognized to occur include 'asphyxial'-type deaths and others where there may be a prolonged congestive phase.

Confusion may arise at postmortem examination in that misinterpretation can occur between this and true cerebral oedema or congestive brain swelling with a raised intracranial pressure. It is well recognized that the cerebellar tonsils descend and ascend within the foramen magnum with normal changes in respiration and blood pressure. It is not uncommon to see prominent cerebellar tonsils at postmortem examination – this cannot necessarily be taken as evidence that there has been a true rise in intracranial pressure. Minor uncal grooming may also be seen at postmortem where there is no cerebral pathology and again care should be taken in the interpretation (Gordon *et al.* 1988).

Histologically, congestion of intracerebral vessels with perivascular haemorrhages (Figure 9.1) may be seen. Whist these may be seen in conditions such as suffocation or overlying in an infant, as well as 'asphyxial' deaths in adults, they should not be taken as diagnostic for these conditions (see also Chapter 11, p. 130–1).

CONGESTIVE BRAIN SWELLING

Congestive brain swelling can occur as a result of an increase in intravascular blood, causing an increase in brain volume. It may occur very rapidly, particularly in children, as well as in patients who have sustained a head

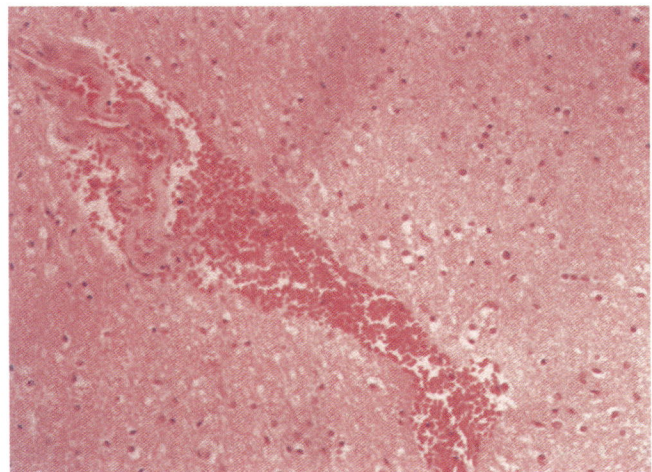

Figure 9.1   Petechial perivascular haemorrhages.

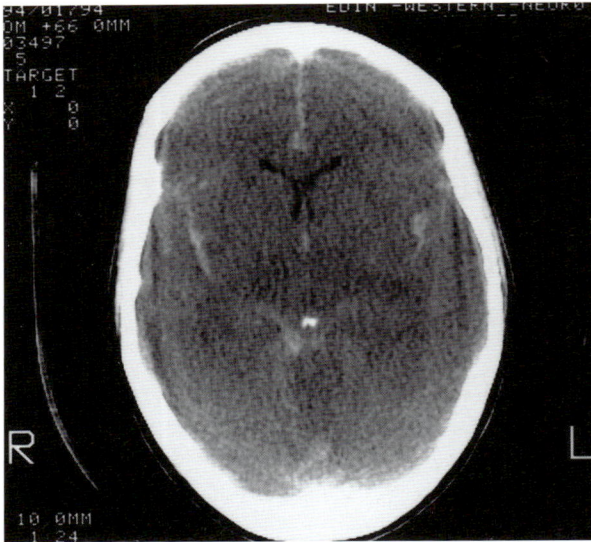

Figure 9.2   Non-contrast – enhanced cranial CT scan showing diffuse brain swelling in a patient with subarachnoid haemorrhage. (Reproduced from Ironside, J.W. and Pickard, J.D. 2002. Raised intracranial pressure, oedema and hydro-cephalus. In Graham, D.I. and Lantos, P.L. (eds), *Greenfield's Neuropathology*, 7th edn. London: Arnold, supplied by Dr R. Gibson.)

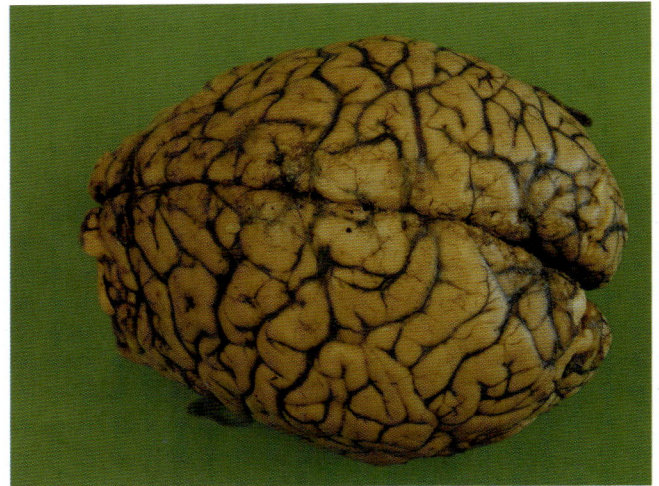

Figure 9.3   Severe brain swelling with gyral flattening.

injury including acute subdural haematomas and aneurys-mal rupture with subarachnoid haemorrhage. On a com-puted tomography (CT) scan obliteration of the basal cisterns and compression of the 3rd ventricle is shown to have occurred (Figure 9.2). Expansion takes place particularly rapidly when the arterial blood pressure is high. It is a diffuse process, affecting one or both hemi-spheres. The increase in blood volume occurs predomi-nantly in the capillary- and post-capillary bed. It is recognized that it may subside leaving only specific fea-tures of previous raised intracranial pressure such as evidence of necrosis in either or both parahippocampal gyri. If congestive brain swelling is prolonged, vasogenic cerebral oedema supervenes with water accumulation in the extracellular spaces.

Pathology

At postmortem examination the brain appears swollen with extremely prominent surface vessels and marked gyral flattening (Figure 9.3). On section there is promi-nence of intracerebral vessels. Histological examination reveals congested vessels with perivascular haemorrhages in areas particularly within the cortex.

CEREBRAL OEDEMA

The two most common types of cerebral oedema encoun-tered are vasogenic and cytotoxic. Of the two, vasogenic is found most frequently. Cerebral oedema is defined as an increase in brain tissue volume as a result of increased tissue water content.

Vasogenic oedema

The normal water content of grey matter is 80 per cent wet weight, whereas that of white matter is considerably lower at about 68 per cent wet weight. In cerebral oedema there is greater accumulation of water within the white matter with typical water contents being 81–82 per cent in grey and 76–79 per cent in white matter (Ironside and Pickard 2002). This is caused by the fact that the water can move more freely between fibre bundles, and is apparent on both sectioning of the brain as well as mag-netic resonance imaging (MRI)/CT. In practice the most common type of oedema seen is vasogenic oedema where there is breakdown of the blood – brain barrier. This occurs locally around mass lesions such as intracerebral haematomas (including contusions, abscesses and tumours) and also more diffusely in the cerebral white matter. It is unclear what the precise mechanism of the increase in vascular permeability is: in tumours it may be related to structural abnormalities in the capillary wall. In the setting of trauma the rate of brain swelling/oedema is increased by hypertension, hypercapnia and vasodilatory substances (Atkinson *et al.* 1998).

Cytotoxic or cellular oedema

Cytotoxic or cellular oedema is defined as cellular swelling associated with a reduced extracellular space but, initially at least, an intact blood–brain barrier (Klatzo 1967). This type of oedema is seen most commonly in association with hypoxia and ischaemic conditions (including in association with cerebral contusions), although generally it is eventually accompanied by a vasogenic type of oedema, with breakdown of the blood–brain barrier (Bullock *et al.* 1991). As the metabolically more active cellular components are in grey matter, this type of oedema is more prominent in this location (Ironside and Pickard 2002).

Other forms of oedema include:

- Hydrostatic oedema, where the blood–brain barrier is intact but there is a sudden rise in intravascular pressure causing water to be driven through the capillary bed into the brain tissue. This occurs particularly during craniotomy when the arterial pressure becomes suddenly increased. This may be seen following bony decompression in patients with a high intracranial pressure leading to herniation of the brain.
- Interstitial oedema, where there is increase in water content in the periventricular white matter as a result of acute obstructive hydrocephalus. Fluid passes through the ependyma into the white matter around the ventricular horns.

Pathology

Typically cerebral oedema produces gyral flattening with compression of the sulci and close application of the swollen brain to the dura. Variable internal herniation may be seen depending on whether or not a mass lesion is present. The brain will be increased in weight if it is in the fresh state. It should be recognized, however that tables referring to normal ranges can give only an indication. Such tables are given in Appendices 1, 2 and 3 of Chapter 2.

In the case of a mass lesion, on section oedema may be found to be localized or diffuse. In diffuse brain swelling there will be compression of the ventricles and no midline shift, but bilateral tentorial herniations may occur. The brain tissue oozes fluid and appears soft. Fixation may be generally poor as formalin is slow to penetrate oedematous tissue. If systemic jaundice is present the tissue may have a greenish discolouration.

Histology shows a spongy appearance of the affected areas, appearing rather pale on haematoxylin and eosin (H&E) stain (Figure 9.4). There is pallor of myelin staining that often appears rather vacuolated, swollen and less packed than in normal white matter. After a day or so the

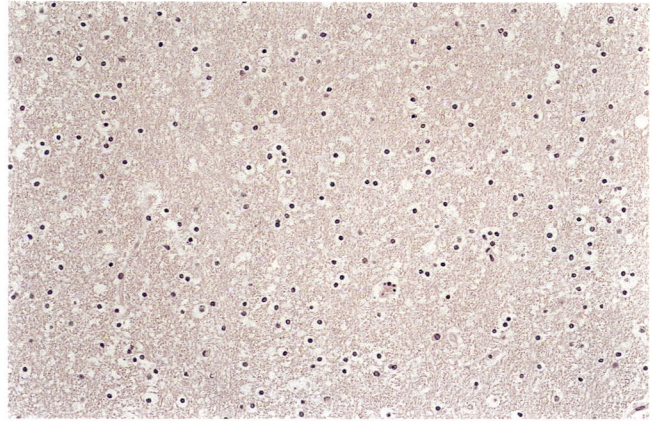

Figure 9.4 Histological appearances of oedema (H&E) – particularly prominent in white matter.

astrocytes appear swollen with later gliosis occurring. If oedema persists, macrophages infiltrate with breakdown of myelin. Small cystic spaces in the white matter may be the end result.

DIFFUSE BRAIN SWELLING

Diffuse brain swelling appears to be a more common complication of head injury in the younger age group (Graham *et al.* 1989). Other factors such as age and sex have also been implicated in the severity of brain swelling after head injury (Farin *et al.* 2003). The reasons for this are complex and ill-understood. It is well recognized that rapid neurological deterioration can follow within minutes or be delayed after head injury (Bruce 1984). No expanding intracranial lesion is usually found but a proportion progress to coma and death. The pathology shows diffuse generalized brain swelling with little evidence of brain injury. It appears that the diffuse swelling may, in a number of these cases, be related to vascular congestion with hyperaemia rather than true cerebral oedema (Bruce *et al.* 1981, Sakas *et al.* 1997). A number of these cases follow a benign course (Snoek *et al.* 1984) and the syndrome can occur with relatively trivial trauma. More recently, it has been reported that delayed cerebral oedema can occur in the familial hemiplegic migraine condition (Kors *et al.* 2001), and with apolipoprotein E deficiency in the experimental situation (Lynch *et al.* 2002).

Pathology of raised intracranial pressure

A mass lesion within the brain causes the brain to shift to accommodate it. Initially the adjacent brain distorts to accommodate, and subsequently a reduction in

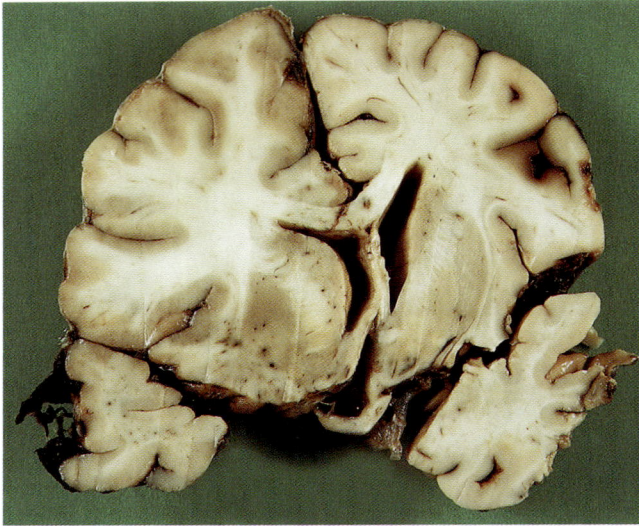

Figure 9.5   Marked left-to-right shift with subfalcine herniation.

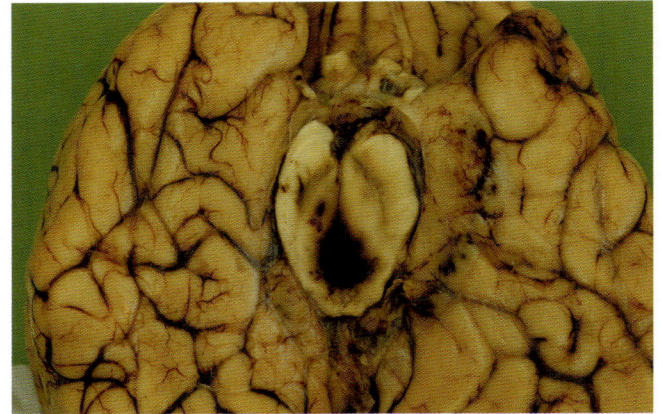

Figure 9.6   Left uncal herniation viewed from base of brain.

cerebrospinal fluid (CSF) volume occurs within the ventricles and the subarachnoid space. With progression of a mass lesion, displacement of the brain tissue from one intracranial compartment to another occurs creating internal hernias. The development of such hernias depends on the location of the primary mass.

SUPRATENTORIAL MASS LESIONS

The classic picture of a mass lesion involving one or other hemisphere shows compression of adjacent brain structures with expansion of the hemisphere. This is followed by compression of the ipsilateral ventricle, reduction in size of the third ventricle and contralateral displacement of the midline structures. This is followed by development of herniation as described below, although in infants the typical picture may not be seen before the closure of sutures.

Subfalcine herniation

In this situation the ipsilateral cingulate gyrus herniates under the falx as the result of a mass lesion in one cerebral hemisphere (Figure 9.5). This may lead to infarction in the parasagittal cortex as a result of compromising circulation of the pericallosal arteries. A wedge of necrosis where the gyrus is in contact with the falx may occur. This is most commonly seen in frontal or parietal regions because the posterior falx is tethered. Unless the herniation is relatively minor it is usually accompanied by other hernias, for example uncal or tonsillar.

Tentorial (uncal) herniation

A supratentorial mass may produce herniation of the ipsilateral uncus of the temporal lobe as well as the medial part of the parahippocampal gyrus (Figure 9.6). This is displaced through the tentorial opening. It is also known as lateral transtentorial herniation. Dilatation of the pupil is one of the earliest signs of tentorial herniation and is recognized to occur before any impairment of consciousness. This is due to compression of the oculomotor nerve. Necrosis may occur along the parahippocampal gyrus. Compression of the anterior choroidal arteries may cause infarction in the medial part of the globus pallidus, internal capsule and optic tract. A common complication is compression of the posterior cerebral artery leading to infarction in the posterior inferior temporal lobe including the hippocampus, together with the medial and inferior surfaces of the occipital cortex. This is most commonly haemorrhagic in nature (Figure 9.7). Cerebellar infarction may also occur as a result of compression of a superior cerebellar artery. These infarctions are most commonly unilateral on the side of the mass lesion but can be bilateral.

Central transtentorial herniation

This may complicate lateral tentorial herniation and produces a bilateral or ring herniation most prominent in the posterior aspect. It occurs particularly with bilateral expanding lesions in the hemisphere. Focal infarction may occur in mamillary bodies. There is elongation of the thalamus and oculomotor nerves. In addition infarctions in the territories, as with tentorial herniation, frequently occur (see above).

Both tentorial and central transtentorial herniation are commonly associated with centrally placed haemorrhagic

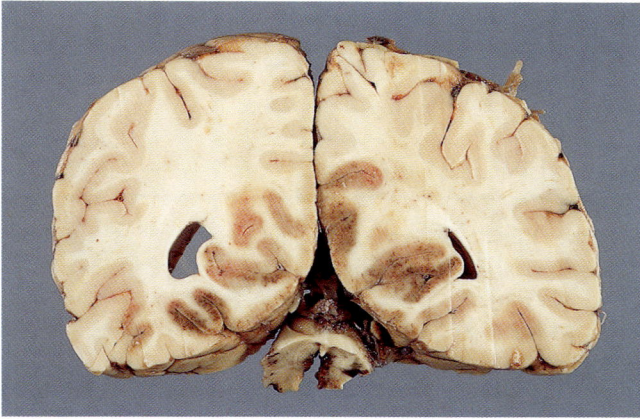

Figure 9.7 Posterior cerebral artery territory infarction.

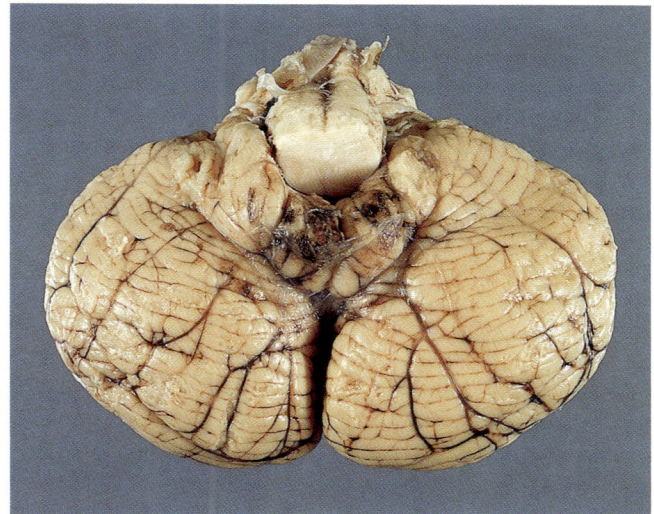

Figure 9.8 Cerebellar tonsillar herniation showing haemorrhagic necrosis of the tonsils.

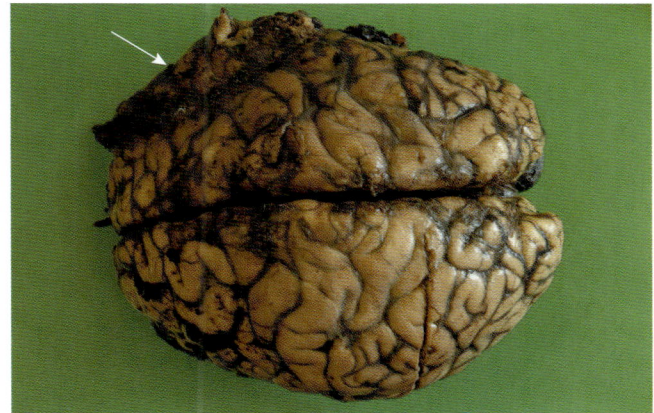

Figure 9.9 Herniation of cortex in relation to craniotomy site.

lesions in the midbrain and pons (also known as Duret haemorrhages). As discussed in Chapter 7, these should be differentiated from primary traumatic haemorrhages. The probable mechanism is progressive displacement, with the perforating branches of the basilar artery supplying the brain stem becoming stretched and narrowed, leading to ischaemia with haemorrhage (Hassler 1967). Microscopically both areas of haemorrhage and ischaemia may be seen.

INFRATENTORIAL MASS LESIONS

Tonsillar herniation (cerebellar cone)

Initially a mass lesion in the posterior fossa produces hydrocephalus with dilatation of the lateral and third ventricles. If unilateral, there is lateral shift with compression of the aqueduct and fourth ventricle. Masses in the posterior fossa produce downward displacement of the cerebellar tonsils through the foramen magnum. The pathological hallmark of this type of herniation is haemorrhagic necrosis of the tips of the cerebellar tonsils, together with grooving of the ventral medullary surface as a result of compression against the anterior border of the foramen magnum (Figure 9.8). If this is very severe, occlusion of branches of the posterior inferior cerebellar artery may occur. Tonsillar herniation also occurs with supratentorial lesions.

Upward tentorial herniation (reversed tentorial herniation)

This occurs as a result of herniation of the superior cerebellar surface through the tentorial opening caused by a posterior fossa mass. Conditions that may produce this include cerebellar tumours, haemorrhage or other space occupying cerebellar lesions, as well as posterior fossa subdural or extradural haemorrhage.

EXTERNAL CEREBRAL HERNIATIONS

Brain tissue may herniate externally through any defect – these may be traumatic or surgical. Small amounts of cortex may protrude through burr holes (see Chapter 2, Figure 2.7) – large areas through decompression sites (Figure 9.9). Oedema with haemorrhagic pressure necrosis at the edges occurs with the development of larger areas of cortical infarction.

NON-PERFUSED (RESPIRATOR) BRAIN

This occurs when perfusion to the brain ceases due to the mean arterial blood pressure being insufficient to overcome the intracranial pressure and allow blood to flow through the brain. The clinical aspects are covered in Chapter 15, p. 189. Blood stagnates in the cerebral circulation. Macroscopically there is swelling of the brain with a dark-grey dusky discoloration and evidence of uncal or tonsillar herniation and, depending on the circumstances, other types of herniation. Fixation is poor and central parts of the brain, such as the basal ganglia, may disintegrate. There is blurring of the grey–white junction of the cortex including the cerebellar cortex (Figure 9.10a). Fragments of cerebellar tissue may be found around the spinal cord.

Histologically there is general tissue pallor, lysed red cells with stagnation and perineuronal vacuolation and neuronal pallor (Figure 9.10b). Tissue reaction is lacking. This may make interpretation of timing of injuries difficult. These changes take around 12 h to develop, becoming more obvious with the passage of time (Black 1978). Bacterial overgrowth, seen in postmortem autolysis, is not usually found.

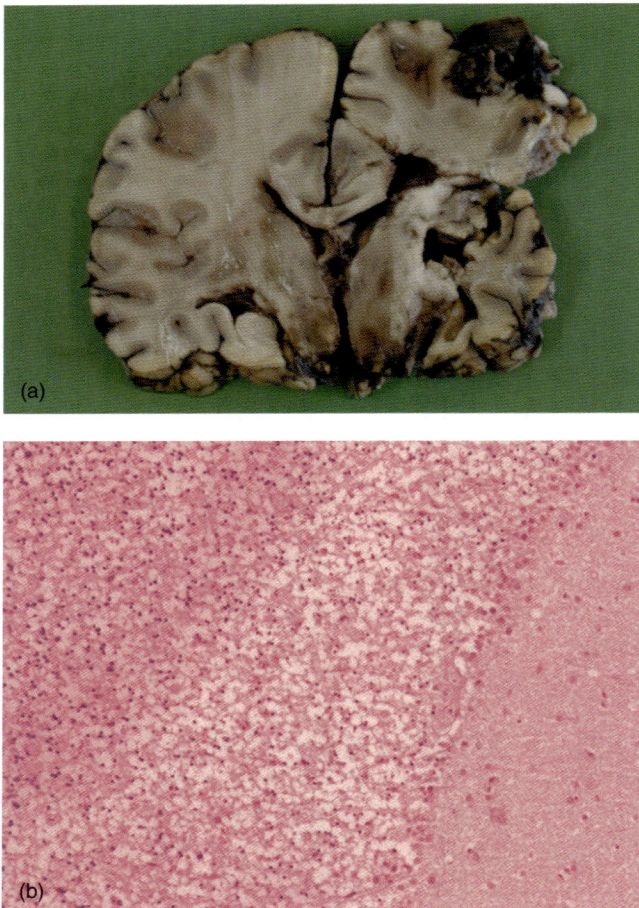

Figure 9.10   (a) Non-perfused brain showing grey cortical discoloration. (b) Lysis of cerebellar granule cells in non-perfused brain.

REFERENCES

Atkinson, J.L.D., Anderson, R.E. and Murray, M.H. 1998. The early critical phase of severe head injury: importance of apnea and dysfunctional respiration. *J Trauma*, **45**, 941–5.

Black, P.Mc.L. 1978. Brain death. *N Engl J Med*, **299**, 338–44, 393–401.

Bruce, D.A. 1984. Delayed deterioration of consciousness after trivial head injury in childhood. *BMJ*, **289**, 715–16.

Bruce, D.A., Alavi, A., Bilaniuk, B. *et al.* 1981. Diffuse cerebral swelling following head injuries in children: the syndrome of 'malignant brain edema'. *J Neurosurg*, **54**, 170–8.

Bullock, R., Maxwell, W.L., Graham, D.I., Teasdale, G.M. and Adams, J.H. 1994. Glial swelling following human cerebral contusion: an ultrastructural study. *J Neurol Neurosurg Psychiatry*, **54**, 427–34.

Farin, A., Deutsch, R., Biegon, A. and Marshall, L.F. 2003. Sex-related differences in patients with severe head injury: greater susceptibility to brain swelling in female patients 50 years of age and younger. *J Neurosurg*, **98**, 32–6.

Gordon, I., Shapiro, H.A. and Berson, S.D. 1988. Forensic Medicine: a guide to principles, 3rd edition. Edinburgh: Churchill-Livingstone.

Graham, D.I., Ford, I., Adams, J.H. *et al.* 1989. Fatal head injury in children. *J Clin Pathol*, **42**, 18–22.

Hassler, O. 1967. Arterial pattern of human brain stem. Normal appearance and deformation in expanding supratentorial conditions. *Neurology*, **17**, 368–75.

Ironside, J.W. and Pickard, J.D. 2002. Raised intracranial pressure, oedema and hydrocephalus. In Graham, D.I. and Lantos, P.L. (eds), *Greenfield's Neuropathology*, 7th edn. London: Arnold, Chapter 4.

Klatzo, I. 1967. Neuropathological aspects of brain edema. *J Neuropathol Exp Neurol*, **26**, 1–14.

Kors, E.E., Terwindt, G.M., Vermeulen, F.L. *et al.* 2001. Delayed cerebral edema and fatal coma after minor head trauma; role of the CACNA1A calcium channel subunit gene and relationship with familial hemiplegic migraine. *Ann Neurol*, **49**, 753–60.

Lynch, J.R., Pineda, J.A., Morgan, D. *et al.* 2002. Apolipoprotein E affects the central nervous system response to injury and the development of cerebral edema. *Ann Neurol*, **51**, 113–17.

Sakas, D.E., Whittaker, K.W., Whitwell, H.L. and Singounas, E.G. 1997. Syndromes of post-traumatic neurological deterioration in children with no focal lesions revealed by cerebral imaging: Evidence for a trigeminovascular pathophysiology. *Neurosurgery*, **41**, 661–7.

Snoek, J.W., Minderhoud, J.M. and Wilmink, J.T. 1984. Delayed deterioration following mild head injury in children. *Brain*, **107**, 15–36.

10

Spinal injuries

Damianos E Sakas and Helen L Whitwell

EPIDEMIOLOGY – CAUSE

Fractures and subluxations of the spinal column associated with damage of the spinal cord are not common. It has been estimated that, in a population of 100 000, 4 serious spinal injuries with cord injury occur every year. In the USA, 10 000 new cases of permanent spinal cord injuries are documented every year. Of these, 50 per cent involve the cervical region and about 10 per cent will result in quadriplegia. Spinal cord injuries are more frequently seen in young adult males, with the most common causes being road traffic incidents, falls, injury during sports such as diving, skiing and rugby, and gunshot injury. Of the patients with a serious spine injury, about 20 per cent have an additional spinal injury and other injuries such as chest trauma or arterial dissections as well as cerebral injury (Sherk 1989).

TYPES AND MECHANISMS OF INJURY: ACUTE AND LONGER-TERM EFFECTS

The proper evaluation of the spinal injuries should be based on a thorough understanding of their classification into numerous types according to specific criteria (Allen *et al.* 1982, Harris *et al.* 1986).

Classification of injuries

CLASSIFICATION INTO STABLE AND UNSTABLE INJURIES

The application of perpendicular force (fall of an object on the head or landing on the ground) causes a compressed or explosive fracture. This may be stable if there is no damage of the ligaments, discs or fracture of the facet joints, or unstable if there is disruption of the posterior longitudinal ligament or fracture of the laminae. A stable fracture results from vertical compression (an object falling on to the head, or jumping from a height) or hinge injury (a weight falling on to the back or a blow to the forehead). An unstable fracture results from a shearing injury (a fall from a height or a road traffic accident) often in association with a rotational force (Torrens and Dickson 1991) (Figure 10.1).

CLASSIFICATION INTO COMPLETE AND INCOMPLETE INJURIES

For therapeutic decisions and prognosis, spinal injuries are classified according to the completeness of the lesion.

An injury is described as complete when it is associated with no preservation of any motor and/or sensory function more than three segments below the level of the injury. In these cases, the persistence of a complete neurological deficit for more than 24 h indicates that no distal function will recover.

An injury can be incomplete, i.e. associated with residual motor or sensory function more than three segments below the level of the injury; it manifests itself with residual sensation or voluntary limb movement, sensation around the anus, voluntary rectal sphincter contraction or voluntary toe flexion. The most common types of incomplete lesions are: central cord syndrome, Brown–Séquard syndrome, and the anterior and posterior cord syndromes.

CLASSIFICATION INTO MINOR AND MAJOR INJURIES

The minor injuries involve a part of the vertebra, do not lead to acute instability and include the following types:

- fracture of transverse process – the patients are usually neurologically intact with the exception of L4–L5 lumbosacral plexus injuries which may be associated

Figure 10.1 Cervical spine injury. Mechanism of hyperflexion–hyperextension after head-on (frontal) collision of motor vehicles. Modified from Rea, E.L. and Muller, C.A. (eds) 1993. *Spinal Trauma*. Rolling Meadows, IL: AANS, with permission.

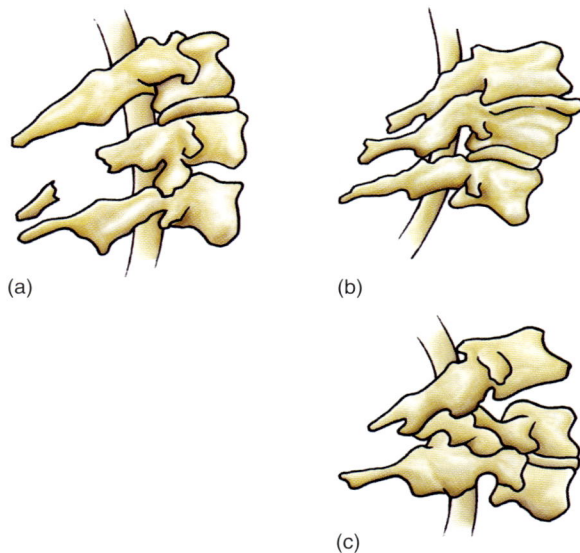

Figure 10.2 Mechanisms of spinal fractures (selected cases). (a) Extension and creation of a bone fragment; (b) extension and vertebral compression; and (c) extension and catastrophic compression of the vertebra. Modified from Rea, E.L. and Muller, C.A. (eds) 1993. *Spinal Trauma*. Rolling Meadows, IL: AANS, with permission.

with renal injuries and T1–T2 injuries which may be associated with branchial plexus injuries
• fracture of articular process
• isolated fractures of the spinous process (Figure 10.2).

The major injuries can be classified into the following four main types:

1. Compression fracture involving only the anterior vertebral region while the middle region remains intact.
2. Compression fracture as a result of severe axial load causing a burst of the vertebral body in both anterior and middle regions.
3. Compression fracture with rotation and shear.
4. Compression fracture with dislocation and distraction.

Mechanisms of spinal cord injury

It is useful for the injuries of the spinal cord to be classified by mechanism of trauma. These mechanisms include:

• direct compression by bone, ligament or disc
• interruption of the vascular supply
• traction.

The spinal cord damage can be caused by direct force applied to the cord, haemorrhage in and around the cord, or vascular injury causing ischaemia. The following types of spinal injury, which may be associated with spinal cord injury, deserve a more extensive description.

COMPRESSION DAMAGE

The height of the vertebral body is decreased and the posterior aspect of it may extend into the spinal canal, compressing the cord. These fractures are usually stable because frequently the posterior longitudinal ligament is intact. They most commonly affect the level C5–C6. A special type is the 'tear-drop' fracture, a separation of a small anteroinferior fragment with displacement of the posterior part into the spinal canal, which may occur when a rotation force in flexion has been applied. About 50 per cent of these cause incomplete damage affecting the anterior cord, with the remainder causing complete neurological deficit below the level of the damage.

HYPEREXTENSION DAMAGE

These lesions most frequently afflict patients who either have degenerative spinal canal stenosis or are elderly. The cord damage is caused by the hyperextension and a spinal fracture is not always present. The cord damage results because of cord compression between the degenerative body and the disc anteriorly and the hypertrophic ligamentum flavum posteriorly (Figure 10.3). Most of these injuries are stable but result in permanent neurological impairment (Epstein et al. 1980).

FLEXION AND FLEXION–ROTATION DAMAGE

In the cervical spine, the flexion and flexion–rotation injuries are the most frequent, with the C5–C6 level being the most commonly affected. One or both of the posterior facets may be subluxed or dislocated. The associated extensive posterior ligament damage makes these injuries unstable. The spinal cord may be compressed and injured,

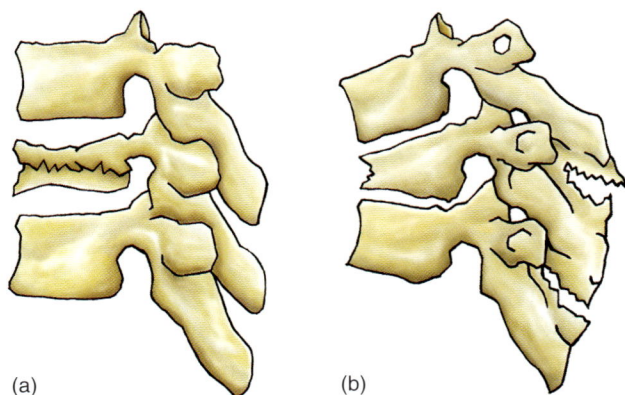

Figure 10.3 Cervical fractures. (a) Stable; (b) unstable, associated with rupture of the ligament over the spinous processes. Modified from Rea, E.L. and Muller, C.A. (eds) 1993. *Spinal Trauma*. Rolling Meadows, IL: AANS, with permission.

from either direct pressure or vascular compromise of the anastomotic segmental feeding vessels.

Distraction–flexion injuries

A further category of spinal injuries is the distraction–flexion injuries. The range includes hyperflexion sprain, minor subluxation and locked facets.

HYPERFLEXION SPRAIN

The injury involves disruption of the posterior ligament without bony injury. It can be missed on plain lateral C-spine radiographs if they are obtained in normal alignment, and, hence, flexion–extension views are required. Spasm of the cervical paraspinal muscles splints the neck and prevents true flexion. Instability may go unnoticed when films are obtained shortly after the injury, thus, if the pain persists, the radiographs should be repeated one to two weeks later (Webb *et al.* 1976).

SUBLUXATION

It has been shown that a horizontal subluxation of more than 3.5 mm of one vertebral body on another, or an angulation of more than 11° of one vertebral body relative to the next, indicates ligamentous instability. If a subluxation of less than 3.5 mm is seen on plain films, and there is no neurological deficit, flexion–extension films should be obtained.

LOCKED FACETS

'Locked facets' are caused by severe flexion injuries with reversal of the normal 'shingled' relationship between facets (normally the inferior facet of the level above is posterior to the superior facet of the level below) and disruption of the facet capsule. Bilaterally locked facets occur after disruption of the ligaments of the apophyseal joints, the ligamentum flavum, longitudinal and interspinous ligaments, and the annulus. Usually, they present with cervical spinal cord and root injury. Facets that have not completely locked may have had significant ligamentous disruption (Sonntag 1981). In the cervical spine, where the apophyseal joints lie almost horizontally, dislocation may occur without a fracture but at other sites a fracture and/or a dislocation is always present.

Types of neurological impairment

MOTOR DEFICIT

The injuries of the spinal cord cause upper motor neuron paralysis with loss of voluntary function, increased muscle tone and hyperreflexia. Lumbar spine lesions with damage of the cauda equina cause lower motor neuron paralysis which presents with reduced muscle tone, muscle wasting and with loss of reflexes. Thoracolumbar injuries affecting the conus medullaris and cauda equina can cause a combination of upper and lower motor neuron lesions.

SENSORY DEFICIT

In complete lesions, the afferent sensory tracts become disrupted at the level of the lesion, resulting in sensory loss below the lesion. This manifests as impairment of sensation, such as loss of feeling and numbness as well as reduced sensitivity to temperature (both hot and cold). Position sense may also be affected, with lack of positional awareness of a foot, for example. Visceral sensation abnormalities involving the bladder or rectum are common.

AUTONOMIC DEFICIT

The most common problems involve:

- vasomotor control – spinal cord lesions above T5 may cause hypotension, and lesions affecting the sympathetic splanchnic vasomotor control will also cause severe postural hypotension
- temperature control – complete spinal lesions impair the autonomic mechanisms for vasoconstriction and vasodilatation, and thermal regulation (there may be hypo- or hyperthermia).

Spinal shock

The term 'spinal shock' refers to the sudden transient depression of neurological function of the segments caudal to the cord lesion, resulting from sudden withdrawal of facilitating or excitatory influences from supraspinal centres. In most patients, this is essentially an areflexic flaccid paralysis. It lasts from three to four days to six to eight weeks. This term also describes the hypotension and bradycardia that follow spinal cord injury after interruption of the sympathetic influence (with relatively unopposed parasympathetic influence) and the loss of muscle tone caused by skeletal muscle paralysis below the level of injury (Atkinson and Atkinson 1996).

Complete lesions

These are the most serious spinal lesions. The patient has complete transverse myelopathy and all neurological function below the level of the lesion, including bladder and bowel (depending on the level) has ceased, resulting in paraplegia or quadriplegia and impairment of autonomic function. A special type of complete lesion is the bulbar–cervical dissociation. This results from spinal cord injury at or above C3 (in cases of atlanto-occipital and atlantoaxial dislocation) and leads to immediate pulmonary and cardiac arrest and death if cardiopulmonary resuscitation (CPR) is not immediately available (Figure 10.4). The patients are quadriplegic and ventilator dependent.

Incomplete lesions

The most common types are described below.

SPINAL CORD CONCUSSION

This is a temporary impairment of spinal cord neuronal function that may follow injury such as blows to the spinal column or whiplash injury. Paralysis occurs immediately and is flaccid in nature. Function usually returns within 6–48 h. It is essentially a clinical diagnosis: the precise mechanism is unclear.

ANTERIOR CERVICAL SPINAL CORD SYNDROME

This consists of cord infarction in the territory of the anterior spinal artery and may result from occlusion of the anterior spinal artery, anterior cord compression (for example by dislocated bone fragment), or traumatic herniated disc. Compression of the anterior aspect of the cord damages the corticospinal and spinothalamic tracts, resulting in motor paralysis and dissociated sensory loss below the level of the lesion, i.e. loss of pain, temperature and touch sensation, but relative preservation of light

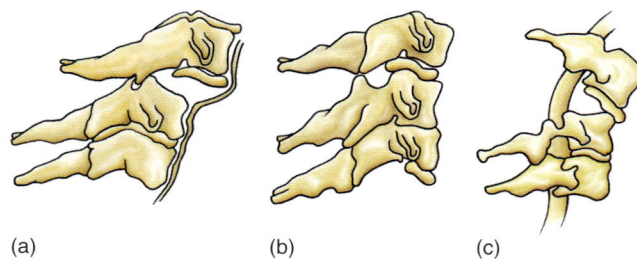

Figure 10.4 (a) Subluxation of the cervical spine with vertebral dislocation and compression of the spinal cord. (b) Subluxation associated with intervertebral disc hernia, which narrows the diameter of the spinal canal and exerts pressure on the spinal cord. (c) Subluxation with large cephalad dislocation of the vertebral body (upwards), detachment of the intervertebral disc and compression of the spinal cord. Modified from Rea, E.L. and Muller, C.A. (eds) 1993. *Spinal Trauma*. Rolling Meadows, IL: AANS, with permission.

touch, proprioception and position sense, which are carried via the posterior columns. With respect to management, it is vital to differentiate a non-surgical case, such as an anterior spinal artery occlusion, from a surgical one, such as pressure from an anterior bone fragment or disc. The latter requires urgent surgical decompression. The prognosis is not good, with only 10–20 per cent recovering functional motor control. There may, however, be sufficient recovery of sensation to help prevent burn injuries.

BROWN–SÉQUARD SYNDROME

This is essentially a hemisection of the spinal cord from a penetrating trauma, stab wound or blunt injury. It may also occur as a result of radiation myelopathy, cord compression by spinal epidural haematoma, large cervical disc herniation, spinal cord tumours, spinal arteriovenous malformations and cervical spondylosis. The limbs below the lesion are paralysed. On the opposite side of the body, the sensation of pain, temperature and touch is lost one to two segments below the lesion's level but light touch is preserved, because of the existence of redundant ipsilateral and contralateral pathways (dissociated sensory loss). The posterior columns deficit is not significant. The prognosis is relatively good with 90 per cent of the patients regaining sphincter control and the ability to ambulate independently.

CENTRAL SPINAL CORD SYNDROME

This usually results from hyperextension of the cervical spine with compression of the spinal cord between the degenerative intervertebral disc anteriorly and the thickened ligamentum flavum posteriorly. Long tract fibres passing through the cervical spinal cord are somatotopically organized. The cervical fibres are located more centrally than the fibres serving the lower extremities. The cord damage is located centrally and is most severe to the centrally lying cervical tracts, which supply the upper limbs, resulting in disproportionate weakness in the upper compared with the lower limbs. The pathophysiology relates to the fact that the central region of the spinal cord is a vascular watershed zone and, hence, more susceptible to injury from swelling.

The central cord syndrome is the most common type of incomplete spinal cord injury. It is sometimes superimposed on congenital spinal stenosis. This often occurs during a motor vehicle accident or after a forward fall causing a blow to the forehead. It may occur without cervical fracture or dislocation in rheumatoid arthritis. The clinical presentation resembles that of syringomyelia with motor weakness, mainly of upper rather than lower extremities, with varying degrees of sensory disturbance below the level of the lesion and sphincter dysfunction.

POSTERIOR CORD SYNDROME

This produces pain and paraesthesias in the neck and upper arms. It is relatively rare.

SPINAL TRAUMA

After an accident, it is vital to ensure that the spine is stable because an unstable lesion may aggravate the spinal cord injury and the pre-existing damage. In cases of suspected or proven cervical spine instability, nasotracheal intubation and cricothyroidotomy are preferred over endotracheal intubation and cervical manipulation. It is also important to remember that the bony or ligamentous damage may or may not be associated with spinal cord damage. The extent of the vertebral column damage does not always correlate with the degree of spinal cord injury and neurological damage. The proper evaluation of the severity of the injury should be primarily based on imaging investigations.

A radiological investigation immediately after a spinal cord injury shows the vertebral alignment at that time, but it reveals neither the degree of disruption that may have occurred at the moment of injury nor the degree of ligamentous damage. To obtain sufficient information, patients with spine injuries should be investigated as follows:

- In the cervical spine, an investigation with anteroposterior, lateral and open-mouth odontoid radiological views will detect all unstable fractures. In patients who are severely injured, anteroposterior and lateral views from the craniocervical junction down to C7–T1 are sufficient for standard evaluation. Evidence of soft tissue swelling between the pharynx and the vertebrae, malalignment of the anterior or posterior margins of the vertebral body or of the lamina, i.e. subluxation, undue widening of the interspinous distance or the disc space, damage to the vertebral body, apophyseal joints, lamina or spinous process, anterior wedge collapse or 'burst' fracture, should be looked for. The following additional studies may be indicated: oblique views to demonstrate the neural foramina, and assess the integrity of the articular masses and lamina; flexion–extension views; polytomograms; computed tomography; and magnetic resonance imaging (MRI). Disruption of the foraminal outline suggests malalignment. In the anteroposterior view the alignment and the width of the apophyseal joints should be noted as well as the presence of vertical fracture lines. If doubt remains, oblique views should be done to demonstrate the intervertebral foramina. In addition, the anteroposterior 'open

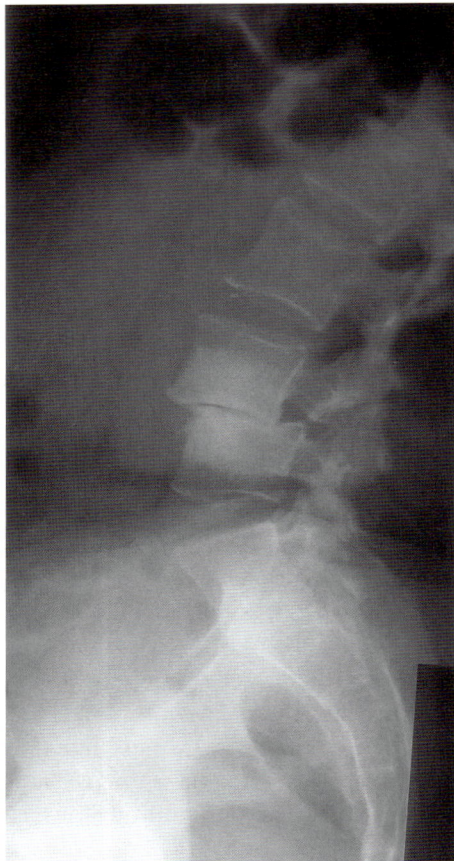

Figure 10.5 Complex severe lumbar spine injury which consists of multiple fractures of the laminae, subluxation of L5 vertebra and displacement upwards and posteriorly, sponylolisthesis at the L4–L5 levels and elimination of the L4–L5 intervertebral space.

mouth' view may be required to demonstrate a fracture of the odontoid peg. The purpose of flexion–extension cervical spine radiographs is to disclose occult instability. It is possible for posterior ligamentous complex injury to occur without any bony fracture. Lateral flexion–extension views detect these injuries.

- In the upper thoracic spine, only tomography may satisfactorily demonstrate the lateral view.
- In the lumbar spine, a 'Scotty dog' appearance suggests a fracture and/or a dislocation (Figure 10.5).

A bone scan may be helpful in distinguishing an old injury from an acute one. Difficulties arise whenever the patients are unconscious, are unable to describe symptoms, have altered mental status or have an unknown mechanism of injury. Finally, it should be remembered that there are several contraindications to an extensive radiological investigation. These include the following: subluxation about 3.5 mm at any level, neurological disability and lack of patient cooperation.

CERVICAL TRAUMA: UPPER AND LOWER

The most common types of cervical spine trauma are described below.

Atlanto-occipital dislocation

This is probably underdiagnosed, because it is found in 8–19 per cent of fatal cervical spine injuries at postmortem examination. It is more commonly seen in children than in adults. The cruciate ligament and apical dentate ligament are insufficient to maintain spinal stability. The tectorial membrane and alar ligaments are the most important structures in maintaining atlanto-occipital stability (Powers *et al.* 1979). The picture may include bulbar–cervical dissociation, lower cranial nerve deficits and worsening neurological deficit on application of traction. It is important to remember that the patients may be neurologically intact, and therefore investigation for the injury must be undertaken.

Atlantoaxial dislocation

The rotatory atlantoaxial subluxation may occur spontaneously in rheumatoid arthritis, other arthritides, or following major or minor trauma. It may also occur as a complication of underlying congenital abnormality. If the transverse ligament is intact, the rotation is not associated with anterior displacement. The neurological deficit may include torticollis and reduced range of motion. If the transverse ligament has weakened, there may be anterior displacement. This is much more serious and one-third of the patients have neurological deficit or die. CT and MRI are essential in evaluating the transverse ligament (Alker and Leslie 1978).

Jefferson's fracture

This injury involves bilateral fractures or a three- or four-point fracture of the arch of the atlas, usually from axial load. The mechanism involves a direct vertical blow to the head and pressure from the head down on the spinal column. The atlas is squeezed between the occipital condyles above and the axis below. The arches of the atlas are weakest at the grooves of the vertebral artery and after the fracture the fragments burst outwards ('blow-out' fracture). It is unstable but usually with no neurological deficit because of the tendency of fragments to burst outwards. It is associated with a C2 fracture in 40 per cent of the patients. Thin-section high-resolution CT from C1 to C3 is the diagnostic procedure of choice.

Axis fractures

Fractures of the axis or C2 constitute approximately 20 per cent of cervical spine fractures and include: odontoid fractures of the axis, hangman's fracture, and other miscellaneous fractures. Neurological injury occurs in less than 10 per cent of the cases and most are treated by immobilization (Hadley *et al.* 1989).

Odontoid fractures

These can be caused after significant force has been applied during a road traffic accident, a fall from a height or a skiing accident, although in older patients (over 70 years old) a fall and head injury may cause such a fracture. The mechanism of injury includes flexion, with resultant anterior displacement of C1 on C2 (atlantoaxial subluxation). They constitute 10–15 per cent of all cervical fractures. They can be missed when other associated injuries are significant (Apuzzo *et al.* 1978). The immediate mortality from odontoid fractures has been estimated at around 25–40 per cent. The symptoms include high posterior cervical pain, paraspinal muscle spasm, tenderness, reduced range of cervical movements, upper extremity paraesthesias and exaggeration of muscle stretch reflexes. Patients affected may tend to support their heads with their hands (Dickman *et al.* 1989).

These fractures occur either at the tip through the base of the dens, or at the base and into the adjacent C2 vertebral body. They may compromise the blood supply to the dens, resulting in pseudoarthrosis of the dens, i.e. non-union of the fracture. Most odontoid fractures are treated by immobilization in a firm brace or halo for four months. In cases of non-union, a C1–C2 fusion is indicated (Bohler 1982).

Hangman's fracture

The lesion consists of the avulsion of the laminar arches of C2 with dislocation of the C2 vertebral body on C3. There may also be fracture of the odontoid process. The term 'hangman's fracture' was coined by Schneider *et al.* (1965) because it was sustained in judicial hangings (where submental placement of the knot results in hyperextension and distraction). However, a detailed study by James and Nasmyth-Jones (1992) demonstrated a relative paucity of cervical fractures – in this series only 3 of 34 were the typical 'hangman's fracture', named 'Type A' by James and Nasmyth-Jones, (see Figure 10.6) whereas in a further three cases involving C2 were of an asymmetrical nature (named 'Type B'). A later anthropological review of six cases showed a transverse cervical process fracturing of C1, C2, C3 and C5 as well as fracture of the C2 vertebral body (Spence *et al.* 1999). It appears that the effects of judicial hanging are variable, although even without bone or cord injury death usually is rapid as a result of compression of the carotid arteries, reflex cardiac arrest, or venous or airway obstruction (James and Nasmyth-Jones 1992). Radiological investigation has also been performed with postmortem examination revealing soft tissue haemorrhage of the neck, avulsion of intravertebral ligaments and tears of the vertebral arteries (Reay *et al.* 1994).

The mechanism of hangman's fractures in cases other than hanging is that of hyperextension and axial loading from motor vehicle incidents, falls, or diving accidents, which are now the most common causes of such injury. It is a bilateral fracture through the pars interarticularis

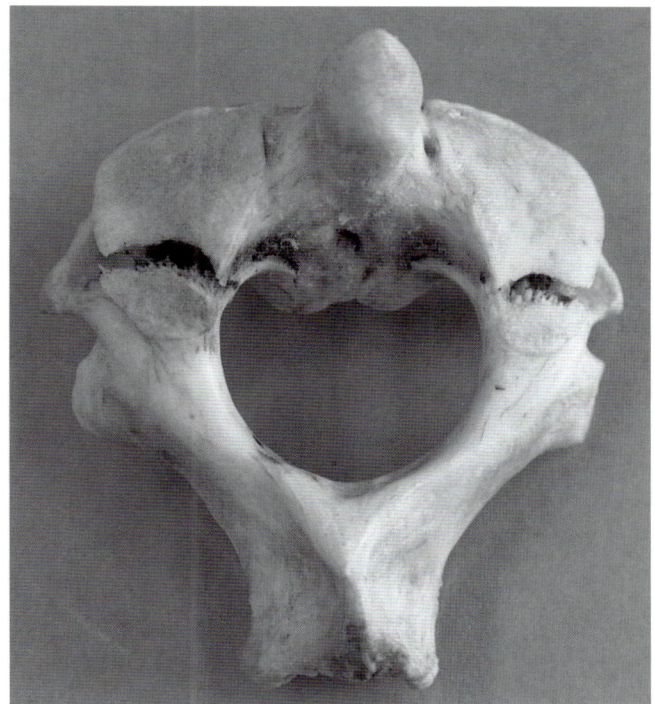

Figure 10.6 Hangman's fracture with fracture across the pedicle of C2.

(isthmus) of the pedicle of C2 often associated with anterior subluxation of C2 on C3. It is usually stable with no neurological deficit. The presentation may include cervical and occipital pain. The hyperextending and axial force usually causes injury to the face and head as well. About 90 per cent heal with immobilization and operative fusion is not needed (Levine and Edwards 1985).

THORACIC AND LUMBAR TRAUMA

Thoracolumbar spine fractures

Of all spine fractures, 60 per cent occur at the thoracolumbar junction, usually at the T12–L1 junction, and 70 per cent of these occur without immediate neurological injury.

Compression injuries

These are common lesions and are easily recognized by the decreased height of the vertebral body. Compression injuries may be seen when injury occurs whilst seated, for example, in air crash survivors. They are stable fractures and most of them are not associated with neurological damage.

Flexion–rotation injuries

These lesions occur most frequently at the T12–L1 level and result in anterior dislocation of the T12 on the L1 vertebral body with disruption of the posterior longitudinal ligament and posterior bony elements. The inferior vertebral body can undergo compression and an antero-superior wedge fracture. They are unstable injuries and result in complete neurological deficit.

Hyperextension injury

This involves rupture of the anterior longitudinal ligament and the intervertebral disc, and also a fracture through the anterior aspect of the vertebral body. This fracture is unstable, usually associated with severe cord damage, but fortunately very uncommon at the thoracolumbar spine.

ROLE OF UNDERLYING DISEASE

Osteoporosis is the main underlying disease that can predispose to serious injury and osteoporotic fractures. Osteoporosis is defined as skeletal fragility because of low bone mass and microarchitectural deterioration of bone. It is found primarily in elderly white women after the menopause. The lifetime risk of symptomatic vertebral body osteoporotic compression fractures is 16 per cent for women and 5 per cent for men. These patients often present with back pain after a minor fall and are found to have significant vertebral body compression fractures with bone retropulsed into the canal. Risk factors include excess weight, history of vertebral body fracture in a first-degree relative, drugs, steroids, warfarin, cigarette smoking, heavy alcohol consumption, physical inactivity and low calcium intake. Other underlying disease conditions may also play a role, for example, osteoarthritis, rheumatoid arthritis and osteomyelitis, metabolic bone disease, congenital spinal abnormalities, as well as degenerative disc disease and spondylitis.

NEUROPATHOLOGY OF SPINAL CORD INJURY

Acute injury: macroscopic findings

In deaths that occur immediately, the clue to any injury may be in the adjacent structures: fracture/dislocation or ligamentous tears (see Figure 10.7a). The spinal cord itself may show minimal findings other than petechial intramedullary haemorrhages and there may be associated subarachnoid haemorrhage. Central haematomyelia that comprises a large space-occupying haemorrhage with distension of the cord may very rarely be seen. In cases where there is significant trauma the cord may show disruption or lacerational injury (for example, as a result of penetrating injury). Swelling and softening of the cord develops during the first 24 hours and haemorrhagic necrosis with loss of grey-white differentiation may be seen.

Histology

Within 24 hours, necrosis of tissue is visible with variable haemorrhage. This begins centrally with extension into white matter – there is usually subpial sparing. Axonal swellings are seen (see Figures 10.7b and c). This is most extensive at the site of damage but extends distally and proximally. Acute inflammation with leucocytes begins, followed by macrophages peaking at three to seven days with removal of debris. Reactive astrocytes with gliosis over the several days or weeks supervenes.

(a)

(b)

(c)

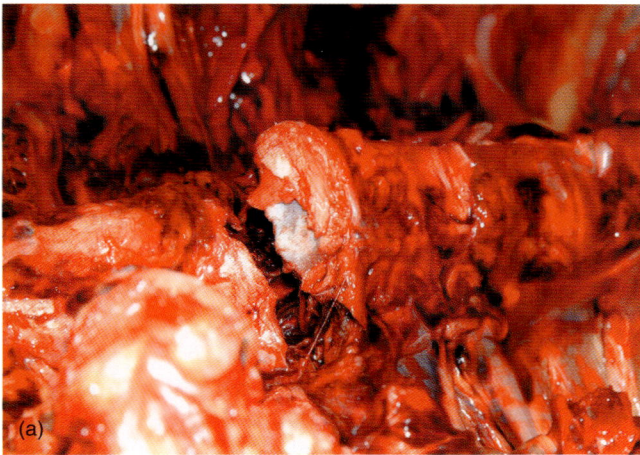

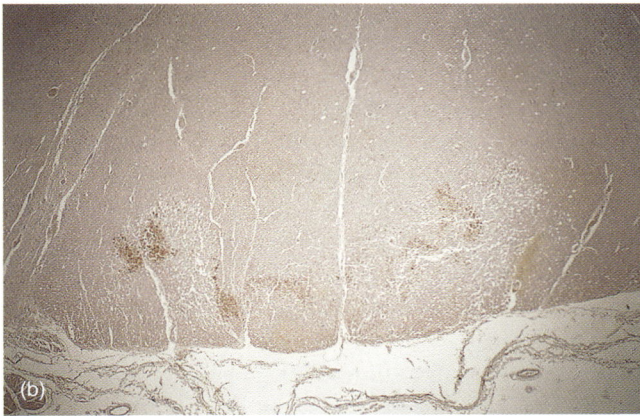

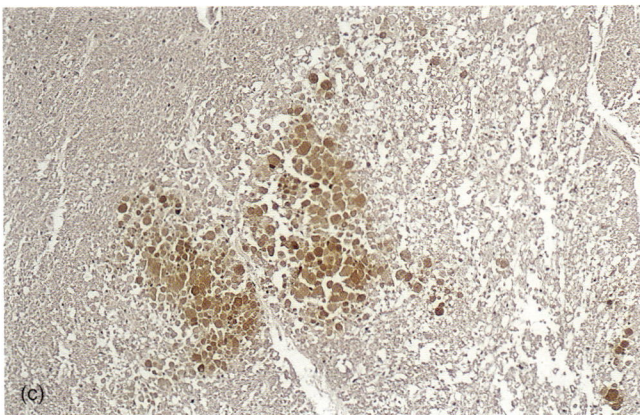

Figure 10.7 (a) Macroscopic appearance of acute spinal cord injury with fracture dislocation T2 following a fall from a height. (b) and (c) Axonal swellings in a high cervical fracture of a few days survival.

Figure 10.8 Post traumatic syringomyelia at T2 level – cavitation ascended from an injury at T11. (Reproduced from De Girolami, U., Frosch, M.P and Tator, C.H. 2002. Regional neuropathology: diseases of the spinal cord and vertebral column. In *Graham, D.I. and Lantos, P.L. (eds), Greenfield's Neuropathology, 7th edn. London: Arnold*)

In cases where there has been extensive cord disruption with breach of the dura, connective tissue scarring, which is frequently extensive, occurs. In the absence of dural laceration intramedullary scarring may be minimal.

The grey matter at the level of injury may become cavitated with hyaline thickening of blood vessels and collagenous fibrosis. Cavitation may extend rostrally and caudally into uninvolved cord to give rise to post-traumatic syringomyelia (see Figure 10.8).

Other long-term complications include traumatic neuromas in the regions of injured nerve roots with proliferation of Schawann cells.

Spinal arachnoiditis with extensive meningeal fibrosis may occur following trauma as well as in association with surgery, infection and spinal subarachnoid therapeutic procedures.

In deaths remote from the time of injury, where survival may be for months or years, the primary cause of death may be as a complication of the injury, such as renal tract infection with or without renal failure, chest infection or pulmonary emboli. It may be necessary to show a direct link between the injury and the cause of death for criminal or civil proceedings. In these situations careful evaluation of other disease, for example the presence of ischaemic heart disease, is important.

Later changes

Rostral and caudal segments show Wallerian degeneration of tracts damaged at the injury site with axonal degeneration and demyelination.

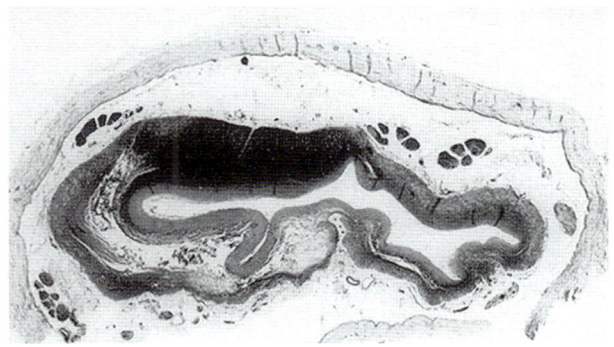

Incomplete lesions

The pathology depends on the location and the extent of injury. These are less often seen at postmortem examination – findings will depend on the anatomical location.

OTHER SPINAL INJURIES

Included under the term 'cervical cord neurapraxia' are the post-traumatic sensory changes, for example numbness, tingling or burning and motor symptoms, that usually involve all four extremities and last from 15 min to 48 hours. After resolution, recurrence rate is high particularly among those with narrower canals. 'Stinger' or 'burner' syndrome consists of burning dysaesthetic pain radiating from the shoulder down one arm and weakness in the C5 or C6 nerve roots. It results from direct nerve root compression in the neural foramina or downward traction on the upper trunk of the brachial plexus. Evaluation must include a cervical spine MRI.

PENETRATING INJURY

Penetrating injury to the spinal cord and surrounding tissues can include injury by gunshot and stabbing or other sharp force implement. The general features of gunshot wounds are covered in Chapter 3, together with the ballistic characteristics. Injury may be caused by a bullet directly to the spinal cord, or may occur at a distance (with high energy transfer wounds) as a result of shock waves or temporary cavitation. In addition, the level of cord injury may not correspond to any external wound due to changes in the position of the spinal cord with posture. Spinal shock may occur.

Radiological assessment should aid identification of bullets or shot, although multiple views may be necessary. Findings at postmortem examination will depend on the time since the event: early deaths may show focalised haematoma formation as well as cord disruption and haemorrhage. Later deaths show the axonal degeneration and demyelination as described above. Additional damage may occur as a result of vascular involvement with the formation of haematomas or infarction.

Penetrating injury due to sharp force implements, such as a knife, cause localised damage: either direct to the spinal cord, or as a result of vascular damage. Other weapons include screwdrivers and machetes. Penetration by a sharp object usually requires considerable force to injure the spinal cord due to the vertebra, although lesser force is required to pass between the lamina. Vascular injury may also occur in particular in relation to vessels in the neck such as the common carotid artery. Radiological assessment may help to identify not only any bony injury, but also any broken portion of a weapon. Embolisation of pellets may also occur (Coghill and Sullivan 1995).

MANAGEMENT OF SPINAL INJURY

The management of spinal cord injury depends on the characters of the lesion such as type, location and severity.

Simple contusion of paraspinal soft tissue especially in the cervical region is the most common form of spinal trauma. This presents with local tenderness and restriction of spinal movements. In cervical injuries, the symptoms include headache or vertigo. The best treatment is rest, analgesia and use of a cervical collar for three to four weeks. In many cases the symptoms become chronic and are described as 'post-traumatic cervical syndrome' or 'whiplash injury' syndrome. This is difficult to treat, especially if there is pre-existing spondylosis or premorbid personality.

In fractures or subluxations, the choice of conservative versus surgical treatment depends on the stability of the spine and the severity of the underlying spinal cord lesion. An unstable fracture may cause further spinal cord damage and, hence, needs to be stabilized with spinal traction, collar or surgical spinal fixation (Sonntag and Hadley 1988). In cases of a compressed fracture with fragments compressing the cord, surgical decompression may be indicated (Schneider et al. 1972). The same applies to the cases of intracanalicular haematoma.

The indications for surgery for gunshot wounds include cauda equina injury, neurological deterioration resulting from spinal epidural haematoma, compression of a nerve root, spinal instability, cerebrospinal fluid (CSF) leak, débridement to reduce the risk of infection and removal of a copper-jacketed bullet because copper can cause intense local reaction.

ACKNOWLEDGEMENTS

Figures 10.1, 10.2, 10.3 and 10.4 are modified and redrawn from Rea, G.L. and Miller, C.A. (eds) 1993. *Spinal trauma: current evaluation and management*. Rolling Meadows, IL: American Association of Neurological Surgeons.

REFERENCES

Alker, G.J. and Leslie, E.V. 1978. High cervical spine and craniocervical junction injuries in fatal traffic accidents: a radiological study. *Orthop Clin North Am*, 9, 1003–10.

Allen, B.L., Ferguson, R.L., Lehmann, T.R. *et al.* 1982. A mechanistic classification of closed, indirect fractures and dislocations of the lower cervical spine. *Spine*, 7, 1–27.

Apuzzo, M.L.J., Heiden, J.S., Weiss, M.H. *et al.* 1978. Acute fractures of the odontoid process. An analysis of 45 cases. *J Neurosurg*, **48**, 85–91.

Atkinson, P.P. and Atkinson, J.L.D. 1996. Spinal shock. *Mayo Clin Proc*, **71**, 384–9.

Bohler, J. 1982. Anterior stabilization for acute fractures and non-unions of the dens. *J Bone Joint Surg*, **64**, 18–27.

Coghill, T.H. and Sullivan, H.G. 1995. Carotid artery pseudoaneurysm and pellet embolism to the middle cerebral artery following a shotgun wound of the neck. *J Trauma*, **39**, 763–7.

Dickman, C.A., Hadley, M.N., Brower, C. *et al.* 1989. Neurosurgical management of acute atlas–axis combination fractures. *J Neurosurg*, **70**, 45–9.

Epstein, N., Epstein, J.A., Benjamin, B. *et al.* 1980. Traumatic myelopathy in patients with cervical spinal stenosis without fracture or dislocation: methods of diagnosis, management and prognosis. *Spine*, **5**, 489–96.

Hadley, M.N., Dickman, C.A., Brower, C.M. *et al.* 1989. Acute axis fractures: a review of 229 cases. *J Neurosurg*, **71**, 642–7.

Harris, J.H., Edeiken-Monroe, B. and Kopaniky, D.R. 1986. A practical classification of acute cervical spine injuries. *Orthop Clin N Am*, **17**, 15–30.

James, R. and Nasmyth-Jones, R. 1992. The occurrence of cervical fractures in victims of judicial hanging. *Forensic Sci Int*, **54**, 81–91.

Levine, A.M. and Edwards, C.C. 1985. The management of traumatic spondylolisthesis of the axis. *J Bone Joint Surg*, **67A**, 217–26.

Powers, B., Miller, M.D., Krammer, R.S. *et al.* 1979. Traumatic anterior atlanto-occipital dislocation. *Neurosurgery*, **4**, 12–17.

Reay, D.T., Cohen, W. and Ames, S. 1994. Injuries produced by judicial hanging. *Am J Forensic Med Pathol*, **15**, 183–6.

Schneider, R.C., Livingston, K.E., Cave, A.J.E. *et al.* 1965. 'Hangman's fracture' of the cervical spine. *J Neurosurg*, **22**, 141–54.

Schneider, R.C., Crosby, E.C., Rosso, R.H. *et al.* 1972. Traumatic spinal cord syndromes and their management. *Clin Neurosurg*, **20**, 424–32.

Sherk, H.H. 1989. *The Cervical Spine*. New York: Lippincott Co.

Sonntag, V.K.H. 1981. Management of bilateral locked facets of the cervical spine. *Neurosurgery*, **8**, 150–2.

Sonntag, V.K.H. and Hadley, M.N. 1988. Nonoperative management of cervical spine injuries. *Clin Neurosurg*, **34**, 630–49.

Spence, M.W., Shkrum, J., Ariss, A. and Regan, J. 1999. Craniocervical injuries in judicial hangings: An anthropologic analysis of six cases. *Am J Forensic Med Pathol*, **20**, 309–22.

Torrens, M.J. and Dickson, R.A. 1991. *Operative Spinal Surgery*. Edinburgh: Churchill Livingstone.

Webb, J.K., Broughton, R.B.K., McSweeney, T. *et al.* 1976. Hidden flexion injury of the cervical spine. *J Bone Joint Surg*, **58B**, 322–7.

11

Difficult areas in forensic neuropathology: homicide, suicide or accident

Christopher M Milroy and Helen L Whitwell

In examining the neuropathological aspects of a case, the pathologist will often be asked to give an opinion on the findings in the brain in the context of other pathological evidence and the circumstances of a case. Such scenarios include falls, falls verus kicks, deaths in young children, sporting deaths and patterns of pathology in motor vehicle collisions. Inevitably certain areas of practice present more problems to the pathologist than others. The role of the pathologist is to assist the judicial system in determining what the manner of the death is. In some jurisdictions, notably the North American medical examiners' systems, the determination of manner of death as well as cause of death lies with the pathologist. In other jurisdictions, such as the English coronial system or the Scottish system with the procurator fiscal, the pathologist does not decide manner of death, although clearly pathological evidence plays a pivotal role. Therefore, the identification of patterns of death that indicate accident, suicide or homicide is central to any death investigation. This chapter deals with areas that often cause problems for the pathologist. The extent to which a pathologist will determine the cause and the manner of the death will depend on the information provided, and the full circumstances of a case may not be revealed until the legal proceedings have been completed. The detail of the information provided often depends on external agencies such as the police or health and safety executive, and most pathologists have a limited role in obtaining evidence not directly related to the postmortem examination. In such cases caution should be exercised before coming to a firm conclusion. The pathologist is often asked to consider different scenarios and one set of pathological facts may be consistent with more than one scenario.

PENETRATING INJURIES

Injuries may be self-inflicted with the intention of killing oneself, for psychiatric reasons or for other gain. The latter pattern of injuries are often seen where there is a claim that another person attacked the alleged victim. These injuries are typically caused by a sharp-bladed implement and are superficial incised wounds or abrasions. They often involve the face, but not sensitive areas such as the eyes or lips. The injuries are frequently parallel and inflicted by the dominant hand in a direction that would be expected with a weapon held in that hand (Cordner 2003, Saukko and Knight 2004).

Suicide by penetrating weapons other than firearms rarely involves the head although, with some psychiatrically disordered people, bizarre injuries or methods of self-destruction may be used and case reports describing such events appear from time to time. Most non-firearms deaths caused by penetrating injuries are going to be homicidal or accidental. The surrounding circumstances should help differentiate these cases, even if the pathology does not.

The differentiation of self-inflicted gunshot wounds from homicide is a central role of the postmortem examination. In the USA firearms are the most common method of suicide for both men and women. In contrast, in the UK, suicide by firearm is a less common method of choice, and women very rarely kill themselves by firearms (Chapman and Milroy 1992). Any weapon may be used, but handguns are the most popular weapons in the USA. Typically the gunshot wound will be a contact wound (Figure 11.1), but intermediate wounds will occasionally be encountered. In considering whether or not a gunshot

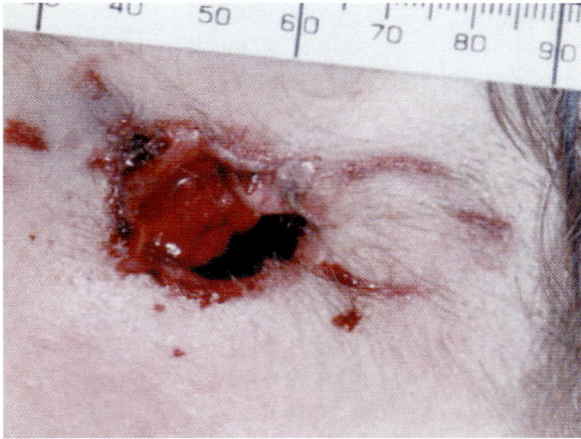

Figure 11.1   Contact shotgun wound to the temple.

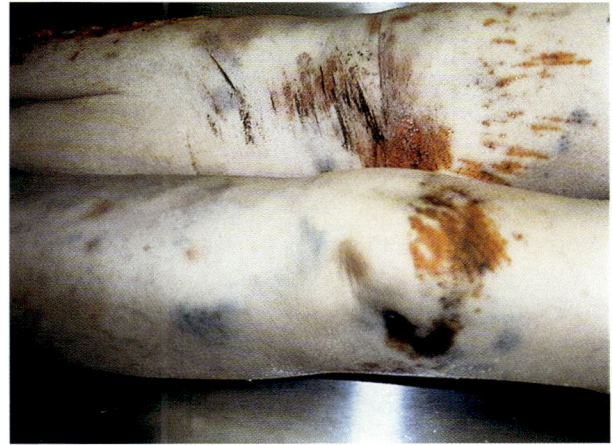

Figure 11.3   Bumper injuries to lower limbs. (Reproduced from Mason, J.K. and Purdue, B.N. 2000. *Pathology of Trauma*. London: Arnold)

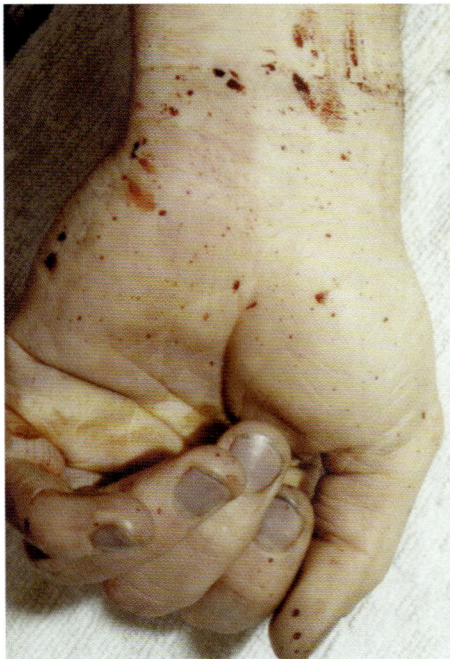

Figure 11.2   Back spatter of blood in firearm suicide.

significant cause of severe head injuries in modern society, accounting for over 50 per cent of severe head injuries in clinical and pathological practice in the UK. Many transportation deaths will present no significant problems in determining the cause of death and in reconstruction of the incident. However, in some cases the pattern of injury is important in reconstructing the incident. This is most commonly encountered where the victim is a pedestrian and there are questions about what position the person was in when struck by a vehicle. A second problem case is where a body is found on the roadside and the question raised is whether the person has been struck by a vehicle or died by some other means. Another problem is the run-over victim. Usually deaths of occupants of a vehicle pose no particular pathological questions, although one question that is occasionally raised is which of the occupants was driving when someone has been expelled from the car after the collision. Before commencement of the post-mortem examination, the clothing should be carefully retained and documented for forensic examination should it become necessary.

wound is self-inflicted, ancillary investigations can be very important. As well as examining for gunshot residue, blood spatter analysis may be helpful (Figure 11.2). In self-inflicted wounds blood pattern analysis of the hands may provide important evidence (Yen *et al.* 2003). Multiple self-inflicted gunshot wounds to the head are well recorded and do not exclude suicide (Kury *et al.* 2000).

ROAD TRAFFIC DEATHS

Deaths associated with transportation form an important part of forensic pathology practice, and they are the most

Pedestrians

Most pedestrians are struck by the front of the car (Clark and Milroy 2000). Most of the other impacts involve a pedestrian colliding with the side of a vehicle and a small percentage of incidents involve reversing. When the front of a car strikes an upright pedestrian, there is damage to the impact site, which is usually on the legs, the bumper striking the lower limbs (Figure 11.3). These injuries are known as the primary impact site. The impact may cause bruising, abrasion or laceration of the skin. This is characteristically around the knee area, but will obviously depend on the height of the victim and the car frontage. If the car is braking when it strikes the victim the car bumper will be at a lower height than when stationary. If the impact is with sufficient force, the underlying bones of the leg may fracture. The injuries may be seen on the presenting leg, when the person has been struck side on. The other leg may have injury at a higher level as he or she walks or runs into the path of the vehicle.

Where there are injuries to both legs at the same height, it supports the person being struck while facing forwards or being stuck from behind when the injuries are to the back of the legs. Deep dissection should be undertaken because bruising may be evident in the deeper tissues in the absence of surface injuries. When the tibia or femur is fractured, a wedge-shaped fracture site may occur, with the base indicating the site of impact and the front the direction of travel. Rotational movement may result in spiral fractures distant from the site of impact. It is important that the height from the heel of any injuries on the leg is measured, because this will allow comparison with any implicated vehicle. In a child the primary impact site may be higher on the leg or at the level of the pelvis. Patterned abrasions or bruising may be encountered but, with modern car design, badges, mirrors and number plates are structured not to cause damage.

Once struck the body will be propelled on to the bonnet (hood) of the car, causing secondary injuries. These injuries may occur with impact on the bonnet, front pillars of the car windscreen ('A' pillar) or the windscreen itself, or by the body travelling on to the roof of the car. It is during this movement that the head and trunk are injured. The body of the victim will then typically fall to the ground, causing tertiary injuries, which may be more severe than the secondary injuries. External injuries to the head include abrasions, bruising and lacerations. The skull may be fractured, with the most common pattern of skull fracture in fatalities being a transverse hinge fracture caused when the head is impacted sideways. Linear vault fractures may occur and the facial skeleton may be fractured. Localized depressed fractures may be seen. Brain injury patterns will correspond to the degree of force applied to the head. Vehicular collisions are a major cause of diffuse axonal injury. Extradural, subdural and subarachnoid haemorrhage may occur, as may coup and contre-coup contusions.

In an analysis of 150 fatal pedestrian collisions, Ryan and colleagues (1994) identified 31 victims with evidence of a single head impact with the vehicle. Using sector-mapping techniques, they found that in lateral impacts (18 cases) there was a greater degree of damage than in occipital impacts (13 cases). In lateral impacts, greater cortical damage was seen in the inferior regions than in the superior or parasagittal regions of both left and right cerebral hemispheres. In the corpus callosum, more than 40 per cent of sectors were injured, as a consequence of shearing forces around the inferior edge of the falx cerebri and in the parasagittal regions. In occipital impacts, the distribution of cortical and white matter injury was symmetrical about the midline. There was less corpus callosum injury. Local effects caused by indentation of bone were not observed in this series of 31 victims. The frontal and temporal regions appear to be more sensitive to injury than other areas of the brain, explaining coup and contre-coup injuries in impacts in the sagittal plane (see also Chapter 7, p. 85).

The most common neck injury is fracture dislocation of the atlanto-occipital junction. This injury is inevitably fatal and injury is most frequently encountered in motorcyclists, but is seen in pedestrian fatalities. Other damage to the cervical vertebrae may be seen (see also Chapter 10, p. 119).

Chest injuries are the most significant injuries in pedestrians after head injuries. Externally the same constellation of abrasions, bruising and lacerations may be seen. Abrasions to the trunk may be caused when the tertiary impact takes place and may consist of broad sheets of abrasion as the body moves along the road. These abrasions provide evidence that the body has travelled in a particular direction. Internally there may be more damage than is expected from the external injuries present. Rib fractures are common and may indicate the side of impact. They may puncture underlying lung, which may also be contused. The heart may be contused, with tears in the atria. The aorta may be ruptured, typically at the start of the descending thoracic aorta, although this injury is seen more often in vehicle occupants. Abdominal injuries are less common in pedestrian fatalities, the most common injury being lacerations of the liver. The spleen, kidney and mesentery may be damaged. The thoracic and lumber spine may be fractured and dislocated.

Where a high-fronted vehicle, such as a van or lorry, strikes the pedestrian, he or she will be pushed forward. If remaining in the path of the vehicle the person will be run over, presenting a further pattern of injuries. Pedestrians who have been struck by a car and suffered primary, secondary and tertiary injuries may also then be run over by another vehicle.

Runover injuries

Pedestrians may be run over because they have already been struck by another vehicle, or because they were lying in the road when initially struck. The pattern of injuries is different in runover cases although if the victim has already been struck or run over by more than one vehicle this may complicate the picture. With all road traffic deaths, the

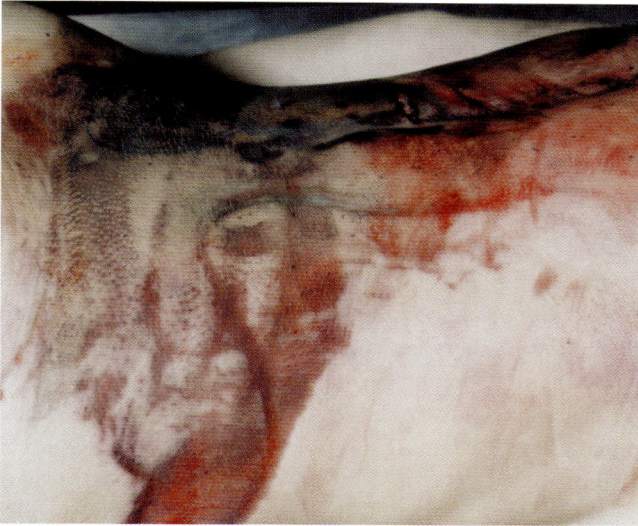

Figure 11.4 Sheet abrasions and grease to the lower back from being dragged underneath a vehicle.

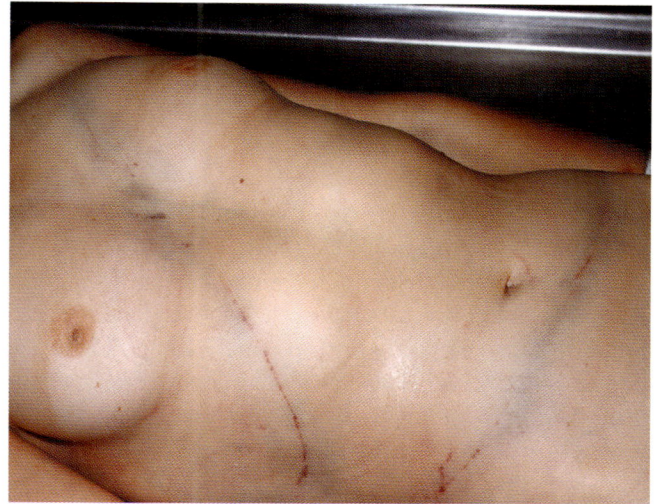

Figure 11.5 Seat-belt marks to car passenger (right-hand drive).

postmortem findings should be correlated with any vehicle examination. It is often the case that an examination of the car that has allegedly struck someone will show damage indicating the position of the victim. Furthermore, blood, tissue and other trace evidence may be recovered from the car to indicate whether the victim has been run over or struck in another position.

Where a person has been lying in the road when struck, he or she is typically found to be intoxicated with alcohol or some other drug, but it is important to try to exclude natural disease or a previous attack by another person. If there is a possibility that the person was already dead when hit, a useful test is to perform a fat stain on the lung. The presence of fat embolism supports an active circulation at the time that the runover injuries were inflicted.

External examination typically reveals broad areas of abrasion indicating dragging of the body (Figure 11.4). There may be oil and grease from the underside of the car. Damage from the under-surface structures of the car may cause further abrasion, bruising and laceration. Tyre imprints may be present. Where no dragging has taken place, external injury may be slight.

When the wheels of a vehicle run over the body, the head may not be significantly damaged, the principal areas of injury being to the chest and abdomen. If the head is run over it will show crush injury patterns. Abrasions, bruising and lacerations may occur and the skull will be fractured. The brain may show direct damage, but not the pattern of injury seen in secondary or tertiary impact injuries. Extensive thoracic and abdominal injury may indicate severe crushing, with extensive damage to the chest wall, lungs, heart and major vessels. The abdominal organs

may be lacerated, with pulping of the liver and spleen seen in some cases. The vertebral column may show extensive damage, as may the pelvis. The bladder may be ruptured. The mesentery is often lacerated.

Vehicle occupants

Postmortem examinations on occupants of cars usually present less controversy than those on pedestrians although, where occupants have been ejected from the car, questions over who was driving may be important. In these cases seat-belt injuries (Figure 11.5) can be useful in determining the seating positions (Milroy and Clark 2000). Injuries to the driver may be caused by the steering wheel and foot injuries may be found where the driver has been braking hard. The front-seat passenger may be injured when he or she strikes the front dashboard. Engine intrusion may occur in high-speed collisions, although modern car design structures the car so that crumple zones absorb the energy of an impact to decrease this event and increase occupant safety. Pathologists should be aware that injury to front-seat passengers may result from unrestrained rear-seat passengers moving forwards after a collision. As well as injury patterns forensic examination of the vehicle is likely to provide important evidence as to who the driver was. The use of seat-belts and air bags has done much to reduce the death rate following vehicular collisions, but head injury remains an important cause of death in car occupants. Airbags themselves produce injuries to the head, including facial and neck abrasions as well as eye injuries. Children are particularly prone to head and neck injuries if placed in the rear position on the passenger side

(Milroy and Clark 2000). In addition, fatalities are also reported where a child is restrained using an adult seat belt (Cooper *et al.* 1998) and where occupants are unrestrained or out of position (Boyd 2002).

Motorcycles and pedal cycles

Head and spinal injuries are a common cause of death in this group. Motorcyclists have a higher incidence of skull fractures than car occupants, including linear, hinge and rarely ring fractures. The pattern of injury in both types will be modified depending on the circumstances, including the presence or otherwise of bicycle helmets (Macpherson and Macarthur 2002, Wardlaw 2002).

Vehicular collision investigation

The following is a question that may be asked of the pathologist: can a correlation be made between the injuries present and the speed of the vehicle? Although there is a general rule that the higher the speed the greater the likelihood of significant injury, low-speed impacts may be associated with a fatal outcome. Where a limb is amputated or transection of the torso occurs, Zivot and DiMaio (1993) found speeds were above 88 km/h (55 miles/h or mph). In a study of 47 standing or walking pedestrian deaths, age range 15–86 years, Karger and colleagues (2000) found that brain damage did not depend on impact velocity. In their study they found 64 spinal fractures in 38 victims. Nearly half of all fractures were in the cervical spine. Spinal fractures were seen with impact speeds above 27.5 km/h, with 36 of 38 victims having them at 45 km/h and in every case above 67.5 km/h. They concluded that, if there was no spinal fracture, the speed was below 70 km/h (43.5 mph) and probably below 50 km/h (31.25 mph). They also showed that aortic and inguinal ruptures and dismemberment correlated with higher-impact speeds. However, in view of the variation in possible speeds with possible injuries, dogmatic statements on speed should not be made. Police traffic investigators are often very accurate in determining speed.

FALLS

Falls are a common problem in forensic practice. The pathologist may need to assess whether there are suspicious findings in an individual who is found at the bottom of a flight of stairs or at the foot of a building; assessment

of possible defence injuries, injuries caused by hanging on to a ledge, as well as those that could have been caused by an attack, should be done. A history of underlying disease such as ischaemic heart disease or epilepsy may also aid interpretation, as will evidence of use of drugs or alcohol.

Falls from standing

Falls leading to head injury are a common occurrence and differentiation between accidental and deliberate infliction is important. The presence of more than one laceration to the head should raise suspicion although there may be an explanation, e.g. multiple falls in an intoxicated individual. A clinical analysis of 189 elderly patients (over 60 years of age) admitted to hospital (Nagurney *et al.* 1998) revealed that falls from a standing height (76 per cent) were more common than falls on stairs (19 per cent) or from a height (5 per cent). However, abnormalities on computed tomography (CT), identified in 16 per cent, were more common in stair falls (42 per cent) and from a height (40 per cent), with the most common patterns of pathology being cerebral contusions (38 per cent) and subdural haematoma (33 per cent). In fatal falls from a standing height the typical pathology seen is after a fall on to the back of the head with an occipital impact site, contre-coup pattern of contusions and subdural haematoma formation.

Falls from a height

High falls are a common method of suicide in some countries. In Singapore, where falls from heights are common, an analysis of 603 postmortem examinations performed on falls from more than 3 m, mostly from 20–40 m, revealed the head and face as the primary impact site in 67 cases (11.1 per cent) (Lau *et al.* 1998). Over one-third of victims had cerebral lacerations, with a subset of about 20 per cent showing massive craniocerebral destruction (Lau *et al.* 2003). Subarachnoid and subdural haemorrhages were seen in a third of these cases. Extradural and intracerebral haemorrhages were uncommon, with cerebral contusions seen in 10 per cent of cases, presumably because of the rapidity of death in these cases. Beale and colleagues (2000) examined 341 people who had suffered a fall from more than 2 m in Scotland. Of these, 89 died, with mortality significantly higher in falls of greater than 10 m. Skull fractures were found in 59 victims with cranial pathology consisting of extradural haematoma, subdural haematoma, subarachnoid haemorrhage, cerebral contusions and intracerebral haematoma. The assessment of traumatic axonal injury, taken in isolation, does not show any significant difference between individuals falling from their standing height or those falling from a greater height (Abou-Hamden *et al.* 1997).

Homicidal falls from a height are a rarity, as published in a recent series from Singapore (Lau 2004). He describes five cases of pure homicide and nine episodes of dyadic death from a total of 533 homicides and 3963 fatal falls from a height. The majority of victims were children.

Falls down stairs

Falls down a flight of stairs are an important cause of head injury (Ragg *et al.* 2000). In an analysis of 51 deaths after falls down stairs by Wyatt and colleagues (1999), the peak age range was 71–80 years, with most victims being over 50 years of age. Males only slightly numbered females; 28 victims had alcohol in the blood, with 20 victims having a blood alcohol in excess of 80 mg/100 ml; 35 had brain or brain-stem damage, and 8 cervical injuries. The findings of head injury and an association with alcohol intoxication is also shown in a review by Preuß *et al.* (2004) in their series of 59 cases over 11 years.

An unpublished analysis of 57 fatalities from the authors' department revealed a bimodal age distribution of victims. One group was 30–60 years of age and there was a second peak of 70–90 years of age. Natural disease was common in this latter group. Of these 57 cases, 15 died of natural disease and 42 of injuries caused in the fall. Of this latter group, 30 died of brain injury and 8 of neck injuries. In the younger group alcohol intoxication was a significant finding. Evidence of frontal impact was seen in 42 per cent, temporoparietal impact in 46.5 per cent and occipital in 30 per cent of cases. Lacerations were seen in half the cases. The most common site for lacerations was the occiput (Figure 11.6), with other sites being the frontal and parietal scalp and the face, including the supraorbital ridges. Most lacerations were single although in four cases there were two lacerations and in one there was three lacerations. More than one area of scalp damage may therefore be present, caused as the victim tumbles down the stairs. Of the victims 27 had a skull fracture, 25 of these consisting of a linear non-displaced fracture. No depressed skull fractures were seen. The temporoparietal skull and occiput were the most common fracture sites. Frontal fractures were unusual. Subdural and subarachnoid were the most common type of intracranial haemorrhage. The principal pattern of brain injury was fronto-temporal contusions with only one case of occipital contusions. The cervical spine was fractured in nine cases, the thoracic spine in five cases and the lumbar spine in one case. Rib fractures were identified in 12 cases.

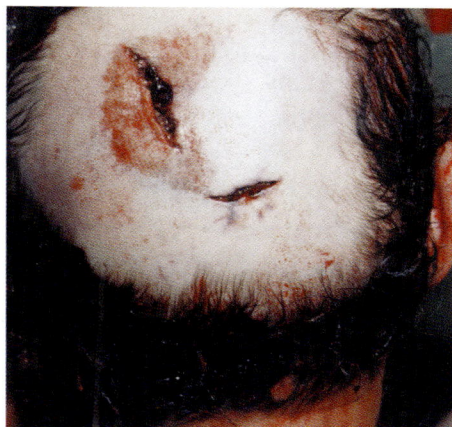

Figure 11.6 Occipito–parietal lacerations from a fall down stairs.

The position in which the body ends up at the bottom of the stairs may be very variable and caution must be exercised in interpretation of the final position, in relation to the fall down the stairs. The victim may still have some capacity to move after the fall, and the tumbling body can come to rest in unusual positions. An apparently unusual position should not lead to the automatic conclusion that the person could not have fallen down a flight of stairs. However, multiple lacerations should raise concerns of an attack with a blunt weapon. Elsewhere on the body, bruising may be found on the trunk and limbs, often with an abrading appearance – the skin overlying the bruise showing an abrasion in the direction of travel. Bruising with abrasion is commonly over body prominences, such as the iliac crest, and over the cervical prominence. However, these injuries to the trunk may not be as substantial as one might expect. Again, caution must be exercised in over-interpreting the relative absence of injuries.

Kicking versus falling

In the medico–legal setting, the pathologist is often asked to distinguish between different types of assault by different assailants. For example, a not uncommon scenario is one individual punching the victim, causing the victim to fall to the ground, and another individual kicking the victim. Diffuse axonal injury (DAI) is most likely to occur biomechanically when the moving head is rapidly decelerated on the ground. Whether DAI can be produced solely by kicking without impact from a fall is unclear, although lesser degrees of traumatic axonal injury do occur. There are limited fully documented cases in the literature as the exact circumstances are frequently unclear and detailed neuropathological studies few. It may be difficult to be confident in any given situation as to the relative contribution of differing assaults to the neuropathological findings (Geddes *et al.* 2000). Correlation with external and scalp injuries may aid the analysis.

SUDDEN DEATH IN HEAD INJURY

Sudden death in the context of head injury is relatively uncommon contrary to the general perception of both non-medical and at times medical personnel. Clearly, where there is massive craniocerebral disruption such as after significant fall from a substantial height or a gunshot injury, death is likely to be instantaneous. With less severe injury there is usually a period of survival with, for example, diffuse axonal injury. Death occurring rapidly in this context is usually related to diffuse vascular injury (see Chapter 8, p. 94). Sudden immediate death may also occur in the forensic setting in the context of basal traumatic subarachnoid haemorrhage – this is covered in detail in Chapter 6, p. 71. Sudden death has been reported after a blow to the back of the neck (Davis and Glass 2001), where the likely mechanism of death was neurogenic shock caused by concussion of the particularly sensitive area at the junction of the medulla and spinal cord. This has also previously been reported by Freytag (1963) who reviewed the postmortem findings in head injuries caused by blunt forces in 1367 deaths. In her series 6 per cent of cases were concluded to result from 'concussion'.

In addition, a number of cases of sudden death with head injury in association with alcohol intoxication have now been reported. The postmortem findings vary in these in that some show evidence of significant trauma. However, in other cases evidence of traumatic injury such as skull fractures or brain injury is minimal (Ramsay and Shkrum 1995, Milanovich and DiMaio 1999, authors' personal experience). A proposed mechanism is concussive brain injury, which may rarely produce prolonged post-injury apnoea in combination with alcohol (Zink and Feustel 1995). Pathological examination in these cases may reveal small haemorrhages, particularly in the corpus callosum and sometimes in the periventricular region or brain stem (Voigt 1981, author's personal experience) (Figure 11.7). As a result of the short time interval between injury and death amyloid precursor protein (APP) staining and other markers are negative.

SIDS AND SUFFOCATION

This is a difficult area in forensic pathology with serious implications in the medico–legal field. The term 'sudden

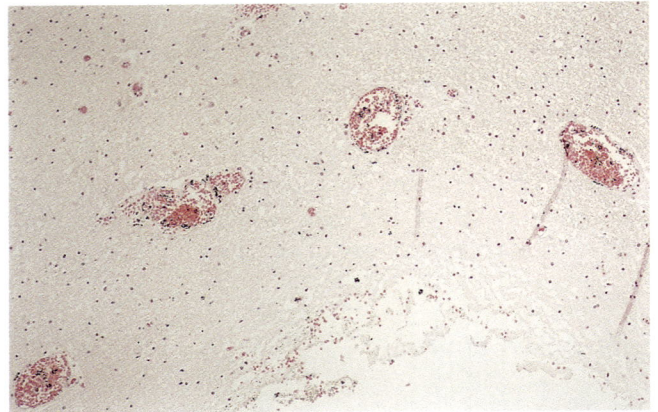

Figure 11.7 Microhaemorrhages in the corpus callosum in a case of sudden death in head injury.

infant death syndrome' became widely accepted in the 1970s and was defined as the sudden death of an infant or young child which is unexpected by history and in whom a thorough postmortem examination fails to demonstrate an adequate cause of death (Saukko and Knight 2004). The incidence of true SIDS has fallen, partly as a result of the 'Back to sleep' campaign. There is now an increased risk of sudden death in association with co-sleeping, sleeping on a sofa, and where there is exposure to tobacco smoke (Blair *et al.* 1996, 1999).

The major difficulty for the pathologist is distinguishing a true SIDS from upper airway obstruction, as in suffocation. This topic is covered extensively in the literature (Byard 2004).

Suspicion should be raised by facial petechiae, including conjunctival (although these can also be seen in SIDS and may not be evident in cases of suffocation) and facial or mouth bruising and/or abrasions. Attempts at resuscitation may complicate the issue (Hanzlick 2001, Leadbeatter 2001). The presence of extensive areas of intra-alveolar haemorrhage in the acute situation has been described (Yukawa *et al.* 1999). Bleeding from the nose is not infrequently seen. Substantial numbers of haemosiderin-containing macrophages have been described in repeated upper airway obstruction (Beecroft and Lockert 1997, Hanzlick and Delaney 2000, Milroy 1999). It appears that they are an uncommon finding in true SIDS and should lead to a search for their cause, which includes infections, pulmonary haemosiderosis and bleeding disorders. It should be recognised, however, that the above findings are non-diagnostic in themselves. It is essential to have a full evaluation of the history and scene, as well as detailed pathology. Increasingly metabolic or cardiac defects are being identified in such cases.

Developmental abnormalities of the brain stem have been described in a number of studies, particularly of the arcuate nucleus (Matturri *et al.* 2000, 2002), which shows hypoplasia in a proportion. However, there is considerable variability in the architecture of the arcuate nucleus. A simplified procedure for examination of the brain stem in infant deaths has recently been described (Matturri *et al.* 2004), although further assessment of the significance of the findings needs evaluating.

Suffocation in the adult as a cause of death is relatively uncommon. Most cases are in the elderly or infirm who not infrequently have other disease such as carcinoma. Drugs such as morphine may be involved. Cases of homicide, for example in head injury, may on rare occasions be complicated by terminal suffocation.

Neuropathological findings of suffocation in both the infant and adult are not well described. Survivors of attempted upper airway obstruction may show changes of hypoxic-ischaemic damage of the global type with widespread cortical involvement.

Acute changes are non-specific. The weight of the brain may be increased in infants, presumably as a result of congestive swelling. This may not be the case in adults however, as examination of organ weights in cases of asphyxiation as compared with trauma cases does not show any significant differences in brain weight across the various groups (Hadley and Fowler 2003). However, in a further series, brain weights of suicidal hangings were significantly higher than in those dying of overdose (Hamilton and McMahon 2002).

Petechial haemorrhages may be seen histologically, often perivascular in the cortex, however they cannot be taken as diagnostic of upper airway obstruction in the absence of other evidence as they may also be a feature of other causes of sudden death.

CONTACT SPORTS

Boxing has raised controversy over many years, in relation not only to acute injury, but also to the long-term chronic effects. In addition there is now considerable literature concerning repetitive mild head injury such as soccer, American football, rugby and ice hockey (Kirkendall *et al.* 2001, Rabadi and Jordan 2001, Marchie and Cusimano 2003, Guskiewicz *et al.* 2003). Second impact syndrome is also recognised to occur following a second injury giving rise to cerebral swelling which may lead to death (see Chapter 15, p. 187). However, the long-term effects with regard to permanent damage are still being evaluated (McCrory 2003). Widespread neocortical neurofibrillary tangles have been identified in cases of chronic repetitive head injury in young individuals in the perivascular location. This is thought to

represent an early consequence of repetitive head injury (Geddes *et al.* 1999). Genetic influences may also play a role as *ApoEε*4 is recognized to be associated with a poorer outcome after traumatic brain injury (Graham *et al.* 2002).

Boxing

DEATH AS A RESULT OF ACUTE INJURY IN BOXERS

This is rare but well recognized (Ross and Ochsner 1999). The most common cause is acute intracranial haemorrhage, most commonly acute subdural haematoma, and in the UK this has led to a re-evaluation of neurosurgical facilities at boxing events (Sawauchi *et al.* 1996, Garfield 2002) Traumatic subarachnoid haemorrhage is rare in boxing, probably as a result of the protective influence of well-developed neck muscles and lack of alcohol intoxication in the fighters. Traumatic subarachnoid haemorrhage has been suggested as the mechanism of death in a historic case (Plant and Butt 1993).

Examination of the brain in these cases shows the typical pathology of acute subdural haemorrhage. There may also be evidence of previous injury such as old contusions.

CHRONIC BRAIN DAMAGE

Dementia pugilistica, commonly called the 'punch drunk' syndrome has been described in both amateur and professional boxers. It develops in boxers who have competed in many bouts over a long period of time. Intellectual deterioration develops in association with Parkinson-type symptoms, as well ataxia. In a large neuropathological series of ex-boxers (Corsellis *et al.* 1973) the characteristic macroscopic features were identified, including:

- Enlargement of the cavum of the septum pellucidum and fenestration of its leaves and thinning of the fornices and corpus callosum (Figure 11.8). This was seen in 77 per cent of cases.
- Cerebellar atrophy in relation to the folia adjacent to the foramen magnum with gliosis and Purkinje cell loss. This is thought to result from repetitive transient tonsillar herniation.
- Depigmentation with neuronal loss of the substantia nigra and locus ceruleus with neurofibrillary tangles. Lewy bodies are not a feature.
- Cortical pathology with neurofibrillary tangles identified on silver staining in the absence of senile plaques was described by Corsellis and colleagues (Figure 11.9). This is most prominent in the medial temporal cortex but also seen elsewhere in the cerebral cortex and brain stem. Later Aβ protein staining demonstrated immunoreactive plaques (Roberts *et al.* 1990).

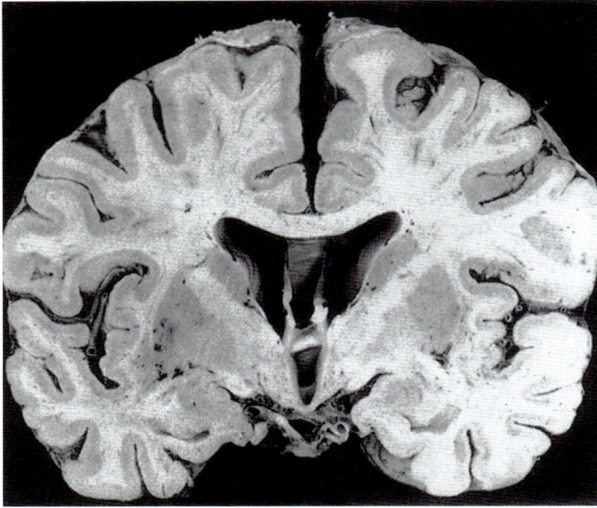

Figure 11.8 Cerebral atrophy, enlargement of the ventricular system with a wide cavum with torn walls in the interventricular septum. (Reproduced from Adams, J. H. 1992. Head Injury. In Adams, J. H. and Duchern, L. W. (eds), *Greenfield's Neuropathology*, 5th edn. London: Edward Arnold)

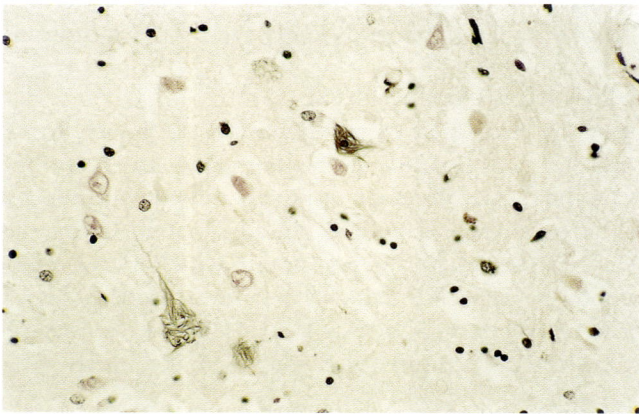

Figure 11.9 Neurofibrillary tangles in a case of long-standing damage caused by boxing.

OTHER SPORTS

Head and neck injuries are seen in a variety of other sporting activities. In general, unlike the issues relating to boxing, the injuries themselves are non-specific and are in most cases the picture of blunt injury. Golf and horse riding were identified as the most common cause of head injuries requiring hospital stay (Lindsay *et al.* 1980). Golf injuries may be as a result of contact with the ball or club. Skiing injuries are most commonly as a result of a fall followed by collision with an object such as tree (Myles *et al.* 1992). Sailing accidents may cause severe head injury when the boom swings across, striking the head. Blunt head injury in equestrian

accidents may arise as a result of falling or contact with the horse or other object such as a branch. Occasionally, horse-shoe abrasions (which may be incomplete) may be identified (Lee 2000, author's personal experience).

Aside from injuries, assessment of natural disease is essential in interpreting sequence of events. A number of sports-related deaths result from natural causes such as ischaemic heart disease with injuries being secondary. In the young age group congenital heart disease such as hypertrophic cardiomyopathy may be present. The assessment of injury and its relationship to natural disease may be difficult in some cases. In addition toxicological assessment is necessary, particularly with the increased use of recreational and illicit drugs.

Diving

Investigation of diving deaths is complex. There has been an increase in deaths occurring during recreational diving with a downturn in occupational deaths. Detailed discussion about the investigation of these is covered in detail by Ross in *The Pathology of Trauma* (Ross and Grieve 2000). Decompression illness is seen in divers who fail to control their rate of ascent. It comprises pulmonary barotrauma and cerebral gas embolism. Bubbles of inert gas come out of solution during decompression. Cerebral symptoms range from dizziness to unconsciousness, with spinal cord involvement manifesting as paraesthesias/numbness. The lower spinal cord is particularly affected and damage may be permanent. Pathology shows acute infarcts typically in lateral and dorsal columns (Calder 1986) (Figure 11.10). Chronic damage may also be seen in the absence of overt illness (Palmer *et al.* 1987).

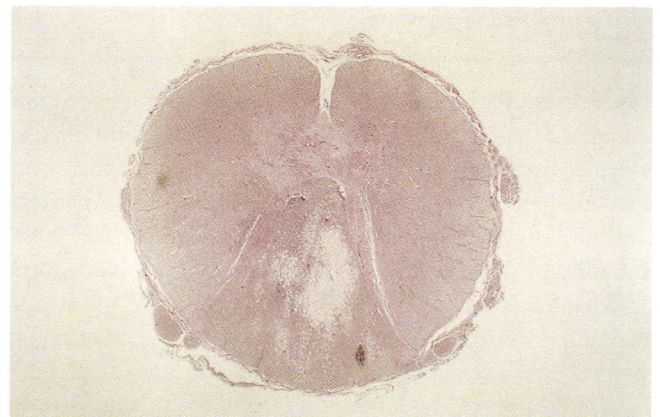

Figure 11.10 Spinal cord in a case of acute decompression sickness with dorsal column involvement. (Courtesy of Dr I M Calder.)

REFERENCES

Abou-Hamden, A., Blumbergs, P.C., Scott, G. *et al.* 1997. Axonal injury in falls. *J Neurotrauma*, **14**, 699–713.

Beale, J.P., Wyatt, J.P., Beard, D., Busuttil, A. and Graham, C.A. 2000. A five year study of high falls in Edinburgh. *Injury*, **31**, 503–8.

Beecroft, D.M. and Lockett, B.K. 1997. Intrapulmonary siderophages in sudden infant death: a marker for previous imposed suffocation. *Pathology*, **29**, 60–3.

Blair, P.S., Flemming, P.J., Bensley, D. *et al.* 1996. Smoking and the sudden infant death syndrome: results from 1993–1995 case-control study for confidential inquiry into stillbirths and deaths in infancy. *BMJ* 313, 195–8.

Blair, P.S., Flemming, P.J., Smith I.J. *et al.* 1999. Babies sleeping with parents: case-control study of factors influencing the risk of sudden infant death syndrome. *BMJ*, **319**, 1457–62.

Boyd, B.C. 2002. Automobile supplemental restraint system-induced injuries. *Oral Surg Oral Med Oral Pathol*, **94**, 143–8.

Byard, R.W. 2004. *Sudden Death in Infancy, Childhood and Adolescence*, 2nd Edn. Cambridge: Cambridge University Press.

Calder, I.M. 1986. Dysbarism. A review. *Forensic Sci Int*, **30**, 237–66.

Chapman, J. and Milroy, C. 1992. Firearm Deaths in Yorkshire and Humberside. *Forensic Sci Int*, **57**, 181–91.

Clark, J.C. and Milroy, C.M. 2000. Injuries and deaths of pedestrians. In Mason, J.K. and Purdue, B.N. (eds), *Pathology of Trauma*. London: Arnold, pp17–29.

Cooper, J., Balding, L. and Jordan, F. 1998. Airbag mediated death of a two year-old child wearing a shoulder/lap belt. *J Forensic Sci*, **43**, 1077–81.

Cordner, S. 2003. Suicide, accident or natural death. In Payne-James, J., Busuttil, A. and Smock, W. (eds), *Forensic Medicine. Clinical and pathological aspects*. London: Greenwich Medical Media, pp135–47.

Corsellis, J.A.N., Bruton, C.J. and Freeman Browne, D. 1973. The aftermath of boxing. *Psychol Med*, **3**, 270–303.

Davis, G.G. and Glass, J.M. 2001. Case report of sudden death after a blow to the back of the neck. *Am J Forensic Med Pathol*, **22**, 13–18.

Freytag, E. 1963. Autopsy findings in head injuries from blunt forces: statistical evaluation of 1,367 cases. *Arch Pathol*, **75**, 402–13

Garfield, J. 2002. Acute subdural haematoma in a boxer. *Br J Neurosurg*, **16**(2), 96–9.

Geddes, J.F., Vowles, G.H., Nicoll, J.A.R. and Revesz, T. 1999. Neuronal cytoskeletal changes are an early consequence of repetitive head injury. *Acta Neuropathol*, **98**, 171–8.

Geddes J.F., Whitwell H.L. and Graham D.I. 2000. Traumatic axonal injury: practical issues for diagnosis in medicolegal cases. *Neuropathol Appl Neurobiol*, **26**, 105–16.

Graham, D.I., Gennarelli, T.A. and McIntosh, T.K. 2002. Trauma. In Graham, D.I. and Lantos, P.L. (eds), *Greenfield's Neuropathology*, 7th edn. London: Arnold, pp831.

Guskiewicz, K.M., McCrea, M., Marshall, S.W. *et al.* 2003. Cumulative effects associated with recurrent concussion in collegiate football players. *JAMA*, **290**, 2549–55.

Hadley, J.A. and Fowler, D.R. 2003. Organ weight effects of drowning and asphyxiation on the lungs, liver, brain, heart, kidneys and spleen. *Forensic Sci Int*, **133**, 190–6.

Hamilton, S.J. and McMahon, R.F. 2002. Sudden death and suicide: a comparison of brain weight. *Br J Psychiatry*, **181**, 72–5.

Hanzlick, R. 2001. Pulmonary haemorrhages in deceased infants. *Am J Forensic Med and Pathol*, **22**, 188–92.

Hanzlick, R. and Delaney, K. 2000. Pulmonary haemosiderin in deceased infants. *Am J Forensic Med and Pathol*, **21**, 319–22.

Karger, B., Teige, K., Bühren, W. and DuChesne, A. 2000. Relationship between impact velocity and injuries in fatal pedestrian-car collisions. *Int J Legal Med*, **113**, 89–97.

Kirkendall, D.T., Jordan, S.E. and Garrett, W.E. 2001. Heading and head injuries in soccer. *Sports Med*, **31**, 369–86.

Kury, G., Weiner, J. and Duval, J. 2000. Multiple self-inflicted gunshot wounds to the head: report of a case and review of the literature. *Am J Forensic Med Pathol*, **21**, 32–5.

Lau, G., Ooi, P.L. and Phoon, B. 1998. Fatal falls from a height: the use of mathematical models to estimate the height of fall from the injuries sustained. *Forensic Sci Int*, **93**, 33–44.

Lau, G., Teo, C.E.S. and Chao, T. 2003. The pathology of trauma and death associated with falls from heights. In Payne-James, J., Busuttil, A. and Smock, W. (eds), *Forensic Medicine. Clinical and pathological aspects*. London: Greenwich Medical Media.

Lau, G. 2004. Homicidal and dyadic falls from a height: rarities in Singapore. *Med Sci Law*, **44**, 93–106.

Leadbeatter, S. 2001. Resuscitation injury. In Rutty, G.N. (ed.), *Essentials of Autopsy Practice*, Vol 1. London: Springer-Verlag.

Lee, K.A. 2000. Injuries caused by animals. In Mason, J.K. and Purdue, B.N. (eds), *Pathology of Trauma*. London: Arnold, pp265–82.

Lindsay, K.W., McLatchie, G. and Jennett, B. 1980. Serious head injury in sport. *BMJ*, **281**, 789–91.

McCrory, P.R. 2003. Brain injury and heading in soccer. *BMJ*, **327**, 351–2.

Macpherson, A.K. and Macarthur, C. 2002. Bicycle helmet legislation: evidence for effectiveness. *Pediatr Res*, **52**, 472.

Marchie, A. and Cusimano, M.D. 2003. Bodychecking and concussions in ice hockey: should our youth pay the price? *CMAJ*, **169**, 124.

Matturri, L., Biondo, B. and Mercurio, P. *et al.* 2000. Severe hypoplasia of medullary arcuate nucleus: quantitative analysis in sudden infant death syndrome. *Acta Neuropathol*, **99**, 371–5.

Matturri, L., Biondo, B. and Suàrez-Mier, M.P. *et al.* 2002. Brainstem lesions in the sudden infant death syndrome: variability in the hypoplasia of the arcuate nucleus. *Acta Neuropathol*, **104**, 12–20.

Matturri, L., Ottaviani, G., Alfonsi, G. et al. 2004. Study of the brainstem, particularly the arcuate nucleus, in sudden infant death syndrome (SIDS) and sudden intrauterine unexplained death (SIUD). *Am J Forensic Med Pathol*, **25**, 44–8.

Milovanovic, A.V. and DiMaio, V.J.M. 1999. Death due to concussion and alcohol. *Am J Forensic Med Pathol*, **20**, 6–9.

Milroy, C.M. 1999. Munchausen syndrome by proxy and intra-alveolar haemosiderin. *Int J Legal Med*, **112**, 309–12.

Milroy, C.M. and Clark, J.C. 2000. Injuries and deaths in vehicle occupants. In Mason, J.K. and Purdue, B.N. (eds), *Pathology of Trauma*. London: Arnold.

Myles, S.T., Mohtadi, N.G. and Schnittker, J. 1992. Injuries to the nervous system and spine in downhill skiing. *Can J Surg*, **35**, 643–8.

Nagurney, J.T., Borczuk, P. and Thomas, S.H. 1998. Elderly patients with closed head trauma after a fall: mechanisms and outcome. *J Emerg Med*, **16**, 709–13.

Palmer, A.C., Calder, I.M. and Hughes, J.T. 1987. Spinal cord degeneration in divers. *Lancet*, ii, 1365–6.

Plant, J.R. and Butt, J.C. 1993. Laceration of the vertebral artery. An historic boxing death. *Am J Forensic Med Pathol*, **14**, 61–4.

Preuß, J., Padosch. S.A. and Dettmeyer R. et al. 2004. Injuries in fatal cases of falls downstairs. *Forensic Sci Int*, **141**, 121–6.

Rabadi, M.H. and Jordan, B.D. 2001. The cumulative effect of repetitive concussion in sports. *Clin J Sport Med*, **11**, 194–8.

Ragg, M., Hwang, S. and Steinhart, B. 2000. Analysis of serious injuries caused by stairway falls. *Emergency Med*, **12**, 45–9.

Ramsay, D.A. and Shkrum, M.J. 1995. Homicidal blunt head trauma, diffuse axonal injury, alcoholic intoxication, and cardiorespiratory arrest: A case report of a forensic syndrome of acute brainstem dysfunction. *Am J Forensic Med Pathol*, **16**, 107–14.

Roberts, G.W., Allsop, D. and Bruton, C. 1990. The occult aftermath of boxing. *J Neurol Neurosurg Psychiatry*, **53**, 373–8.

Ross, R.T. and Ochsner, M.G. Jr. 1999. Acute intracranial boxing-related injuries in U.S. Marine Corps recruits: report of two cases. *Milit Med*, **164**(1), 68–70.

Ross, A.S. and Grieve, J.H.K. 2000. Underwater diving. In Mason, J.K. and Purdue, B.N. (eds), *Pathology of Trauma*. London: Arnold, pp341–62.

Ryan, G.A., McClean, A.J. and Vilenius, A.T.S. et al. 1994. Brain injury patterns in fatally injured pedestrians. *J Trauma*, **36**, 469–76.

Saukko, P. and Knight, B. 2004. *Knight's Forensic Pathology*, 3rd edn. London: Arnold.

Sawauchi, S., Murakami, S., Tani, S. et al. 1996. Acute subdural hematoma caused by professional boxing. *No Shinkei Geka*, **24**, 905–11.

Voigt, G.E. 1981. Small haemorrhage in the brain stem: a sign of injury? *Am J Forensic Med Pathol*, **2**, 115–19.

Wardlaw, M. 2002. Butting heads over bicycle helmets. *JAMA*, **167**, 337–8.

Wyatt, J.P., Beard, D. and Busuttil, A. 1999. Fatal falls down stairs. *Injury*, **30**, 31–4.

Yen, K., Thali, J.M., Kneubuehl, B.P., Peschel, O., Zollinger, U. and Dirnhofer, R. 2003. Blood-spatter patterns. Hands hold clues for the forensic reconstruction of the sequence of events. *Am J Forensic Med Pathol*, **24**, 132–40.

Yukawa, N., Carter, N., Rutty, G. et al. 1999. Intra-alveolar haemorrhage in sudden infant death syndrome: a cause for concern? *J Clin Pathol*, **52**, 581–7.

Zink, B.J. and Feustel, P.J. 1995. Effects of ethanol on respiratory function in traumatic brain injury. *J Neurosurg*, **82**, 822–8.

Zivot, U. and DiMaio, V.J.M. 1993. Motor vehicle-pedestrian accidents in adults: relationship between impact speed, injuries and distance thrown. *Eur Spine J*, **14**, 185–6.

12
Head injury in the child

Helen L Whitwell

Much of this chapter covers the major area of non-accidental injury in the infant and child. It is important to realise, however, that this label may, in some cases (unless the circumstances are known in great detail) be more honestly left to be determined by a jury (Geddes 2003a) as the differentiation between accidental and inflicted injury may be impossible to determine on the medical evidence alone.

ACCIDENTAL HEAD INJURY

The majority of incidences of accidental head injury in the paediatric age group arise as a result of motor vehicle collisions, including occupants as well as pedestrians and cyclists. Other causes of accidental head injury include falls and injuries relating to sports and other recreational activities. The pattern of brain injury, particularly in the older child, resembles that seen in adults. However there is some evidence that the maturing brain in the younger age group responds differently to that of the older child. There is a smaller risk of traumatic intracranial haematomas in children less than five years of age than in adults (Teasdale *et al.* 1990). Diffuse swelling is more common in the paediatric age group – this may be secondary to other pathology such as ischaemia or contusional injury or may have no underling cause (Graham *et al.* 1989). This is covered further in Chapter 9, p. 109.

NON-ACCIDENTAL HEAD INJURY

This is an area that has attained increasing importance both within the medico–legal context and the public eye. Head injury is the major cause of death in cases of non-accidental injury (Hargrave and Earner 1992, Ellis 1997) and is being recognized as a cause of morbidity in the survivors with varying degrees of long-term physical and psychological effects (Caffey 1974, Ewing-Cobbs *et al.* 1998, Barlow and Minns 1999). Estimates of the true incidence of non-accidental head injury are difficult and vary in the UK from between 11.2 per 100 000 children younger than 1 year to 24.6 per 100 000 (Barlow *et al.* 1998, Jayawant *et al.* 1998, Barlow and Minns 2000). The long-term morbidity is estimated at 78 per cent (Barlow and Minns 1999).

It is becoming increasingly apparent that there are significant differences in the neuropathology of injury in the child compared with the adult, as well as between children of different ages. The lack of reliable data has meant that opinions in this area, to a great extent, have been based on knowledge of adult head injury from both the biomechanical and the neuropathological aspects. Assumptions have been made that on close scrutiny lack a firm scientific basis (Donohoe 2003). This is a field where because of the nature of a case a careful unbiased assessment is of paramount importance. This is particularly so in the very young infant where issues relating to birth as well as the relative immaturity of the infant brain are important.

POSTMORTEM EXAMINATION IN A CASE OF SUSPECTED NON-ACCIDENTAL HEAD INJURY

Examination in these cases should follow a standard forensic routine with paediatric input. If a paediatric pathologist commences a postmortem examination and identifies injury or other suspicious findings, he or she should request the participation of a forensic pathologist in a joint examination (Royal College of Pathologists and

Royal College of Paediatrics and Child Health 2004). As well as a detailed external and internal examination with sampling, including histology, virology, bacteriology and frozen tissue for metabolic abnormalities, it is essential to perform full toxicology and collect samples for DNA analysis. DNA analysis may be required for coleration with the scene as well as for establishing parenthood. The Confidential Enquiry into Sudden Deaths in Infancy (CESDI) report (CESDI 2000) includes guidelines for investigation, in particular for underlying natural disease. The joint Royal College of Pathologists and Royal College of Paediatrics and Child Health report also contains postmortem examination protocols as well as additional investigations (Royal College of Pathologists and Royal College of Paediatrics and Child Health 2004).

The postmortem examination should always be preceded by as much history of events as possible. It should include details of the story given by the carer(s) at the time of presentation as well as any events in the preceding days/weeks. Photographs of any scene should always be taken including measurements made of the position and type of furniture and other items within a room. An accurate scale plan drawing may be necessary in some cases. A visit to the scene may be part of the initial examination in a case where the child is found dead, but it is more likely that it will take place at a later date when the account of events can be assessed and compared with injuries identified. When it is possible that injuries may have been caused by falling on to a surface, detailed examination of the surface including coverings such as carpet will be necessary. The crucial question always is 'is the injury consistent with the story?' (Whitwell 2001).

In the young infant the birth history should be obtained and in all cases past medical history. Details of resuscitation procedures should be obtained, including not only those carried out by medical personnel but also attempts by carers or others. Recognized resuscitation injuries to the head area include damage to the frenulum, bruising to the mouth area both internally and externally, as well as scalp bruising. This last may be either in association with neurosurgical procedures or, if in the posterior position, related to the position of the head (Leadbeatter 2001). If the child has survived for a period of time in hospital, full clinical details including radiological results and results of other investigative tests should be sought. The timing of brain-stem death should also be recorded. Kemp (2002) has reviewed diagnostic procedures for investigating subdural haemorrhage in infants in hospital and gives in detail what investigations should be carried out. These include radiological, ophthalmological and laboratory investigations (Kemp 2002). In a number of cases the child will present as a sudden or unexpected death. As with all such cases skeletal survey is mandatory before postmortem examination – if this is done after, identification of true injury may become impossible. This is particularly so with skull and rib fractures. Postmortem magnetic resonance imaging (MRI) has also been used in cases of suspected head trauma (Hart *et al.* 1996, Kahana and Hiss 1999). However, this is at present restricted in availability and its value has not yet been fully evaluated.

Extracranial injury is common in non-accidental injury (NAI). This includes abdominal injury (which is more common in the older child – Geddes *et al.* 2001a) and the second most common cause of death in NAI (Hargrave and Earner 1992, McClain *et al.* 1993, Saukko and Knight 2004). In fatal cases rupture of the liver or spleen, and tearing of the mesentery and pancreatic lesions may be found. Other extracranial injuries include rupture of the small intestine or duodenum, renal lacerations and lung contusions.

Skeletal injuries include long bone fractures, rib fractures and clavicular fractures. The pathology and radiology of these are well covered in other texts (Rao and Carty 1999, Carty and Pierce 2002) and are not discussed further here. However, assessment of extracranial injury may have significance in interpretation of a head injury, including issues such as timing as well as mechanism. This is most classically seen in the infant where rib fractures seen in the paravertebral gutters on one or both sides may indicate a squeezing mechanism. If old they produce a 'string of beads' appearance on a radiograph. Limb fractures may indicate that the child was grasped by the ankles or wrists and propelled against a hard surface.

EXTERNAL FINDINGS IN HEAD INJURY

Photography is essential in recording injuries both external and internal. This should include scales to indicate size and location. In addition serial photographs may be useful to record bruising that may become obvious later, although interpretation of this after a postmortem examination may be difficult. Ultraviolet (UV) photography may be included although interpretation of this may also be difficult (Whitwell 2001). The following may be found.

No external injury

This is not infrequent. Atwal *et al.* (1998), in a series of 24 fatal cases, all younger than 5 years, identified 21 per cent with no visible external injuries. Careful inspection may reveal soft tissue swelling. It is recognized that bruising may develop over a period of time and deep bruising takes longer than superficial bruises to appear (Johnson 1990, Langlois and Gresham 1991). Bruising in dark complexioned children may be masked. It is important within the general postmortem examination to check for deep bruising of the trunk and limbs. There is some merit in delaying the postmortem examination to allow a truer picture to develop, as if it is undertaken too early bruising may develop subsequent to the postmortem, making further examination necessary. It is unusual to see no bruising on internal examination apart from the so-called 'shaken baby syndrome' (see below).

Bruising

This may comprise single or multiple areas and overly any skull fracture that may be present (Figure 12.1). Large diffuse areas are indicative of a broad surface impact, e.g. the floor or wall. It may occasionally be possible to identify patterns within the bruising relating to the surface, e.g. material such as a carpet or object may produce a recognizable pattern, which may also be seen in associated surface abrasion.

Small bruises may be the result of fingers, with larger more irregular areas being caused by slapping (Figure 12.2). Injuries to the ears may be as a result of direct blows or pinching. The classic picture of the battered child is now less often seen, and presents less diagnostic difficulty than the child with a single head injury and few if any external injuries (Whitwell 2001). In an attempt to clarify 'normal' accidental bruising in children, Carpenter (1999) surveyed 177 babies aged 6–12 months. As expected, there was an increase in bruises with mobility. Frequent locations were on the bony prominences of the face and head. Mobility was again highlighted by Sugar *et al.* (1999). Atypical sites include the trunk and buttocks as well as the hands. Worrying sites on the head include the ear lobes and non-prominent areas of the face. It is well recognized that bruises can be difficult to age, both from the naked eye appearance and histologically. Stephenson and Bialas (1996) documented the sequence of colour changes in photographs after accidental bruising in 23 children with a total of 50 bruises. They identified red coloration in bruises less than 1 week old, whereas green or yellow hues suggested that the injury was at least 24 hours old. Several different colours may be present in an individual bruise. Bruises may change colour at different rates in different locations in the same individual. In a review of the available literature

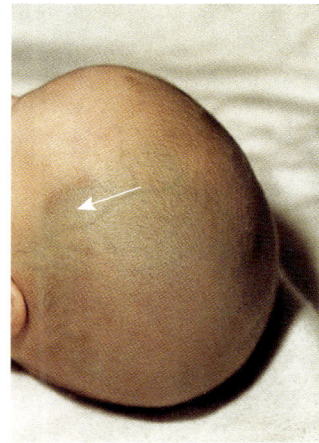

Figure 12.1 Diffuse bruise overlying skull fracture.

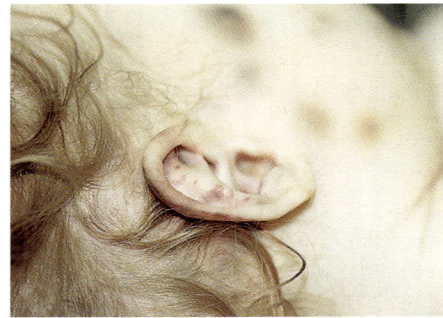

Figure 12.2 Finger-tip bruising to the side of the face and ear.

on estimating the age of a bruise in child abuse by Schwartz and Ricci (1996), they suggested that this should not be the sole criterion for the diagnosis and other factors including history of injury and medical history should be taken into account. Histological assessment with assessment of inflammatory cell infiltrate and haemosiderin deposition is essential (see Chapter 3, p. 51). Haemosiderin deposition suggests an age of probably at least 48 hours. Red cells, both lysed and intact, may persist for some considerable time, particularly if there is a degree of sequestration as may occur on scalp bruises (Perper and Wecht 1980). Care must be taken to differentiate birthmarks from bruises as well as other skin conditions, bleeding disorders and birthmarks (Wheeler and Hobbs 1988, O'Hare and Eden 1994, Stephenson 1995).

Abrasions

These can occur in association with bruising. The features may be non-specific. Care should be taken in interpretation, because children are able to scratch themselves. Patterned abrasions may result when an object such as a ring or belt is involved. Accurate photographs with a scale, at a 90° angle, should be used to match a potential object with the injury.

Injury to the frenulum

This can occur as the result of a direct blow to the mouth (Figure 12.3). It may also occur when a feeding bottle is rammed into the mouth and is, rarely, reported to occur in the process of resuscitation (Leadbeatter 2001). If this is an issue, clearly full details of any such procedures are essential.

Traumatic alopecia

These are irregular shaped areas of hair loss, which occur in some cases. The force may be sufficient to cause an underlying subgaleal haematoma (Kempe 1975).

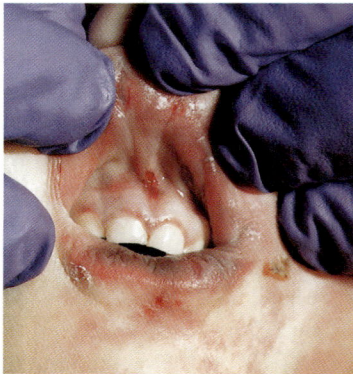

Figure 12.3   Torn frenulum as a result of a direct blow to the mouth.

Ocular injury

Retinal haemorrhages, which are a frequent finding in these cases, do not show external findings. Other ocular findings include vitreous haemorrhage, traumatic retinoschisis, perimacular retinal folds and retinal detachment as well as optic nerve haemorrhage. This is a controversial area and there is debate on the interpretation of retinal haemorrhages and other ocular findings as to the aetiology and pathogenesis. A specialist ocular pathologist should be consulted. Gilliland and Luthert have undertaken a review on the histological aspect of retinal haemorrhages (Gilliland and Luthert 2003). It is, however, becoming increasingly recognised that ocular findings cannot be taken in isolation as diagnostic for non-accidental injury. Other pathological findings should be taken into account together with details of known history, including witness statements, and assessment of circumstances (Lantz *et al.* 2004). Retinal haemorrhages are commonly found where there is evidence of impact, but there is some evidence to suggest they may be related to severe brain swelling rather than shearing injuries to the retina (Ommaya *et al.* 2002).

Rarely, direct trauma to the eyes occurs producing scleral/conjunctival haemorrhage. Examination for petechial haemorrhages in the conjunctivae and skin, suggestive of external pressure to the neck or chest, should be undertaken.

INTERNAL EXAMINATION

Bruising to the subscalp tissues may be extensive or more discreet (Figure 12.4). If a hand has been used, small areas

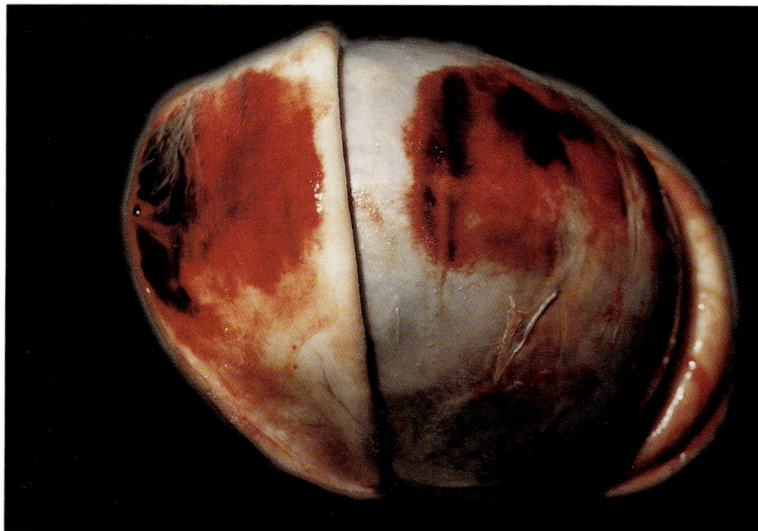

Figure 12.4   Diffuse subscalp bruise caused by impact with a wall.

with a fingertip pattern may be identified. Careful dissection of these is necessary. Bruising is usually seen where there is a recent fracture. The absence of it should alert to the possibility that the fracture may not be genuine, e.g. a naturally occurring fissure. The latter is a well-recognized pitfall in the diagnosis of skull fractures in the infant. They are naturally occurring defects, which fuse as ageing occurs. Identification is usually radiological (Keats 1996, Swischuk 2003). If there is continuing doubt, a sample should be taken for histology, which will show a picture of immature bone merging with mature lamellar bone.

Care should be taken in interpretation of subscalp bruising in the very young infant, particularly with a history of instrumental delivery, which may show cephalo-haematomas or subgaleal haemorrhage (O'Grady et al. 2000).

In a small number of cases there may be no evidence of either external or internal bruising. This is particularly so in the young infant (up to a few months in age). In the series of Geddes, 8 of 53 cases showed no external or internal bruising and no skull fracture (Geddes et al. 2001a). Exploration of the upper cervical region should be undertaken in all suspect NAI cases because it is becoming increasingly apparent that neck injury with either cervical musculature bruising and/or cervical extradural haemorrhage is common and has probably been under-identified in the past (Hadley et al. 1989, Hart et al. 1996, Geddes et al. 2001a). Readers may find helpful the technique for postmortem dissection of the brain and spinal cord that preserves the anatomy of the cranio – cervical junction together with details of histological sections as described by Judkins (Judkins et al. 2004).

SKULL FRACTURES

Skull fractures are a frequent finding in inflicted head injury. They indicate impact injury – whether or not there has been additional shaking may be impossible to say. In the largest series of fatal inflicted head injury in childhood (Geddes et al. 2001a), 36 per cent of cases had one or more skull fractures. This compares with Atwal et al. (1998), where the incidence was 42 per cent, and is generally in accordance with other series.

A number of features relating to characteristics of fractures have been described in order to attempt to differentiate accidental injury from NAI (Hobbs 1984, Rao and Carty 1999). The features that have been identified as being suggestive of NAI are as shown in the box. Most authors, however, emphasize that there is no specific feature diagnostic of NAI and indeed most skull fractures in cases of NAI are linear with the more complex picture less commonly seen (Figure 12.5). The majority of fractures are located in the parietal or occipital regions. Diastatic fractures occur where the skull sutures are not completely fused and fractures may travel along them. Separation of the sutures may also occur in severe cerebral oedema.

Features of skull fractures suspicious of non-accidental injury

Non-parietal fracture
Multiple or complex fractures
Depressed fracture
Diastatic fractures >5 mm
Growing fractures

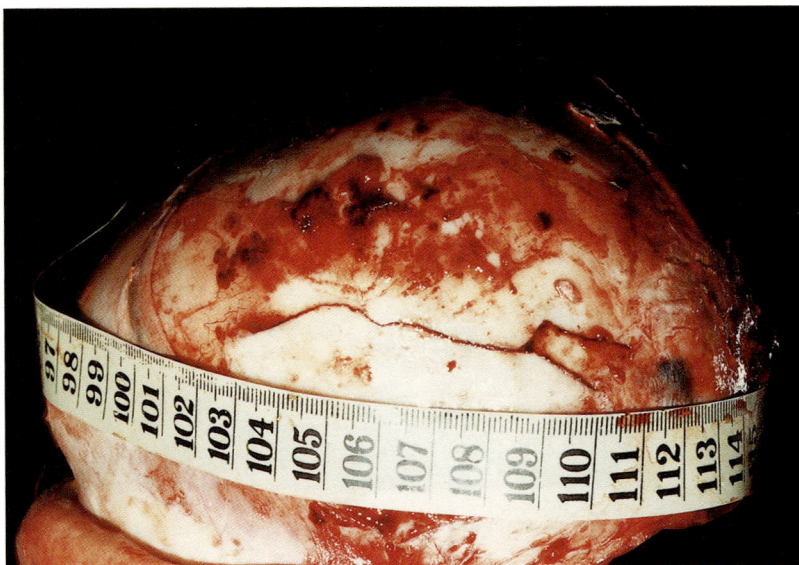

Figure 12.5 Fracture in a child as a result of a blow with a baseball bat.

Underlying abnormalities of connective tissue such as osteogenesis imperfecta should be excluded (Brown and Minns 1993). Fractures of varying ages and of a number of bones may be seen and subdural haematomas have been reported (Tokoro *et al.* 1988, Steiner *et al.* 1996). Some forms present with subtle physical signs. Other metabolic conditions include copper deficiency, which may be associated with pathological fractures (Shaw 1988, Brown and Minns 1993).

There is support in the literature that simple skull fractures have a low risk of intracranial complication (Harwood-Nash *et al.* 1971, Helfer *et al.* 1977). The literature on infants and children relating to falls has been extensively published (Chadwick *et al.* 1991, Williams 1991, Reiber 1993), and there has been a general consensus that low level falls are rarely associated with skull fractures and are usually simple and linear, commonly parietal. However, Weber (1984, 1985), in a series of experiments in which he dropped infants on to a variety of surfaces, found skull fractures in all that were dropped from a height of 82 cm. This appears to be at odds with the clinical situation, although skull fractures may not necessarily indicate intracranial haemorrhage or other pathology. Skull fractures occur in asymptomatic head-injured infants. Greenes and Schutzman, in 1998, reported a retrospective review of 101 infants aged under 2 years admitted to a paediatric hospital with head injury. In 20 per cent of cases there were occult intracranial findings in asymptomatic children and all but one had a skull fracture.

At postmortem examination histology should be taken of any fracture. Details of the histology of skull fractures are given in Chapter 4, p. 59. The findings in infants and children are not well documented in the literature, but assessment of the age of any associated haemorrhage should be done. Radiological dating of fractures is difficult (Rao and Carty 1999).

NEUROPATHOLOGY OF NON-ACCIDENTAL INJURY

The major neuropathological findings are:

- Subdural haemorrhage
- Subarachnoid haemorrhage
- Cerebral swelling
- Intracerebral haemorrhage and contusional tears
- Hypoxic–ischaemic damage
- Axonal injury, both traumatic and ischaemic
- Brain damage in survivors
- Spinal cord and nerve root injury.

Subdural haemorrhages

These are a frequent occurrence – 84 per cent of the infant age group with 81 per cent in the older child (Geddes *et al.* 2001a). In a population-based study by Jayawant *et al.* (1998), in 82 per cent of subdural haemorrhages (SDHs) in infants aged under 2 years a cause was identified but not in the remaining 18 per cent. A lower incidence of 55 per cent was reported in a similar series from Australia (Tzioumi and Oates 1998). The haemorrhages are commonly bilateral and posteriorly sited (Duhaime *et al.* 1998). Particularly in the infant group, as opposed to the older child, they consist of thin films of blood rarely requiring neurosurgical intervention (Figure 12.6). Indeed, in a number of cases SDH is not identifiable radiologically. At postmortem examination caution must be taken not to misinterpret blood that has drained from subdural sinuses, when these are incised, as ante-mortem subdural bleeding. The postmortem appearance classically is of an extremely swollen brain with thin films of acute haemorrhage over the hemispheres, and with not infrequently extension into the posterior interhemispheric region. In the older age group, space-occupying haematomas, which may be unilateral, are also seen, as is the more diffuse type of bleeding (Geddes *et al.* 2001a). The former do not necessarily lie on the same side as a fracture if present. The mechanism of the formation of SDHs is most commonly tearing of the bridging veins crossing the subdural space. Identification of ruptured bridging veins has been reported *post mortem* (Maxeiner 2001). Recently, traumatic tearing of these veins as a mechanism for subdural bleeding has been questioned in the very young with brain swelling, hypoxic damage and no objective evidence of impact. However, at the present time, an

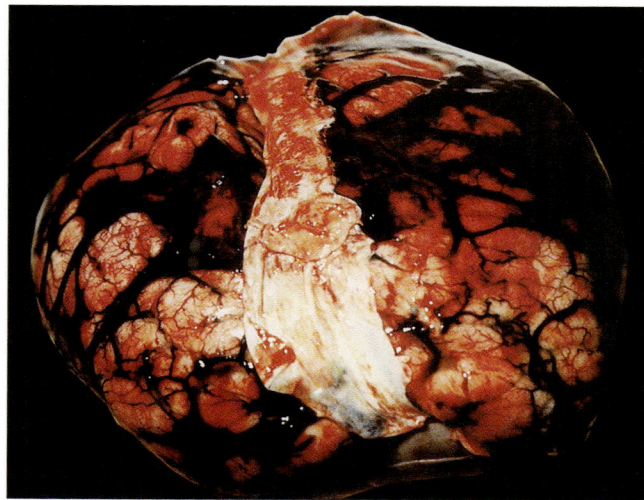

Figure 12.6   Thin films of haemorrhage in an infant with severe brain swelling.

alternative mechanism to trauma needs further evaluation (Geddes 2003c).

Underlying conditions, such as bleeding disorders, vascular malformations, septicaemia and metabolic abnormalities, e.g. glutaric aciduria type I and hypernatraemia, should be sought (Ehrenforth *et al.* 1998, Ng *et al.* 1998, Renzulli *et al.* 1998, Handy *et al.* 1999, Rutty *et al.* 1999, Hartley *et al.* 2001, Menkes 2001). Infants with enlarged extra-axial spaces, such as in shunted hydrocephalus, appear to be at increased risk of subdural bleeding with lesser degrees of trauma (Duhaime *et al.* 1998). The history of the mode of delivery is of extreme importance and ventouse extraction is recognized to have an incidence of SDHs, many of which resolve after birth (Liu *et al.* 1998, Towner *et al.* 1999, O'Grady *et al.* 2000). MRI in routine deliveries identifies asymptomatic SDH in a proportion. These usually resolve by 4 weeks (Whitby *et al.* 2003). Subdural bleeding *in utero*, including frank haematomas, is also described (Barozzino *et al.* 1998).

Not infrequently there is evidence of earlier subdural bleeding histologically (Geddes *et al.* 2001a) with haemosiderin staining of the dura. Evidence of such older haemorrhage should not be taken as certain evidence of previous NAI. The findings in themselves are non-specific and simply indicate the likelihood of previous trauma including birth trauma (Figure 12.7).

It is essential that several specimens of dura be taken to assess the presence of organizing haemorrhage, membrane formation and haemosiderin deposition for the reasons above. This latter is said to take at least 48 h to develop, although it should be said that this is only an estimate. Changes may only be seen microscopically (Geddes *et al.* 2001a). Most of the work on ageing SDH relates to the adult population. This makes dating in the infant very difficult.

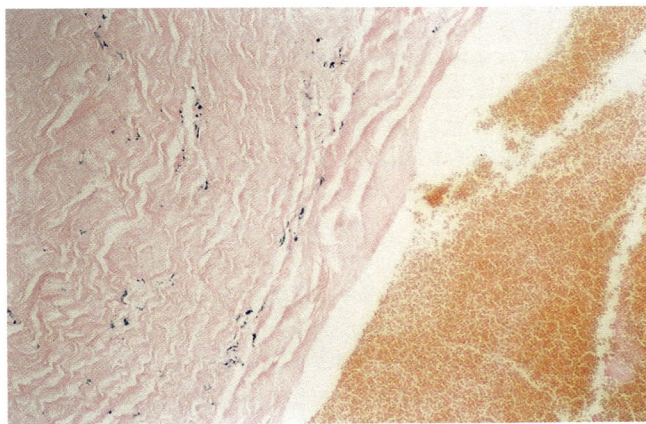

Figure 12.7 Haemosiderin macrophages in the dura – away from acute haemorrhage.

Extra-axial fluid collections, which include chronic SDH, subdural hygroma and subdural effusion, are covered in detail in Chapter 5. It should be noted that extradural haemorrhage is extremely rare in NAI (Duhaime *et al.* 1998).

Subarachnoid haemorrhage

This is present in about half the cases of paediatric head injury (Geddes *et al.* 2001a). It is seen in association with either SDH and/or skull fracture. It may be associated with the fracture site. It is commonly seen in a patchy fashion over the hemisphere *post mortem*, but after the brain has been fixed in formalin it is often less extensive than originally thought. Subarachnoid haemorrhage is not clinically significant and, apart from where it is associated with a fracture, it is not an accurate indicator of location of injury. Very rare cases of true traumatic subarachnoid haemorrhage are reported with rupture of an intracranial vessel (Leestma 1988). This may be delayed as a result of rupture of a traumatic aneurysm, which has rarely been reported in NAI (Lam *et al.* 1996).

Cerebral swelling

One of the major findings in inflicted head injury is cerebral swelling. This is most commonly seen in association with global neuronal hypoxic–ischaemic damage (see below) and is the most common mode of death. The most accurate way to assess this is by weight of the brain compared with the expected size for the age and other growth parameters of the child (Appendix 2, Chapter 2, p. 29). It should be recognized, however, that in the forensic setting congestive cerebral swelling, such as may occur in an 'asphyxial' type of death, may also show an increase in brain weight. The brain should be weighed fresh before suspension using accurate scales. The weight should be re-checked after fixation.

One of the reasons that the infantile brain does not show the same features as the adult is that there is the possibility for the infant's skull to expand. Thus, the various herniations are not usually present as they are in the adult. In infants with severe hypoxic damage, particularly when there has been a period of survival, often with assisted ventilation, the brain is extremely softened and 'pours out' on opening the skull. Artefactual disruption and distortion frequently happen even when extreme care is taken. Removal under water may aid the process.

On section after fixation, macroscopic examination may show only compression of the ventricles. Histological examination demonstrates predominantly white mater oedema with myelin swelling. With increasing survival time additional features such as hypoxic–ischaemic damage are commonly seen. Amyloid precursor protein (APP) staining often shows a vascular pattern (Figure 12.8a). This is usually widespread in the white matter. Focal geographic APP expression, commonly seen in the diencephlon and brain stem, is taken as outlining areas of incipient infarction caused by brain swelling (Geddes *et al.* 2000). Distinguishing between vascular and traumatic damage to axons (Figure 12.8b) is not always easy and great care is needed, particularly in the brain stem.

The importance of this distinction is that, if axonal injury in this location (in the distribution of that seen in diffuse axonal injury rather than the localized corticospinal tract damage – see below) is claimed definitely to be traumatic, this conclusion implies that a significant degree of force has been applied and that there were immediate profound clinical effects at the time of injury. Brain swelling with compression of blood vessels leads to ischaemic white matter damage in similar areas to that seen in traumatic axonal injury (Shannon *et al.* 1998, Geddes *et al.*

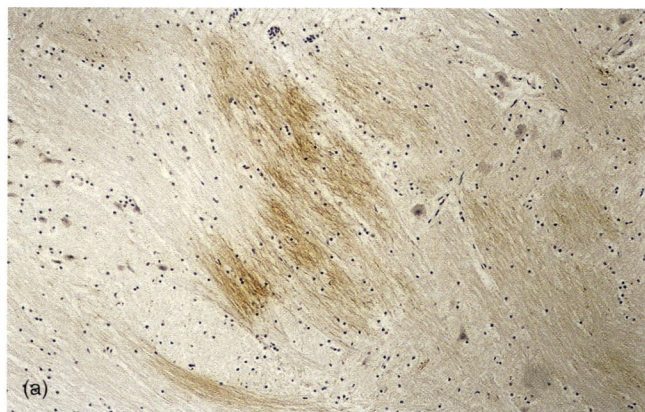

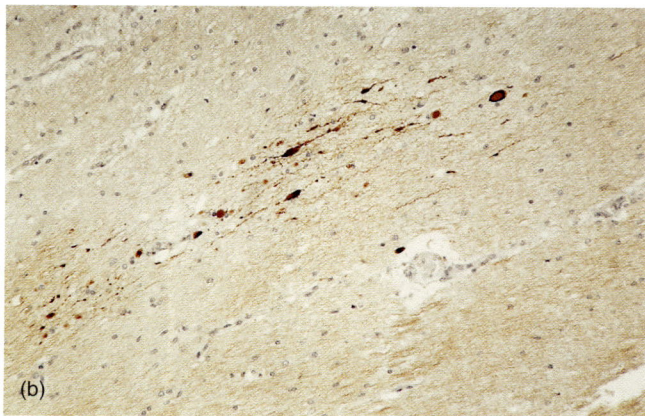

Figure 12.8 (a) Vascular pattern of amyloid precursor protein (APP). (b) Traumatic path of axonal injury (APP).

2000). This is particularly so in cases of short survival which may be impossible to interpret.

In some, albeit infrequent, instances where death occurs rapidly, swelling may not be a feature. If the cause of death is head injury, it is probable that the injury produced traumatic damage incompatible with life – presumably to the brain stem. In the author's experience such a situation occurs most commonly in association with significant impact. The mechanism of acute cerebral swelling after minor head injury is discussed in Chapter 9, p. 109.

Intracerebral haemorrhage

This can be of a number of types.

IN ASSOCIATION WITH CONTUSIONAL INJURY OF ADULT TYPE

With increasing age of the child, the findings are essentially those seen in the adult with surface contusional/lacerating injury. In the younger infant, the 'adult-type' contusional injury is not a common finding, although if present it is most frequently seen in the olfactory bulbs or tracts and gyrus rectus (Rorke 1992). In severe trauma associated with fracture, true coup contusions relating to the fracture site may be seen. The various types of intracerebral haematoma seen are covered in Chapter 5.

CONTUSIONAL TEARS

These relatively uncommon lesions appear to occur only in very young infants (Lindenberg and Freytag 1969, Calder *et al.* 1984). Contusional tears have been described at postmortem examination by a number of authors (Lindenberg and Freytag 1969, Calder *et al.* 1984, Geddes *et al.* 2001a), but can also be identified by transcranial ultrasonography (Jaspan *et al.* 1992). They occur at the junction of the grey and white matter, and are believed to result from shearing at that site. The classic locations are the frontal (beneath the superior frontal convolution and in the orbital gyri), temporal and occipital regions. Although many of these cases occur with impact injury, including severe impact, contusional tears may also be seen where there is minimal or no evidence of impact (Geddes *et al.* 2001a). It is the author's view that these lesions are essentially focal in nature and represent the way that infants' brains (poorly or not myelinated) react, and are not comparable with contusional injury seen in the adult. In this context it is clear that the degree of force required to produce these tissue tears is unknown, and saying that they necessarily require the same amount of force as needed to produce diffuse axonal injury cannot be justified. Diffuse axonal injury requires severe angular or rotational acceleration with or without deceleration, producing widespread axonal damage.

A recent tear can be identified macroscopically as a slit, with variable haemorrhage. Caution is necessary, however, because cracks, slits or apparent tears in the white matter, which may look haemorrhagic, are relatively common findings in paediatric neuropathology, and are not necessarily of any significance. Damage to tissue can occur if the brain is 'roughly' handled, and the author has seen one case where transportation immediately after removal and before adequate fixation resulted in artefactual slits in the tissue that resembled contusional tears. Even after some time in formalin, an infant's brain may crack and fall apart as it is examined, particularly if it is swollen, poorly fixed or infarcted. The only way of proving beyond doubt that a white matter cavity is a contusional tear is by histology, showing haemorrhage or damaged axons around the edges, or a cellular reaction in or around the cavity. For that reason, it is essential that the brain is well fixed and not cut fresh (Figure 12.9a).

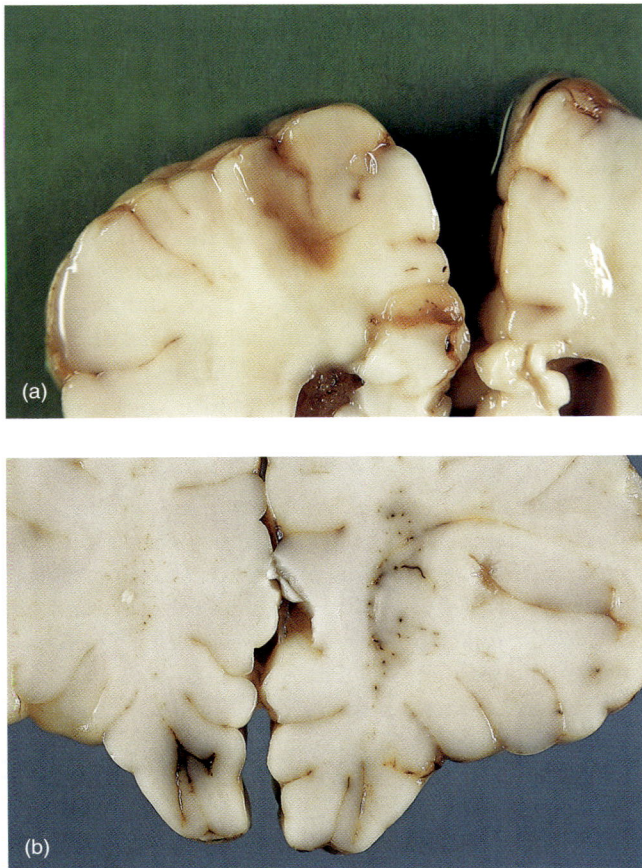

Older tears can be identified macroscopically as slits with yellow–brown discoloration indicative of old haemorrhage (Figure 12.9b). Histologically, a recent tear shows variable recent haemorrhage, with separation of the tissue (Figure 12.10). Depending on the timing after injury APP staining may demonstrate axonal damage, as well as a cellular reaction with neutrophil polymorphs (although these are always scanty). There may be focal shrinkage and 'dark cell' change in adjacent neurons, and after some hours' survival genuine hypoxic–ischaemic changes. With the passage of time macrophages accumulate along with reactive astrocytes and increase in microglial cells.

Haemosiderin is usually identifiable after 48 h. An old tear may be identifiable by haemosiderin macrophages and gliosis. Timing is obviously an issue, but caution should be applied and over-dogmatic opinions are ill-advised.

PETECHIAL HAEMORRHAGES

Not infrequently microscopic petechial haemorrhages, often perivascular, may be seen in the cortex, white matter or deep grey matter. Their significance has to be interpreted with caution because similar lesions may be seen in so-called asphyxial deaths or unexplained deaths in infancy (see also Chapter 11, p. 130).

Hypoxic–ischaemic injury

This has become increasingly recognized as the major pathology in the infant. This is identifiable by the neuropathological criteria of widespread neuronal eosinophilia and shrinkage (Figure 12.11). It is usually widespread throughout the brain. Histologically the classic time for its appearance in adults is 4–6 h. However, it is possible in

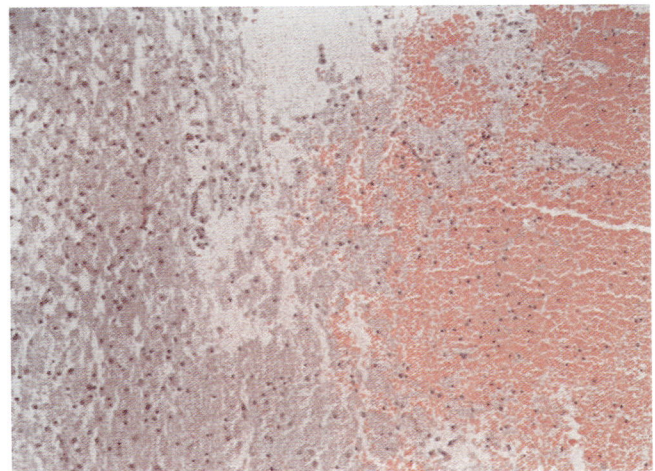

Figure 12.9   (a) Recent contusional tear – superior frontal region. (b) Bilateral older contusional tears – slit-like appearance.

Figure 12.10   Contusional tear – haemorrhage with surrounding macrophages.

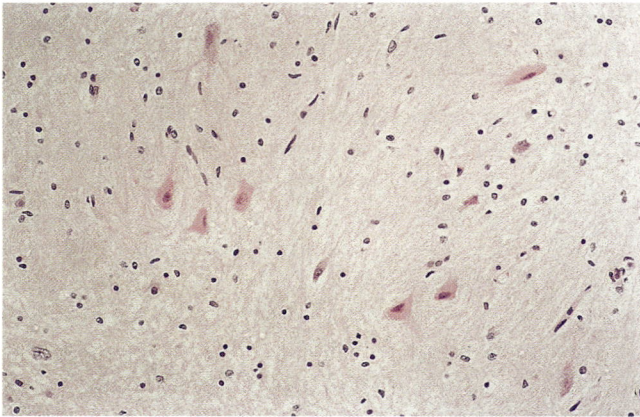

Figure 12.11　Hypoxic–ischaemic neuronal injury.

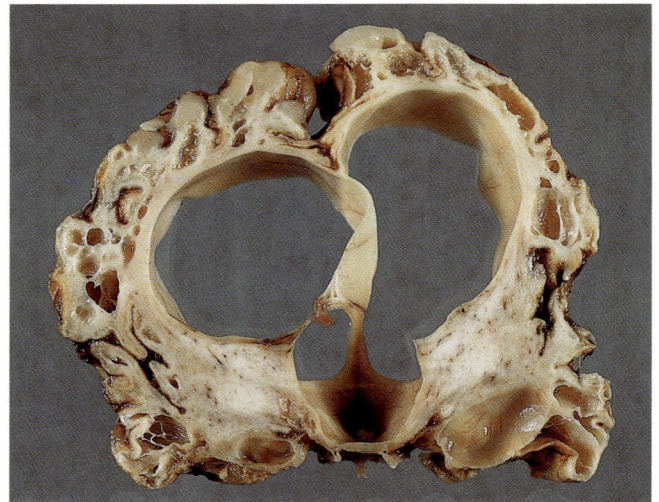

Figure 12.12　Long-term survivor after hypoxic–ischaemic injury.

some cases to identify changes earlier. Differentiation from the so-called dark cell change may be difficult. The usual cause of death is brain swelling related to the hypoxic damage. Apnoea is a frequent presentation in NAI (Johnson *et al.* 1995) and it is thought that this, rather than the mechanism of injury, has the major role in prognosis (Prasad *et al.* 2002).

In most infants (up to 9 months old) this was the most common finding, with 78 per cent of cases showing widespread neuronal hypoxia and a large proportion having a history of apnoea or respiratory arrest (Geddes *et al.* 2001b). Recently, radiological studies using diffusion-weighted MRI in the investigation of non-accidental head injury have identified changes of cerebral ischaemia rather than traumatic lesions as the major abnormality (Suh *et al.* 2001, Biousse *et al.* 2002, Stoodley 2002). In some cases older areas of hypoxic–ischaemic damage may be seen suggesting previous injury, although as in the cases of long-term survivors other causes, such as birth injury, should be sought. In long-term survivors with severe damage gross cerebral atrophy may develop (Figure 12.12).

Axonal injury

Immunohistochemistry has in recent years revealed the spectrum of axonal injury. It has been widely assumed that the brain damage in infants and children in NAI is diffuse axonal injury (DAI). However, detailed neuropathological studies have been few. Earlier studies have claimed that DAI is common. In 1987, Vowles *et al.* published a neuropathological series of ten cases, and reported DAI in six. This study was done before APP and utilized silver stains. However, the cases were inadequately studied by today's standards, and what the authors reported was not what is now understood by DAI (see Geddes *et al.*

2001b for a full discussion). Review of these cases using immunohistochemistry (unpublished data) showed that, in fact, DAI was present in only one case. Most showed lesser degrees of axonal injury, the aetiological nature of which was unclear because of the inadequate nature of the examination and paucity of clinical information. In 1998 Shannon *et al.* demonstrated localized axonal injury in the spinal cervical roots in NAI cases. In addition they found similarities between controls dying of hypoxic–ischaemic brain damage and those in NAI where hypoxic–ischaemic damage was also seen. The APP staining showed similar patterns in both, although these authors felt that additional axonal injury was also present as a result of trauma in some cases. They pointed out that dorsolateral quadrant injury in the brain stem was often not identified – an important criterion for DAI (Geddes *et al.* 2000).

In 1999, Gleckman *et al.* published a series of ten cases of non-accidental craniocerebral trauma in infants and diagnosed DAI in five cases of 'shaken baby syndrome'. However, much of the APP staining appears to indicate the vascular type. In 2001, Geddes and colleagues published the largest series to date of 53 infants and children with inflicted head injury (Geddes *et al.* 2001a). This series also had defined diagnostic criteria for inclusion. In the 37 infants, only two of the cases showed DAI; both of these had evidence of significant impact with skull fractures. The authors concluded that DAI was uncommon in NAI and that the most common pathology related to widespread hypoxic–ischaemic damage and swelling. In the eight cases where there was no impact (no bruising or skull fracture) no DAI was present. In 13 of 37 there was

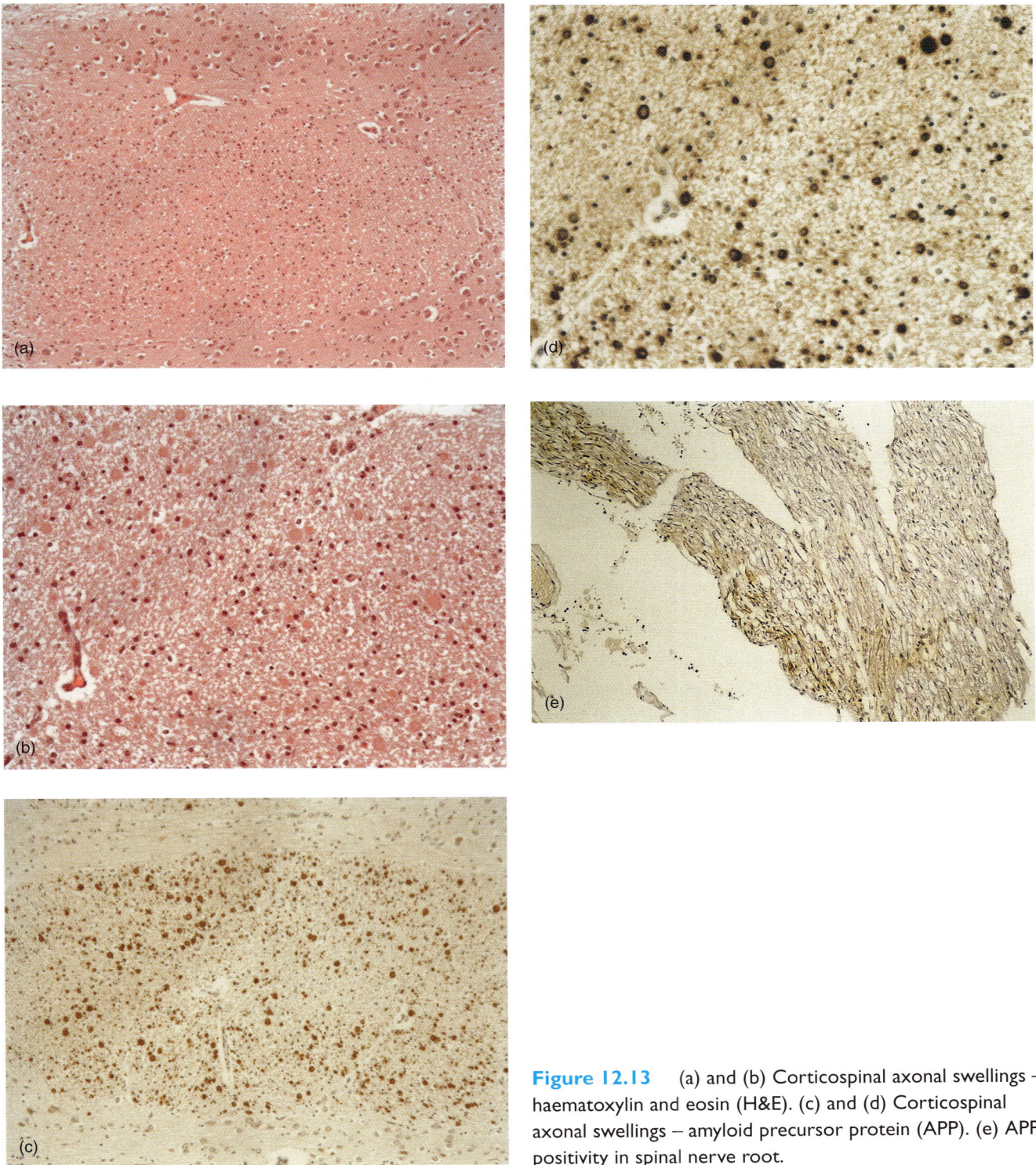

Figure 12.13   (a) and (b) Corticospinal axonal swellings – haematoxylin and eosin (H&E). (c) and (d) Corticospinal axonal swellings – amyloid precursor protein (APP). (e) APP positivity in spinal nerve root.

evidence of either macroscopic or microscopic cervical cord damage, including epidural haemorrhage, APP positivity of cervical nerve roots, and corticospinal axonal damage in the lower pons and medulla (Figure 12.13). This last has been reported in adult hyperextension injury (Lindenberg and Freytag 1970) but not previously in NAI. There was no difference between cases with and those without impact (Geddes *et al.* 2001a).

The experience of the author and her colleagues is that histological evidence of long tract or nerve root damage is more likely to be found if a full craniocervical examination is undertaken, and the whole brain stem and upper cervical cord are embedded for histology. The findings suggest that the significant mechanism of injury is primary brain-stem damage leading to severe apnoea or respiratory abnormalities, which in turn result in raised intracranial pressure from widespread hypoxic–ischaemic damage, and that the diffuse brain damage that occurs in the infants is hypoxic, not traumatic. Geddes *et al.*'s finding that DAI is rare in infants has now been confirmed by Reichard and his colleagues (2003), in a study of 28 cases of abusive head injury which included 10 children aged 12 months or younger. Only one instance of DAI was seen in the subgroup of infants, in a child of 10 months. This is also supported by the work of Duhaime *et al.* (1987) in that the forces generated by shaking are not of sufficient severity to cause DAI without impact. This is further supported by lack of definite pathological evidence in the case of the 'shaken adult' (Geddes and Whitwell 2003b). What is unclear is how much force is required to damage the craniocervical region in the young infant.

Within the diagnostic setting it is important to sample the brain widely along with the spinal cord (see Chapter 2). APP staining may well, if nothing else, give some indication of survival time. Care is needed in the interpretation of axonal injury. Hemispheric injury is seen as either axons or bulbs in the white matter, corpus callosum and internal capsule. Axonal swellings are usually visible from around 12–18 h after injury. Axonal injury varies from scattered foci in the white matter, which is more common than DAI where injury by definition is widespread and includes the rostral brain stem. CD 68 identifies microglial activity – becoming positive a few days after injury – and may be useful in identifying previous episodes of injury, which are not infrequently seen. As commented in the discussion relating to hypoxic–ischaemic damage, interpretation of the changes may be difficult.

Brain damage in survivors

In cases where death occurs at a later stage, a variety of findings may be seen. The major brain pathology results from hypoxic–ischaemic damage and variable grey and white matter damage, either cystic or non-cystic, may be seen (Figure 12.12). Residua of gliding contusions, often in the frontal lobes, may show as well-demarcated cavities. Other findings include old SDH, old surface contusional injury and secondary optic nerve degeneration.

Spinal cord injury

The findings relating to the craniocervical junction are detailed above. Examination is usually best done through a posterior approach. It is essential that the spinal cord in its entirety is retained, fixed and examined. Ideally it should be taken en bloc with the spinal column. This enables additional radiography to be done. Other cervical lesions are occasionally seen, including haematomyelia (Priatt 1995, Parrish 1996).

Other spinal lesions are relatively uncommon in NAI but may be identified on skeletal survey, although their demonstration can be difficult. It is difficult to assess the true incidence of spinal fractures in child abuse. It is said to vary from 0 to 3 per cent (Akbarnia 1999). Many are thought to go unidentified and they may be found without obvious clinical findings – this may have significance in timing (Cramer 1996, Akbarnia 1999). The majority involve the vertebral body with anterior compression fractures to the lower thoracic and upper lumbar areas. In addition rupture of the spinal ligament and vertebral dislocation may be seen. The mechanism of injury varies but includes hyperextension–flexion, compression fractures as a result of compaction to the buttocks or head, and fractures of spinal processes as a result of direct impact. Spinal cord damage may occur. Acute lateral compression has been reported producing extradural and intraspinal haemorrhage without fracture. This is thought to be a result of the flexible cartilaginous spine distorting with direct trauma (Gosnold and Sivaloganathan 1980).

Fracture sites should be examined histologically as should the spinal cord. It may be possible at least to indicate whether an injury is acute or shows ageing features. However, only general comments about timing may be achievable.

A common artefact seen is epidural blood along the length of the cord, resulting from congestion in the epidural vascular fat network (Valdes-Dapena 1975). This should not be mistaken for true trauma. Epidural bleeding localized to the cervical cord region, particularly in association with haemorrhage into the soft tissues of the neck, may be indicative of trauma to the craniocervical junction (Geddes *et al.* 2001a). Recent work suggests that some cases of isolated epidural haemorrhage with no other findings may be postmortem artefact (Rutty *et al.* 2005).

PROBLEM AREAS

Shaking or impact – the controversy

A number of authors over the years have suggested that shaking was the explanation for cases of SDH where no scalp injury was identifiable (Guthkelch 1971, Caffey 1972, Benstead 1983). Caffey (1974) coined the phrase 'whiplash shaken infant syndrome'. The term 'shaken baby syndrome' has now become virtually synonymous with the combination of SDH, retinal haemorrhages and brain injury. The issue of how much trauma is necessary is debatable in this young age group and others have questioned whether unexplained subdurals always equal either abuse or trauma (Fung et al. 2002). However, it is clear from pathological studies that there is commonly evidence of impact post mortem, although this may not be identifiable externally. Shaking alone has, however, been thought to be inadequate to produce sufficient force to generate findings relating to shaken baby syndrome, including SDH. This was identified in the biomechanical paper of Duhaime et al. in 1987, and Ommaya et al. (2002) have extensively reviewed the topic. Recent biomechanical work by Bandak (2005) has indicated that the forces commonly reported in shaken baby syndrome are too great for the infant neck to withstand. Neck injury is not a feature seen in cases of shaken baby syndrome. It is estimated that about 10 per cent of cases will have no evidence of impact and, although in the past this has been attributed to impact with a soft surface, it is being re-questioned in the light of the recent pathological studies, which indicate that diffuse axonal injury occurs only where there is significant impact and that the young infant more usually shows evidence of a focal stretch type injury to the craniocervical junction (Geddes et al. 2001a). Currently, there are no biomechanical data to indicate the degree of force required to produce this localized stretch injury, which, although similar to that seen in adult hyperextension injury (Lindenberg and Freytag 1970, Geddes et al. 2000), may be exceptionally variable in its severity (see Chapter 8, p. 98).

Is it necessary to postulate shaking where there is evidence of impact injury? The answer to this is most probably no. It is recognized that impact produces much greater forces (Duhaime et al. 1987, Ommaya et al. 2002). Impact has been shown to produce subdural bleeding and brain swelling, with various degrees of traumatic axonal injury. The series of Geddes et al. (2001a) showed similar findings in relation to the craniocervical injury in shaking alone as in those cases in which there was evidence of impact. This suggests that flexion and hyperextension of the neck almost certainly occur as part of an impact injury.

Re-bleeding in subdural haematomas

It is well recognized in adults that a process of organization and liquefaction of a subdural haematoma will occur within several days, with membrane formation. It has been said by some authors that re-bleeding in a SDH is not an explanation for symptoms or presentation in the young (David 1999, Case et al. 2001). This is also subject to debate. Evidence of SDHs of varying ages is seen in cases of infant head injury. Work referred to before (Greenes and Schutzman 1998) indicates that scanning reveals intracranial pathology, including SDH in a number of asymptomatic children, and 'disappearing subdurals' have been described (Duhaime et al. 1996). Clinically chronic SDHs are not uncommon and it is logical to assume that they at some point were acute. It does not necessarily mean that the injury was hidden, but rather that the injury was not severe enough to be considered serious at the time (Uscinski 2002). It is well recognized that SDHs occur as the result of birth including normal deliveries. The mechanism of membrane formation with reabsorption of a chronic SDH has never been demonstrated to be different from that in adults (Uscinski 2002). Re-bleeding in SDHs may occur with minimal or no trauma in the adult population. Clearly each case needs to be considered in the light of all available evidence.

Timing of injury

Much of this is within the realms of clinical paediatrics. However, the forensic neuropathologist is often asked to comment on the effect of the injuries identified and the probable clinical symptomatology. Caution is advised in these circumstances, particularly so on the background of previous injury including chronic SDH. In addition, the basis for traumatic unconsciousness was always thought to be that related to diffuse axonal injury. This has now been shown not to be so in most cases. It is likely that the injuries to the craniocervical junction found in a number of infants may lead to breathing abnormalities. In those infants, children who survive for some hours, pathological findings such as APP positivity, axonal retraction balls and other findings, e.g. CD68 positivity may assist in broad terms. It is recognized that the time interval between injury and onset of symptoms in a number of cases remains unclear (Nashelsky and Dix 1995, Gilliland 1998); however, in most cases with no complicating factors it has generally been accepted that an infant in the case of severe injury would not appear normal to a carer (David 1999). However there are reported cases with delay in presentation – these appear to be related to delayed onset of complications such as cerebral swelling or presentation of an intracranial haematoma (Denton and Mileusnic 2003). This is one of the most difficult areas and the forensic neuropathologist would be wise to be limited in opinion.

How far to fall?

This is another difficult area in which to draw conclusions in an individual case. There are many reported series of children with no fatalities from low-level falls (Chadwick and Salerno 1993, Williams 1991, Warrington and Wright 2001, Lyons and Oates 1993). However there have been individual case reports and series of documented genuine injuries including fatalities from low-level falls (Aoki and Masuzawa 1984, Howard *et al.* 1993, Christian *et al.* 1999, Kim *et al.* 2000, Aly-Hamdy *et al.* 2001, Plunkett 2001, Reichelderfer *et al.* 1979, Hall *et al.* 1989), in infants and children with or without skull fractures as well as other intracranial pathology including subdural haemorrhage. Stairway falls are also problematic – deaths are reported to be rare. Joffe and Ludwig (1988) found no correlation with injury and number of stairs, concluding these were less severe than a fall through the same vertical distance. However, reconstruction in the individual situation is difficult.

Detailed assessment of the known circumstances as well as assessment of other injuries is essential. Biomechanical engineering assessment is becoming more frequently used in these cases as is the use of infant models to simulate injury (Prange *et al.* 2003, Bandak 2005).

CONCLUSION

'Non-accidental' head injury in infants and children is highly complex, and probably the most controversial area of forensic neuropathology. No one can argue that protection of children is not paramount. However, in justice to the accused, opinions must be given on the basis of solid scientific findings and careful assessment of all the evidence. The conclusions in a recent paper, that much of what is believed in the field is based on a small database of poor-quality original research without appropriate controls, means that critical evaluation of all the evidence available is essential (Donohoe 2003).

REFERENCES

Akbarnia, B.A. 1999. Pediatric spine fractures. *Orthop Clin N Am*, 30, 521–36.

Aly-Hamdy, N., Childs, A.M., Ferrie, C.D. and Livingstone, J.H. 2001. Subdural haemorrhage with bilateral retinal haemorrhage following accidental household trauma. *Dev Med Child Neurol*, 43 (suppl 90), 24.

Aoki, N. and Masuzawa, H. 1984. Infantile acute subdural haematoma: clinical analysis of 26 cases. *J Neurosurg*, 61, 273–80.

Atwal, G.S., Rutty, G.N., Carter, N. *et al.* 1998. Bruising in non-accidental head injured children: a retrospective study of the prevalence, distribution and pathological associations in 24 cases. *Forensic Sci Int*, 96, 215–30.

Bandak, F.A. 2005. Shaken baby syndrome: a biomechanics analysis of injury mechanisms. *Forensic Sci Int*, 151, 71–9.

Barlow, K.M. and Minns, R.A. 1999. The relation between intracranial pressure and out come in non-accidental head injury. *Dev Med Child Neurol*, 41, 220–5.

Barlow, K.M. and Minns, R.A. 2000. Annual incidence of shaken impact syndrome in young children. *Lancet*, 356, 1571–2.

Barlow, K.M., Milne, S. and Minns, R.A. 1998. A retrospective epidemiological analysis of non-accidental head injury in children in Scotland over the last 15 years. *Scott Med J*, 43, 112–14.

Barozzino, T., Sgro, M., Toi, A. *et al.* 1998. Fetal bilateral subdural haemorrhages. Prenatal diagnosis and spontaneous resolution by time of delivery. *Prenat Diagn*, 18, 496–503.

Benstead, J.G. 1983. Shaking as a culpable cause of subdural haemorrhage in infants. *Med Sci Law*, 23, 242–4.

Biousse, V., Suh, D.Y. and Newman, N.J. 2002. Diffusion weighted magnetic resonance imaging in shaken baby syndrome. *Am J Ophthalmol*, 133, 249–55.

Brown, J.K. and Minns, R.A. 1993. Non-accidental head injury, with particular reference to whiplash shaking injury and medico-legal aspects. *Dev Med Child Neurol*, 35, 849–69.

Caffey, J. 1972. On the theory and practice of shaking infants. Its potential residual effects of permanent brain damage and mental retardation. *Am J Dis Child*, 124, 161–3.

Caffey, J. 1974. The whiplash shaken infant syndrome: manual shaking by the whiplash-induced intracranial and intraocular bleedings, linked with residual damage and mental retardation. *Pediatrics*, 54, 396–403.

Calder, I.M., Hill, I. and Scholtz, C.L. 1984. Primary brain trauma in non-accidental injury. *J Clin Pathol*, 37, 1095–100.

Carpenter, R.F. 1999. The prevalence and distribution of bruising in babies. *Arch Dis Child*, 80, 363–6.

Carty, H. and Pierce, A. 2002. Non-accidental injury: a retrospective analysis of a large cohort. *Eur Radiol*, 12, 2919–25.

Case, M.E., Graham, M.A., Handy, T.C. *et al.* 2001. Position abusive head injuries in infants and young children. National Association of Examiners Ad Hoc Committee on Shaken Baby Syndrome. *Am J Forensic Pathol*, 22, 112–22.

CESDI. 2000. The CESDI SUDI studies. In Fleming, P., Blair, P., Bacon, C. and Berry, P.J. (eds), *Sudden Unexpected Deaths in Infancy*. London: The Stationery Office.

Chadwick, D.L., Salerno, C. 1993. Likelihood of the death of an infant or young child in a short fall of less than 6 vertical feet. *J Trauma*, 35, 968.

Chadwick, D.L., Chin, M., Salerno, C. *et al.* 1991. Deaths from falls in children. How far is fatal? *J Trauma*, **31**, 1353–5.

Christian, C.W., Taylor, A.A., Hertle, R.W. *et al.* 1999. Retinal haemorrhages caused by accidental household trauma. *Paediatrics*, **135**, 125–7.

Cramer, K.E. 1996. Orthopedic aspects of child abuse. *Pediatr Clin N Am*, **43**, 1035–51.

David, T.J. 1999. Shaken baby (shaken impact) syndrome: non-accidental head injury in infancy. *R Soc Med*, **92**, 556–61.

Denton, S., Mileusnic, D. 2003. Delayed sudden death in an infant following an accidental fall. *Am J For Med Pathol*, **24**, 371–6.

Donohoe, M. 2003. Evidence-based medicine and shaken baby syndrome Part 1: Literature review, 1966–98. *Am J Forensic Med Pathol*, **24**, 239–42.

Duhaime, A.C., Gennarelli, T.A., Thibault, L.E. *et al.* 1987. The shaken baby syndrome. A clinical, pathological, and biochemical study. *J Neurosurg*, **66**, 409–15.

Duhaime, A.C., Alario, A.J., Lewander, W.J. *et al.* 1992. Head injury in very young children: mechanisms, injury types, and ophthalmologic findings in 100 hospitalized patients younger than 2 years of age. *Pediatrics*, **90**, 179–85.

Duhaime, A.C., Christian, C., Armonda, R. *et al.* 1996. Disappearing subdural haematomas in children. *Paediatr Neurosurg*, **25**, 116–22.

Duhaime, A.C., Christian, C.W., Rorke, L-B. and Zimmerman, R.A. 1998. Nonaccidental head injury in infants-the 'shaken-baby syndrome'. *N Engl J Med*, **338**, 1822–9.

Ehrenforth, S., Klarmann, D., Zabel, B. *et al.* 1998. Sever factor V deficiency presenting as subdural haematoma in the newborn [letter]. *Eur J Pediatr*, **157**, 1032.

Ellis, P.S.J. 1997. The pathology of fatal child abuse. *Pathology*, **29**, 113–21.

Ewing-Cobbs, L., Kramer, L., Prasad, M. *et al.* 1998. New imaging, physical and developmental findings after inflicted and non-inflicted traumatic brain injury in young children. *Paediatrics*, **102**, 300–7.

Fung, E.L., Sung, R.Y., Nelson, E.A. *et al.* 2002. Unexplained subdural hematoma in young children: is it always child abuse? *Pediatr Int*, **44**, 37–42.

Geddes, J.F. and Whitwell, H.L. 2003a. Neuropathology of fatal infant head injury. *J Neurotrauma*, **20**, 905.

Geddes, J.F. and Whitwell, H.L. 2003b. Shaken adult syndrome revisited. *Am J Forensic Med Pathol*, **24**, 310–11.

Geddes, J.F., Tasker, R.C., Hackshaw, C.D. *et al.* 2003c. Dural haemorrhage in non-traumatic infant deaths: does it explain the bleeding in 'shaken baby syndrome'?. *Neuropathol Appl Neurobiol*, **29**, 14–22.

Geddes, J.F., Hackshaw, A.K., Vowles, G.H., Nickols, C.D. and Whitwell, H.L. 2001a. Neuropathology of inflicted head injury in children. 1. Patterns of brain damage. *Brain*, **124**, 1290–8.

Geddes, J.F., Vowles, G.H., Hackshaw, A.K. *et al.* 2001b. Neuropathology of inflicted head injury in children. 2. Microscopic brain injury in infants. *Brain*, **124**, 1299–306.

Geddes, J.F., Whitwell, H.L. and Graham, D.I. 2000. Traumatic axonal injury: practical issues for diagnosis in medicolegal cases. *Neuropathol Appl Neurobiol*, **26**, 105–16.

Gilliland, M.G.F. 1998. Interval duration between injury and severe symptoms in nonaccidental head trauma in infants and young children. *J Forensic Sci*, **43**, 723–5.

Gilliland, M.G.F. and Luthert, P. 2003. Why do histology on retinal haemorrhages in suspected non-accidental injury? *Histopathology*, **43**, 592–602.

Gleckman, A.M., Bell, M.D., Evans, R.J. and Smith, T.W. 1999. Diffuse axonal injury in infants with non-accidental craniocerebral trauma: enhanced detection by beta-amyloid precursor protein immunohistochemical staining. *Arch Pathol Lab Med*, **123**, 146–51.

Gosnold, J.K. and Sivaloganathan, S. 1980. Spinal cord damage in a case of non-accidental injury in children. *Med Sci Law*, **20**, 54–7.

Graham, D.I., Ford, I., Adams, J.H. *et al.* 1989. Fatal head injury in children. *J Clin Pathol*, **42**, 18–22.

Greenes, D. and Schutzman, S. 1998. Occult intracranial injury in infants. *Ann Emerg Med*, **32**, 680–6.

Guthkelch, A.N. 1971. Infantile subdural haematoma and its relationship to whiplash injuries. *BMJ*, **11**, 430–1.

Hadley, M.N., Sonntag, V.K., Rekate, H.L. and Murphy, A. 1989. The infant whiplash-shake injury syndrome: a clinical and pathological study. *Neurosurgery*, **24**, 536–40.

Hall, J.R., Reyes, H.M., Horvat, M., Meller, J.L. and Stein, R. 1989. The mortality of childhood falls. *J Trauma*, **29**, 1273–5.

Handy, T.C., Hanzlick, R., Shields, L.B.E. *et al.* 1999. Hypernatremia and subdural hematoma in the pediatric age group: Is there a causal relationship? *J Forensic Sci*, **44**, 1114–8.

Hargrave, D.R. and Earner, D.P. 1992. A study of child homicide over two decades. *Med Sci Law*, **32**, 247–50.

Hart, B.L., Dudley, M.H. and Zumwalt, R.E. 1996. Postmortem cranial MRI and Postmortem examination correlation in suspected child abuse. *Am J Forensic Med Pathol*, **17**, 217–24.

Hartley, L.M., Khwaja, O.S. and Verity, C.M. 2001. Glutaric aciduria type 1 and nonaccidental head injury. *Pediatrics*, **107**, 174–6.

Harwood-Nash, D.C., Hendrick, E.B. and Hudson, A.R. 1971. The significance of skull fractures in children. *Radiology*, **101**, 151–6.

Hobbs, C.J. 1984. Skull fracture and the diagnosis of abuse. *Arch Dis Child*, **59**, 246–52.

Helfer, R.E., Slovis, T.L. and Black, M. 1977. Injuries resulting when small children fall out of bed. *Pediatrics*, **60**, 533–5.

Howard, M.A., Bell, B.A. and Uttley, D. 1993. The pathophysiology of infant subdural haematoma. *Br J Neurosurg*, **7**, 355–65.

Jaspan, T., Narborough, G., Punt, J.A. and Lowe, J. 1992. Cerebral contusional tears as a marker of child abuse-detection by cranial sonography. *Pediatr Radiol*, **22**, 237–45.

Jayawant, S., Rawlinson, A., Gibbon, F. *et al.* 1998. Subdural haemorrhages in infants: population based study. *BMJ*, **317**, 1558–61.

Joffe, M. and Ludwig, S. 1988. Stairway injuries in children. *Paeditrics*, **82**, 457–61.

Johnson, C.F. 1990. Inflicted injury versus accidental injury. *Pediatr Clin N Am*, **37**, 791–814.

Johnson, D.L., Boal, D. and Baule, R. 1995. Role of apnea in nonaccidental head injury. *Pediatr Neurosurg*, **23**, 305–10.

Judkins, A.R., Hood, I.G., Mirchandani, H.G. *et al.* 2004. Technical Communication: Rationale and Technique for Examination of Nervous System in Suspected Infant Victims of Abuse. *Am J Forensic Med Patholol*, **25**, 29–32.

Kahana, T. and Hiss, J. 1999. Forensic radiology. Review article. *Br J Radiol*, **72**, 129–33.

Keats, T.E. 1996. *Atlas of Normal Roentgen Variants that May Simulate Disease*, 6th edn. St Louis: Mosby.

Kemp, A.M. 2002. Investigating subdural haemorrhage in infants. *Arch Dis Child*, **86**, 98–102.

Kempe, C.H. 1975. Uncommon manifestations of the battered child syndrome. *Am J Dis Child*, **129**, 1265.

Kim, K.A., Wang, M.Y., Griffith, P.M. *et al.* 2000. Analysis of pediatric head injury from falls. *Neurosurg Focus*, **8**, Article 3.

Lam, C.H., Montes, J., Farmer, J-P. *et al.* 1996. Traumatic aneurysm from shaken baby syndrome: case report. *Neurosurgery*, **39**, 1252–5.

Langlois, N.E.I. and Gresham, G.A. 1991. The aging of bruises: a review and study of the color changes with time. *Forensic Sci Int*, **50**, 227–38.

Lantz, P.E., Sinal, S.H., Stanton, C.A. *et al.* 2004. Perimacular retinal folds from childhood head trauma. *BMJ*, **328**, 754–56.

Leadbeatter, S. 2001. Resuscitation injury. In Rutty, G.N. (ed.), *Essentials of Postmortem Examination Practice*, Vol 1. London: Springer-Verlag.

Leestma, J.E. 1988. *Forensic Neuropathology*. New York: Raven Press.

Lindenberg, R. and Freytag, E. 1969. Morphology of brain lesions from blunt trauma in early infancy. *Arch Pathol*, **87**, 298–305.

Lindenberg, R. and Freytag, E. 1970. Brainstem lesions characteristic of traumatic hyperextension of the head. *Arch Pathol*, **90**, 509–15.

Liu, H.Y., Huang, L.T. and Lui, C.C. 1998. Vacuum extraction delivery complicated with acute subdural haematoma and cerebral infarction: report of one case. *Chung-Hua Min Kuo Hsiao Erh Ko I Hsueh Hui Tsa Chih*, **39**, 119–22.

Lyons, T.J. and Oates, R.K. 1993. Falling out of bed: a relatively benign occurrence. *Pediatrics*, **92**, 125–7.

McClain, P.W., Sacks, J.J., Froehlke, R.G. *et al.* 1993. Child abuse and neglect, United States 1979 through 1988. *Pediatrics*, **91**, 338–43.

Maxeiner, H. 2001. Demonstration and interpretation of bridging vein ruptures in cases of infantile subdural bleedings. *J Forensic Sci*, **46**, 85–93.

Menkes, J.H. 2001. Subdural haematoma, non-accidental injury or …? *Eur J Paediatr Neurol*, **5**, 175–6.

Nashelsky, M.B. and Dix, J.D. 1995. The time interval between lethal infant shaking and onset of symptoms. A review of the shaken baby syndrome literature. *Am J Forensic Med Pathol*, **16**, 154–7.

Ng, P.C., Fok, T.F., Lee, C.H. *et al.* 1998. Massive subdural haematoma: an unusual complication of septicaemia in preterm very low birthweight infants. *J Pediatr Child Health*, **34**, 296–8.

O'Grady, J.P., Pope, C.S. and Patel, S.S. 2000. Vacuum extraction in modern obstetric practice: a review and critique. *Curr Opin Obstet Gynecol*, **12**, 475–80.

O'Hare, A.E. and Eden, O.B. 1994. Bleeding disorders and non-accidental injury. *Arch Dis Child*, **59**, 860–4.

Ommaya, A.K., Goldsmith, W. and Thibault, L. 2002. Biomechanics and neuropathology of adult and paediatric head injury. *Br J Neurosurg*, **16**, 220–42.

Parrish, R. 1996. Isolated cord injury in child abuse. *Pediatr Trauma and Forensic Newsletter Jan*. Letter to the Editor.

Perper, J.A. and Wecht, C.H. (eds) 1980. *Microscopic Diagnosis in Forensic Pathology*. Springfield, IL: Thomas.

Plunkett, J. 2001. Fatal pediatric head injuries caused by short-distance falls. *Am J Forensic Med Pathol*, **22**, 1–12.

Prange, M.T., Coats, B., Duhaime, A.C. and Margulies, S.S. (2003). Anthropomorphic simulations of falls, shakes, and inflicted impacts in infants. *J Neurosurg*, **99**, 143–50.

Prasad, M., Ewing-Cobbs, L., Swank, P.R. and Kramer, L. 2002. Predictors of outcome following traumatic brain injury in young children. *Pediatr Neurosurg*, **36**, 64–74.

Priatt, J.H. 1995. Isolated spinal cord injury as a presentation of child abuse. *Pediatrics*, **96**, 780–2.

Rao, P. and Carty, H. 1999. Non-accidental injury: review of the radiology. *Clin Radiol*, **54**, 11–24.

Reiber, G. 1993. Fatal falls in childhood; how far must children fall to sustain fatal head injury? Report of cases and review of the literature. *Am J Forensic Med Pathol*, **14**, 201–7.

Reichard, R.R., White, C.L., Hladik, C.L. and Dolinak, D. 2003. Beta-amyloid precursor protein staining of nonaccidental central nervous system injury in pediatric autopsies. *J Neurotrauma*, **20**, 347–55.

Reichelderfer, T.E., Overbach, A. and Greensher, J. 1979. Unsafe playgrounds. *Pediatrics*, **64**, 962–3.

Renzulli, P., Tuchschmid, P., Eich, G. *et al.* 1998. Early vitamin K deficiency bleeding after maternal phenobarbital intake: management of massive intracranial haemorrhage by minimal surgical intervention. *Eur J Pediatr*, **157**, 663–5.

Rorke, L.B. 1992. Neuropathology. In Ludwig, S. and Kornberg, A.E. (eds), *Child Abuse: A medical reference*, 2nd edn. New York: Churchill Livingstone, 403–21.

Royal College of Pathologists and Royal College of Paediatrics and Child Health. 2004. *Sudden unexpected death in infancy*. London: RCP and RCPCH.

Rutty, G.N., Smith, C.M. and Malia, R.G. 1999. Late-form hemorrhagic disease of the newborn. A fatal case report with illustration of investigations that may assist in avoiding mistaken diagnosis of child abuse. *Am J Forensic Med Pathol*, **20**, 48–51.

Rutty, G.N., Squier, W.M. and Padfield, C.J. 2005. Epidural haemorrhage of the cervical spinal cord: a post-mortem artefact? *Neuropathol Appl Neurobiol*, **31**, 247–57.

Saukko, P. and Knight, B. 2004. *Knight's Forensic Pathology*, 3rd edn. London: Arnold.

Schwartz, A.J. and Ricci, L.R. 1996. How accurately can bruises be aged in abused children? Literature review and synthesis. *Pediatrics*, **97**, 254–7.

Shannon, P., Smith, C.R., Deck, J. *et al.* 1998. Axonal injury and the neuropathology of shaken baby syndrome. *Acta Neuropathol*, **95**, 625–31.

Shaw, J.C.L. 1988. Copper deficiency and non-accidental. *Arch Dis Child*, **63**, 448–55.

Steiner, R.D., Pepin, M. and Byers, P.H. 1996. Studies of collagen synthesis and structure in the differentiation of child abuse from osteogenesis imperfecta. *J Pediatr*, **128**, 542–7.

Stephenson, T. 1995. Bruising in children. *Curr Paediatr*, **5**, 225–9.

Stephenson, T. and Bialas, Y. 1996. Estimation of the age of bruising. *Arch Dis Child*, **74**, 53–5.

Stoodley, N. 2002. Non-accidental head injury in children: gathering the evidence. *Lancet*, **360**, 271–2.

Sugar, N.F., Taylor, J.A. and Feldman, K.W. 1999. Bruises in infants and toddlers. *Arch Pediatr Adolesc Med*, **153**, 399–403.

Suh, D.Y., Davis, P.C., Hopkins, K.L. *et al.* 2001. Non accidental pediatric head injury: diffusion weighted imaging findings. *Neurosurgery*, **49**, 309–20.

Swischuk, L.E. 2003. *Imaging of the Newborn, Infant and Young Child*. Philadelphia: Lippincott Williams&Wilkins.

Teasdale, G.M., Murray, G., Anderson, E. *et al.* 1990. Risks of acute traumatic intracranial haematoma in children and adults: implications for managing head injuries, *BMJ*, **300**, 363–7.

Tokoro, K., Nakajima, F. and Yamataki, A. 1988. Infantile chronic subdural hematoma with local protrusion of the skull in a case of osteogenesis imperfecta. *Neurosurgery*, **22**, 595–8.

Towner, D., Castro, M.A., Eby-Wilkins, E. *et al.* 1999. Effect of mode of delivery in nulliparous women on neonatal intracranial injury. *New Engl J Med*, **341**, 1709–14.

Tzioumi, D. and Oates, K.R. 1998. Subdural haematomas in children under 2 years accidental or inflicted? A 10 year experience. *Child Abuse Neglect*, **22**, 1105–12.

Uscinski, R. 2002. Shaken baby syndrome: fundamental questions. *Br J Neurosurg*, **16**, 217–19.

Valdes-Dapena, M. 1975. Sudden death in infancy: a report for pathologists. *Perspect Pediatr Pathol*, **2**, 1–14.

Vowles, G.H., Scholtz, C.L. and Cameron, J.M. 1987. Diffuse axonal injury in early infancy. *Clin Pathol*, **40**, 185–9.

Warrington, S.A., Wright, C.M., ALSPAC Study Team. 2001. Accidents and resulting injuries in premobile infants: data from the ALSPAC study. *Arch Dis Child*, **85**, 104–7.

Weber, W. 1984. Experimental studies of skull fractures in infants. *Z Rechtsmed*, **92**, 87–94.

Weber, W. 1985. Biomechanical fragility of the infant skull. *Z Rechtsmed*, **94**, 93–101. [Article in German]

Wheeler, D. and Hobbs, C. 1988. Mistakes in diagnosing non-accidental injury: 10 years' experience. *BMJ*, **296**, 1233–6.

Whitby, E.H., Griffiths, P.D., Rutter, S. *et al.* 2003. Frequency and natural history of subdural haemorrhages in babies and relation to obstetric factors. *The Lancet*, **362**, 846–51.

Whitwell, H.L. 2001. Non-accidental injury in children. In Lowe, D.G. and Underwood, J.C.E. (eds), *Recent Advances in Histopathology*, Vol 19. Edinburgh: Churchill Livingstone.

Williams, R.A. 1991. Injuries in infants and small children resulting from witnessed and corroborated free falls. *J Trauma*, **31**, 1350–3.

13

Non-traumatic neurological conditions in medicolegal work

Colin Smith

Pathology within the central nervous system (CNS) is a common finding at postmortem examination. The intracranial pathology is, however, often an incidental finding and not directly related to the cause of death. This chapter deals with the non-traumatic neurological conditions that have a direct relationship to the cause of death not covered in other chapters and which the forensic pathologist may encounter. The chapter focuses on adult pathology and, therefore, does not cover topics such as sudden infant death syndrome (SIDS) or neonatal/infant ischaemic pathology.

DISORDERS RELATED TO CEREBRAL PERFUSION

Normal function of the brain is dependent on the blood supply delivering adequate levels of both oxygen and glucose. Disruption to the provision of these essential elements will result in neuronal injury, which may be further compounded by the accumulation of neurotoxic metabolites such as lactate (Miyamoto and Auer 2000). In this section the neuropathological changes associated with hypoxia, ischaemia and hypoglycaemia, and how they may impact on the forensic pathologist, are discussed.

Hypoxia refers to the state of low oxygen levels, either in the blood (hypoxaemia) or in the brain tissue. Hypoxia can be produced in a variety of ways (Table 13.1), although the pure forms of hypoxia are rarely seen in forensic practice and produce little pathology (Auer and Sutherland 2002). More frequently hypoxia is seen in association with ischaemia (impairment of blood supply to the brain), such as after cardiorespiratory arrest. Stagnant hypoxia is often

superimposed on other forms of hypoxia, e.g. cardiac arrest after cyanide poisoning. Stagnant hypoxia is descriptive of the tissue conditions in cases of focal or global ischaemia and for practical purposes can be considered to be the same as ischaemia.

In ischaemia, blood flow to the brain fails, resulting not only in failure to deliver oxygen but also in accumulation of toxic metabolites. It is the accumulation of metabolites with resulting low tissue pH that will account for tissue damage in ischaemia (Siesjö 1992). The extent of damage to the brain produced in ischaemia is determined principally by the extent of the ischaemia (focal or global) and its duration. Focal ischaemia is characteristically produced by occlusion of a vessel whereas global ischaemia is a consequence of haemodynamic collapse.

Focal ischaemia and cerebral infarction

The most common form of cerebrovascular pathology encountered at postmortem examination is that of a stroke. Stroke is common, the incidence increasing with age, although cerebrovascular pathology may be seen at any age. In Western countries thromboembolic infarcts account for about 60–80 per cent of strokes, with intracerebral haemorrhages accounting for most of the remaining cases. Rare causes of stroke include a variety of emboli, vasculopathies, haematological disorders and genetic disorders. Vertebral and carotid artery dissection is dealt with in Chapter 6.

Thrombosis, either acute or organizing, may be seen within the major vessels at the base of the brain. More commonly, atherosclerosis is seen in the major branches of the circle of Willis, the carotid arteries (particularly

Table 13.1   Classification of hypoxia

Type of hypoxia	Pathophysiology	Clinical example	Structural changes in the brain
Hypoxic	Reduced oxygen tension in pulmonary alveoli results in hypoxaemia and brain hypoxia	Acute: pulmonary oedema Chronic: interstitial lung disease	None
Anaemic	Hypoxaemia caused by low levels of haemoglobin or by competitive binding of oxygen sites on the haemoglobin molecule	Carbon monoxide poisoning	None if not accompanied by ischaemia
Histotoxic	The neuron is unable to use the oxygen as a result of poisoning of the metabolic pathways within the cell	Cyanide poisoning	None
Stagnant	There is reduced blood flow to the brain caused by either reduced cardiac output or disruption of local perfusion	Reduced cardiac output: cardiac arrest Disrupted local perfusion: thromboembolic infarct	Neuronal necrosis and if prolonged tissue infarction.

Table 13.2   Timing of changes in thromboembolic infarcts

Time since infarct (h)	Macroscopic appearance	Microscopic appearance
6–12	None	Irreversible ischaemic cell change
12–24	Early loss of grey–white matter interface	Neutrophilic infiltrate although may be relatively inconspicuous
48	Cerebral oedema established, with associated mass effect. There may be tissue splitting	Activated macrophages present in damaged tissues
1–3 weeks	Cavitation begins to develop	Gliosis and neovascularization identified
Months	Gliotic scar often golden-brown in colour as a result of haemosiderin staining	Gliotic scar

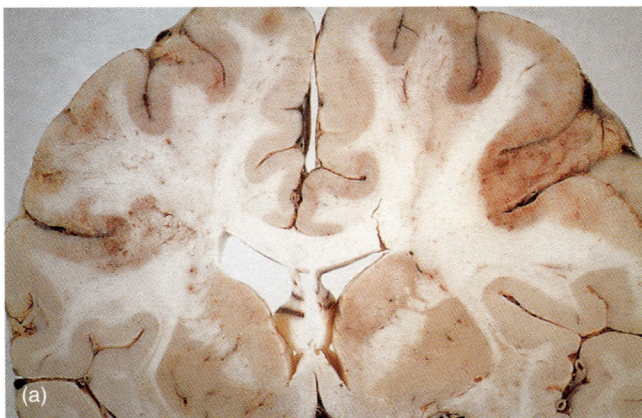

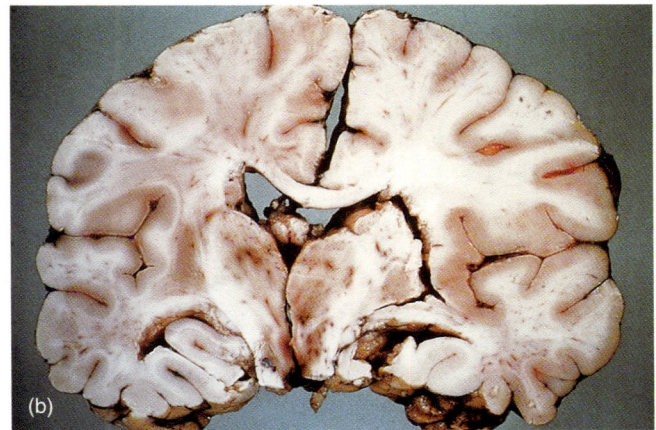

Figure 13.1   Macroscopic appearances of cerebral infarcts at different ages. (a) Recent infarct (24–48 h) with swelling and discoloration of the infarcted tissue. (b) Recent infarct (48+ h) with cerebral swelling and mass effect. The splitting of the tissue is artefact and represents the infarcted tissues separating from viable tissues.

at the bifurcations) and the vertebrobasilar arteries. In rare cases, severe atherosclerosis of the basal vessels may result in severely ectatic vessels. The pathological changes seen at various stages of infarct evolution are detailed in Table 13.2 (Kalimo *et al.* 2002) and illustrated in Figures 13.1 and 13.2.

The following rare causes are associated with specific situations.

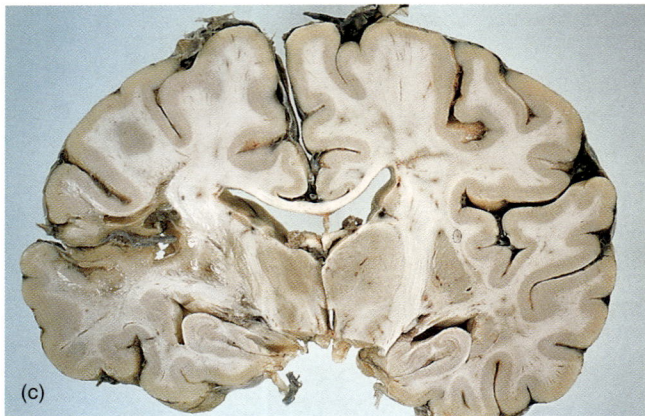

Figure 13.1 (*continued*)   (c) Old infarct (months) with loss of the insular cortical tissue and associated white matter. The discoloration is caused by haemosiderin deposition.

FAT EMBOLI

After traumatic fracture of the long bones or pelvis, the patient, often in an intensive therapy unit (ITU), will develop respiratory and neurological dysfunction, usually 24–48 h after the history of trauma. Macroscopically the brain is swollen and there are widespread petechial haemorrhages (Figure 13.3). Frozen sections of brain tissue should be examined to identify the fat emboli (Kamenar and Burger 1980) (Figure 13.4).

AIR EMBOLI

These may be associated with decompression of deep-sea divers (Caisson's disease) or with cardiac surgery. Microinfarcts may be seen in the spinal cord (Palmer *et al.* 1987) and cerebrum.

FIBROCARTILAGINOUS EMBOLI

Intervertebral disc material has been described as a cause of cerebral infarction after minor trauma (Toro-Gonzalez *et al.* 1993). It is also rarely associated with pulmonary emboli.

Global ischaemia

Global ischaemia may be absolute in situations where cerebral blood flow ceases for a period of time, such as during cardiac arrest, or may be variable in situations where the cerebral blood flow is compromised, such as with systemic hypotension or raised intracranial pressure.

BRAIN DAMAGE AFTER CARDIAC ARREST

In this situation there is cessation of blood flow to the brain. It is important to remember that the pathological changes associated with cardiac arrest require successful

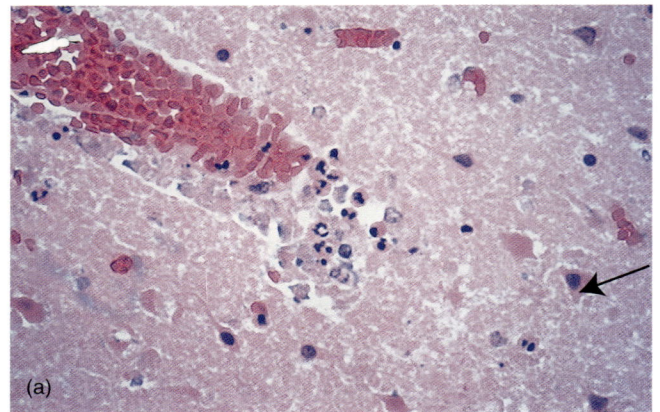

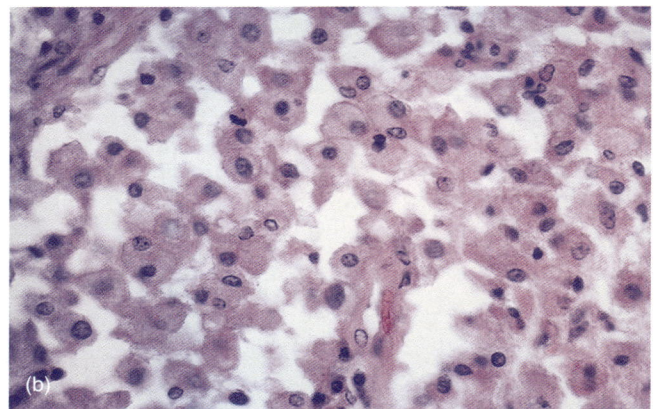

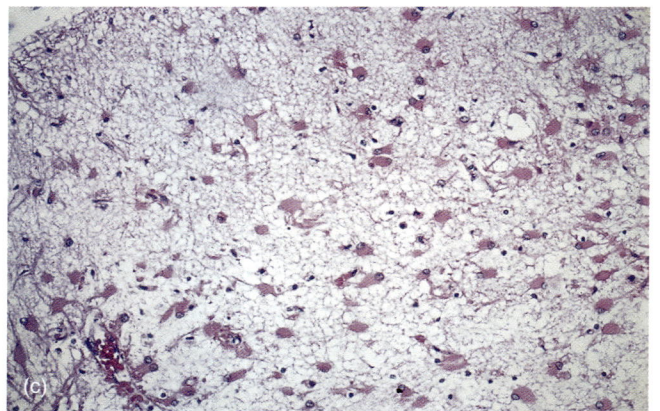

Figure 13.2   Microscopic appearances of cerebral infarcts at different ages. (a) Recent infarct (hours) with neutrophilic infiltration. Note the neuronal eosinophilia consistent with ischaemic neuronal damage (arrow) (H&E, ×40). (b) Recent infarct (48+ h) with macrophage infiltration (H&E, ×40). (c) Old infarct (months) with gliosis at the edge of the area of infarction (H&E, ×40).

resuscitation, with subsequent survival for a period of hours before they can be identified. The period of cessation of cerebral blood flow may last only 5–10 min for irreversible brain damage to develop. Macroscopic identification of

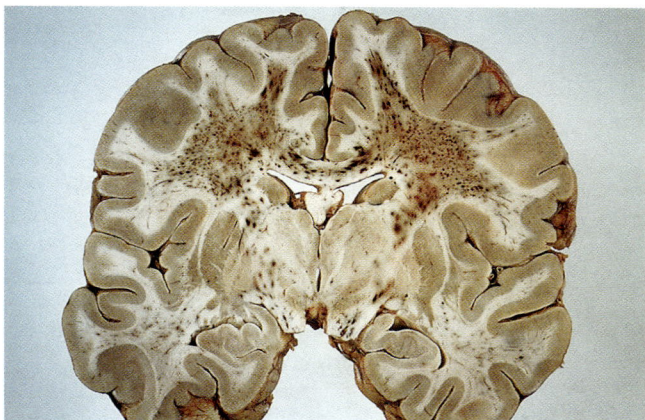

Figure 13.3 Widespread petechial haemorrhages secondary to fat embolism.

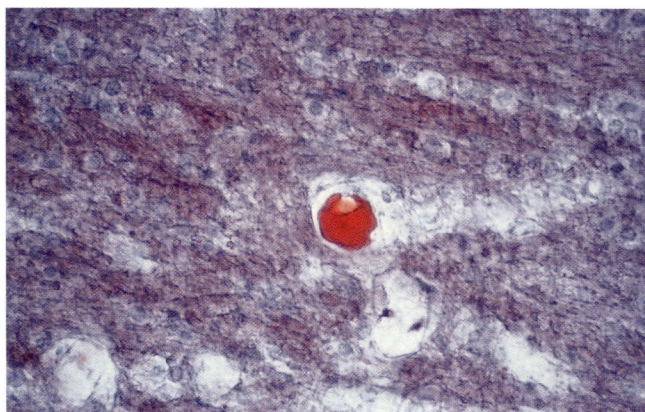

Figure 13.4 Fat embolus seen in a section of frozen brain tissue (Oil red O, ×40).

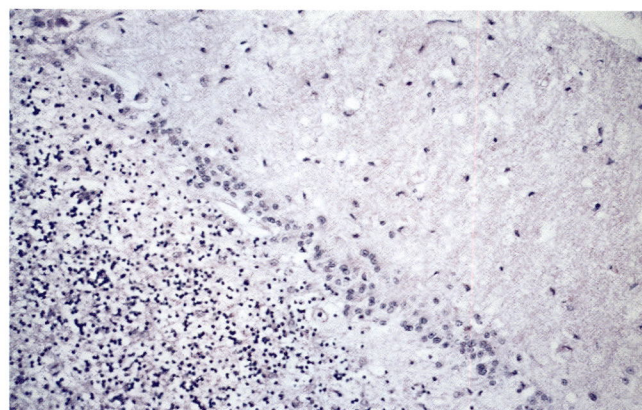

Figure 13.5 Section of cerebellar cortex from a case of case of cardiac arrest with survival of several weeks. There is loss of Purkinje cells with reactive Bergmann gliosis (H&E, ×40).

brain damage, however, requires a survival of approximately 36–48 h. Cortical discoloration may be identified (early laminar necrosis), particularly in the parietal and occipital cortices, or may be accentuated within the depths of gyri. Microscopic diffuse neuronal ischaemic damage can be seen after 5–10 h, initially with a pattern of selective vulnerability. Selective vulnerability refers to the fact that early in global ischaemia neurons in certain regions will be affected before others (Pulsinelli 1997) (Figure 13.5). The selective vulnerability associated with ischaemia differs from that seen in hypoglycaemia, and they are compared in Table 13.3.

In cases of long-term survival, the patient usually being in a vegetative state, the ventricles are enlarged secondary to loss of grey and white matter.

BRAIN DAMAGE AFTER HYPOTENSION

In this situation there is reduced perfusion of the brain. Ischaemic damage is accentuated at boundary zones with damage to both neurons and surrounding brain parenchyma (watershed infarcts) being seen at the posteriorly placed triple watershed zone in mild hypotension (Figure 13.6). The entire cerebral cortex may be involved; if the hypotension is severe and/or prolonged the situation begins to mimic that seen in cardiac arrest. Watershed infarcts may be unilateral as a result of the asymmetry of the circle of Willis (e.g. caused by atherosclerosis).

The cerebellum also has boundary zones, the most commonly involved being between the superior cerebellar (SCA) and the posteroinferior cerebellar (PICA) arteries at the dorsal angle. If hypotension is mild or quickly reversed (within several minutes), damage may be limited to Purkinje cell ischaemic damage or focal infarction within the boundary zones. If more severe or prolonged, the injury is extensive within the cerebellar cortex. Guidelines for sampling the brain in cases with ischaemic damage are covered in Chapter 2, Appendix 5.

Venous infarction

This is most commonly associated with cerebral venous thrombosis. This is more common than previously suspected, with a fatal outcome in a relatively small number of cases. Underlying infection remains an important cause of cerebral venous thrombosis, although the hormonal changes produced by oral contraceptives and pregnancy are probably the most important aetiological factors (Ameri and Bousser 1992). The superior sagittal sinus is most commonly involved, resulting in parasagittal congestion and parenchymal haemorrhage involving both cortex and underlying white matter. Microscopically the appearances

Table 13.3 The microscopic distribution of neuronal damage in ischaemia and hypoglycaemia

Anatomical region	Ischaemia	Hypoglycaemia
Hippocampus	Maximal within sector CA1	Maximal within sector CA1 and within the dentate gyrus
Cerebral cortex	Accentuated within the depths of gyri, particularly at boundary zones, and involving the deeper layers of the cortex	Widespread and most pronounced within the superficial layers
Striatum	Diffuse damage in the striatum, usually medium-sized neurons	Little damage seen in the striatum
Cerebellum	Purkinje cell ischaemic damage or focal infarction within the boundary zones	The cerebellar cortex, and in particular Purkinje cells, show little evidence of injury

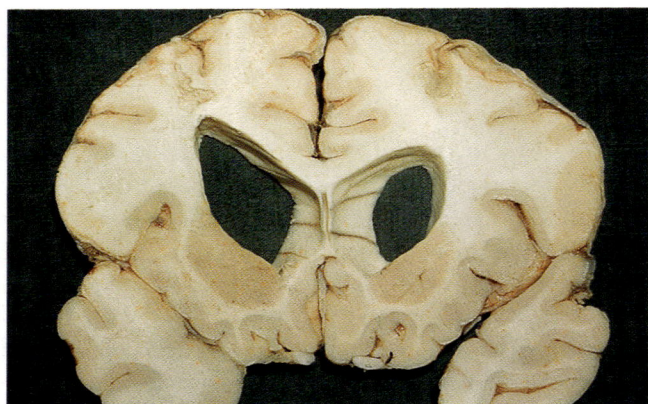

Figure 13.6 Bilateral watershed infarcts secondary to hypotension.

are of haemorrhagic infarction although in venous infarction neutrophils are usually abundant.

HYPOGLYCAEMIA

Hypoglycaemic brain damage will be produced in the adult when blood glucose levels fall below about 1.5 mmol/l. Hypoglycaemia produces a pattern of selective neuronal necrosis which, in its pure form, differs from that seen with ischaemia (Auer and Siesjö 1988). However, hypoglycaemic coma is often accompanied by seizures or cardiorespiratory depression such that ischaemic features may coexist with the pathology associated with hypoglycaemia.

Insulin overdose, resulting in profound hypoglycaemia, may be accidental or intentional (suicide or homicide). The patient with diabetes may accidentally administer an incorrect dose of insulin, or may have difficulty controlling the diabetes, resulting in episodes of hypoglycaemia or hypoglycaemic coma. Rare causes of hypoglycaemia include islet cell tumours of the pancreas (insulinoma).

Macroscopically the brain shows little abnormality with only mild swelling. The microscopic distribution of injury is detailed in Table 13.3.

INFECTION OF THE CNS

Infections of the nervous system remain a considerable cause of morbidity and mortality worldwide. With the increasing ease of world travel, particularly to destinations 'off the beaten track', and an increasing population of immunosuppressed individuals, there is a need for greater awareness of nervous system infections. A nervous system infection may be the cause of sudden unexpected deaths in adults, particularly in elderly and immunocompromised individuals. This section covers some of the more common nervous system infections encountered; for a more detailed review of this subject the interested reader is referred to recent reviews (Davis and Kennedy 2000, Love 2001).

Identification of the underlying organisms requires appropriate tissue samples and swabs to be taken at the time of the postmortem examination, often requiring consultation with the microbiology and virology departments. Guidelines for practice at postmortem examination in such cases are provided in Chapter 2, p. 26.

Sterile cerebrospinal fluid (CSF) can be removed from the brain *in situ* during the postmortem examination by passing a needle into the lateral ventricles after the dura has been removed and may be used for polymerase chain reaction (PCR) identification of viruses. At the postmortem examination the air sinuses, orbit and both the middle and inner ears should be examined for possible foci of infection. Examination of the heart and other viscera may indicate the site of infection.

Infections of the nervous system can be considered based on their anatomical location or the underlying organism.

BACTERIAL INFECTIONS

Bacteria may enter the nervous system by the haematogenous route (most common) or secondary to osteomyelitis and sinusitis/mastoiditis or compound fracture. Bacterial infections are usually suppurative, although notable exceptions include mycobacteria.

FUNGAL INFECTIONS

These tend to be opportunistic infections, being seen in immunocompromised patients. CNS fungal infection is virtually always associated with systemic infection although infection may be most obvious in the brain (Figure 13.7). Spread is usually haematogenous from a focus within the lungs (Walsh *et al.* 1985). In some cases infection is the result of direct spread from either the air sinuses or the orbit. In particular mucormycosis may extend from paranasal sinuses to the orbit or brain. This is classically seen in diabetic ketoacidosis, although it may be associated with any immunosuppressed state.

VIRAL INFECTIONS

Viruses generally enter the body via the mucous membranes of the respiratory or gastrointestinal tract, although in some cases entry may be traumatic, e.g. after a dog bite that breaks the skin. A number of viruses are neurotrophic and will move towards the nervous system, usually via peripheral nerves (e.g. rabies). Some viruses will show a latent period between primary infection and nervous system disease (e.g. HIV).

PROTOZOAL INFECTIONS

These are disseminated via haematogenous spread and can produce a range of pathological appearances depending on the infecting organism. Protozoal infections are common worldwide although tend to be seen less frequently in Western populations.

MACROSCOPIC APPEARANCE OF INFECTIONS

Pachymeningitis

This refers to a collection of pus in relation to the dura. Empyemas are collections of pus that are usually subdural and found in the supratentorial space, although in the spinal region extradural collections can be seen in association with osteomyelitis. Subdural empyemas are usually associated with open skull fractures or neurosurgical procedures (Krauss and McCormick 1992). Macroscopically the appearances are of a collection of pus in the subdural space, often encapsulated by granulation tissue.

Leptomeningitis

This is more commonly referred to as meningitis. The most common form of meningitis to be encountered in the postmortem examination room is acute purulent (bacterial) meningitis (Figure 13.8). In adults bacterial meningitis is usually secondary to bacteraemia, although skull fracture secondary to trauma or previous neurosurgical procedure may result in recurrent meningitis. The organisms commonly involved in adults are *Streptococcus pneumoniae* and *Neisseria meningitides* (Gray and Alonso 2002). Similar organisms are seen in children with, in addition, *Haemophilus influenzae*. Macroscopically the brain is usually swollen and a purulent membrane is seen overlying it; this is most pronounced over the convexities with *Strep. pneumoniae*, and usually most pronounced basally with other organisms (Love 2001). The infection

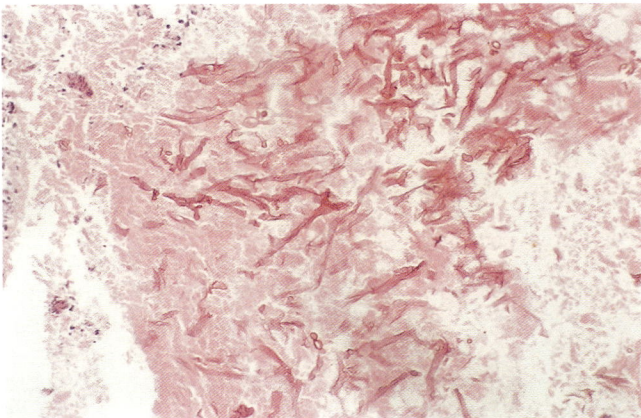

Figure 13.7 Histology of cerebral mucormycosis showing fungal hyphae.

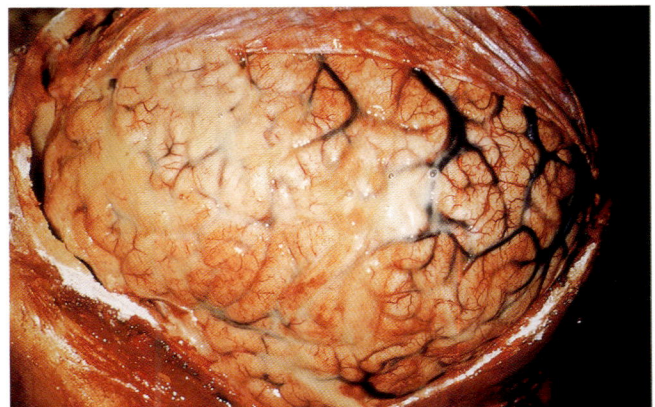

Figure 13.8 Purulent exudate in the subarachnoid space in a case of acute bacterial meningitis. The reflected dura can be seen at the top of the image.

may extend into the ventricles (ventriculitis) and there may be a degree of hydrocephalus secondary to CSF obstruction by pus. Microscopically a florid neutrophilic infiltrate is seen within the subarachnoid space, and there is often thrombosis of cortical vessels with associated superficial cortical infarction.

Granulomatous meningitis

Tuberculous meningitis macroscopically has a rather nodular exudate and is most pronounced basally and laterally, particularly overlying the sylvian fissures (Figure 13.9). Tuberculosis may also present as a parenchymal mass (tuberculoma). The differential diagnosis of granulomatous inflammation of the meninges includes the non-infective condition neurosarcoidosis.

Infections of the brain parenchyma

These may be suppurative or non-suppurative; the suppurative infections will produce an abscess whereas non-suppurative infections tend to produce encephalitis, a non-suppurative diffuse inflammation of the brain parenchyma.

CEREBRAL ABSCESSES
Cerebral abscesses can act as space-occupying lesions resulting in raised intracranial pressure (Figure 13.10). Cerebral abscesses can be solitary or multifocal, determined to a degree by the route of entry. Haematogenous dissemination of the organism tends to produce multifocal cerebral abscesses whereas direct invasion from local infection in the paranasal sinuses, middle ear or maxillary dental roots tends to produce a solitary lesion (Wispelwey *et al.* 1991). For haematogenous dissemination primary sites include cardiac valves and chronic lung infections (bronchiectasis). They also may be seen in intravenous drug abuse. Bacterial organisms are usually responsible for cerebral abscesses in immunocompetent individuals with *Strep. viridans* being the most common organism identified. In immunocompromised individuals fungal organisms, such as *Aspergillus* and *Candida* species, and protozoa, such as *Toxoplasma gondii*, may be seen. Toxoplasmosis, in particular, is associated with HIV infection (Strittmatter *et al.* 1992) (Figure 13.11).

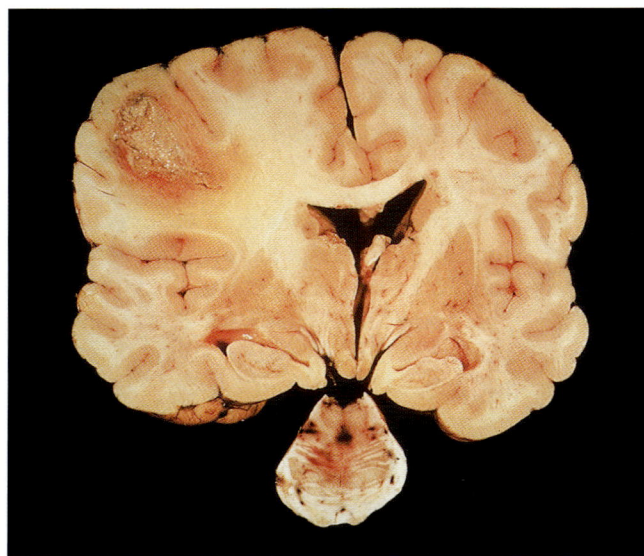

Figure 13.10 A large cerebral abscess acting as a space-occupying lesion with mass effect. There is midline shift and subfalcine herniation. Note the haemorrhages in the brain stem (Duret haemorrhages).

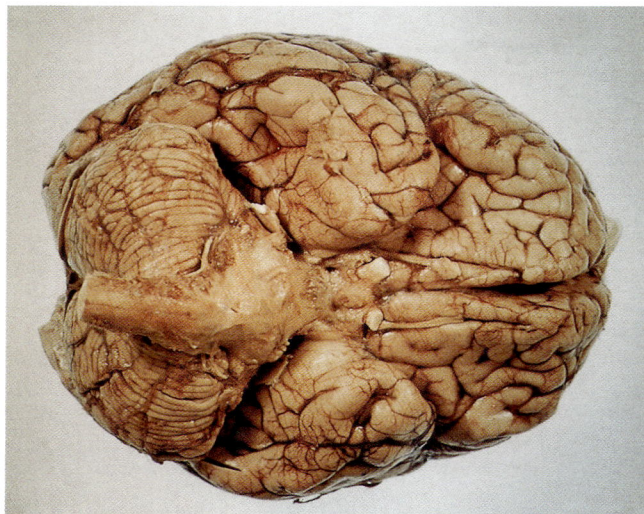

Figure 13.9 Tuberculous meningitis with a dense granulomatous exudate obscuring the base of the brain.

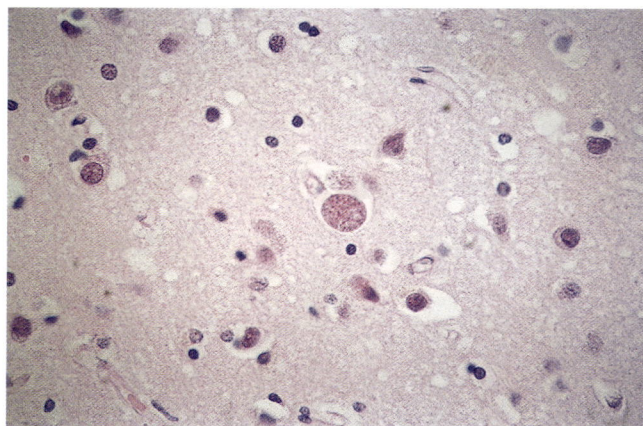

Figure 13.11 A toxoplasmosis cyst seen in the brain of a patient infected with HIV (H&E, ×40).

Macroscopically cerebral abscesses caused by haematogenous dissemination are classically said to develop at the cortical grey–white matter interface, although in some cases they are widely distributed throughout the brain (Figure 13.12). Abscesses associated with paranasal air sinus and maxillary dental root infections are located in the adjacent frontal lobe, whereas those associated with middle-ear infections are usually found in the temporal lobe or, rarely, the cerebellum (Love 2001). Abscesses initially develop as an area of focal swelling and discoloration, often being haemorrhagic. By about 2 weeks the abscess will have a purulent necrotic centre and a thick capsule.

ENCEPHALITIS

There are some general features of encephalitis that are not dependent on the infecting virus. Macroscopically, the brain is often swollen with congestion of parenchymal vessels. There is usually a perivascular lymphocytic infiltrate with both microglial and astrocytic (gliosis) activation. Cell necrosis is variable ranging from individual neurons in poliomyelitis to extensive tissue infarction (necrotizing encephalitis) in herpes simplex encephalitis. Inclusion bodies may be seen in neurons or glial cells, and this may help in identification of the infecting virus (Love and Wiley 2002). Specific causes of encephalitis that are of importance are considered below.

HIV infection

HIV infection is widespread throughout the world and remains a major clinical problem. In forensic practice, it is important to remember that a high proportion of the intravenous drug abuser (IVDA) population are HIV positive. HIV infection results in progressive depletion of CD4 T lymphocytes and subsequent immunosuppression causing acquired immune deficiency syndrome (AIDS). Neuropathology is seen in 70–90 per cent of AIDS cases although much of this is caused by opportunistic infections (Bell 1998). HIV itself can cause nervous system pathology, with neuropathology directly attributable to HIV being seen in 20–30 per cent of cases (Navia *et al.* 1986). HIV encephalopathy (HIVE) is a white matter disease characterized by multinucleated giant cells. HIVE has been shown to be more prevalent in IVDAs than in homosexuals (Bell *et al.* 1995). The HIV related-protein gp41 can be demonstrated immunohistochemically in relation to many multinucleated giant cells. Macroscopically, the HIVE brain often appears normal; microscopically, there are foci of perivascular lymphocytes and microglial nodules often lying adjacent to multinucleated giant cells (Budka *et al.* 1991). Vacuolar myelopathy is a relatively common condition in AIDS patients, which resembles subacute combined degeneration of the cord. The degeneration develops in the posterior and lateral columns of the spinal cord and there is axonal degeneration (Budka *et al.* 1991). In AIDS patients the disorder is not associated with vitamin B12 deficiency.

Herpes simplex encephalitis

This is the most common necrotizing encephalitis and is usually caused by herpes simplex virus (HSV) type 1. The macroscopic appearances of established HSV encephalitis are characteristic, with asymmetrical necrosis of the anterior temporal lobes, cingulate gyri and insular cortices (Figure 13.13). In early stages of the infection, however, the changes can be very mild with only temporal lobe congestion suggestive of infection. Microscopic examination of established cases reveals perivascular inflammation, necrosis with foamy macrophage infiltration and

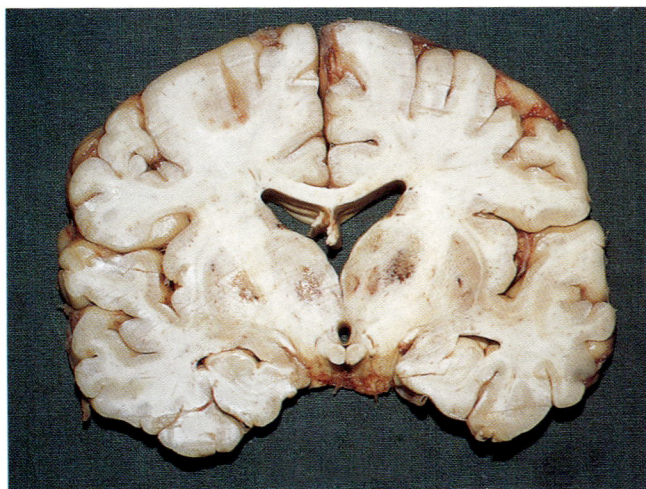

Figure 13.12 Multiple cerebral abscesses: in this example the abscesses were caused by toxoplasmosis.

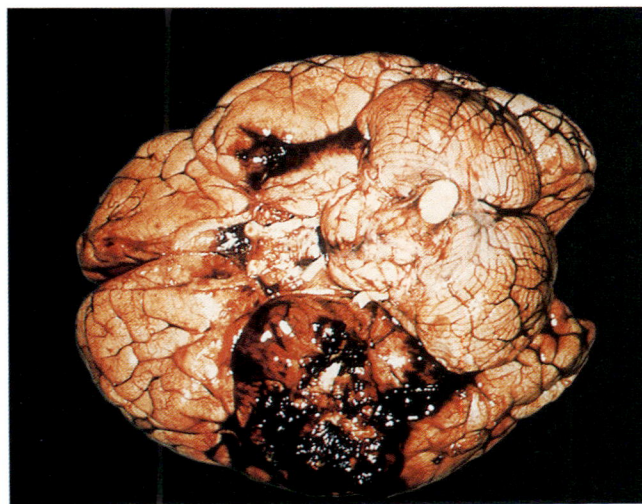

Figure 13.13 Herpes simplex encephalitis: there is haemorrhagic necrosis predominantly involving the temporal lobes.

eosinophilic neuronal inclusions, which are predominantly nuclear (Adams and Milled 1973). Immunohistochemistry can detect the virus up to 3 weeks from the start of infection.

Malaria

Malaria is a protozoal infection caused by *Plasmodium* species. Cerebral malaria may complicate *Plasmodium falciparum* (in up to 10 per cent of infections), the predominant type seen in West Africa but also encountered throughout Africa, Asia and South America. It should always be considered in cases of sudden death in travellers through these regions. Encephalopathy and death can develop rapidly. Macroscopically, petechial haemorrhages are usually widely distributed throughout the brain (Figure 13.14). Microscopically, the capillaries are engorged and malaria parasites or granules of pigment are seen in many of the red cells. Inflammation is usually sparse, although occasional aggregates of neutrophils, lymphocytes and macrophages (Dürck granuloma) may be seen (Day and Turner 2000).

NEOPLASIA INVOLVING THE CNS

A detailed description of tumours of the CNS is outwith the scope of this book and the interested reader is referred to recent specialist accounts (Kleihues and Cavenee 2001, Ironside *et al.* 2002). However, CNS tumours may be encountered at postmortem examination. The tumours may be either primary or secondary; metastatic tumours to the CNS are common, whereas primary intrinsic tumours are rare. The epithelial tumours that metastasize most commonly to the nervous system are of the bronchus and breast, followed by melanoma (Delattre *et al.* 1988). In general, prostatic and female genital tract tumours rarely metastasize to the brain or meninges, although prostatic carcinomas do frequently cause neurological complications secondary to cord compression after vertebral body metastasis and collapse. Diffuse infiltration of the leptomeninges may be seen with haematological malignancies (lymphoma and leukaemia) and epithelial malignancies such as breast and bronchial carcinomas (meningeal carcinomatosis) (Posner 1995).

Of the intrinsic primary brain tumours gliomas are the most common. There are a large number of histological subtypes that reside under the blanket term 'glioma'. More specific terms such as astrocytoma, oligodendroglioma, ependymoma, choroid plexus tumour or mixed glial tumour should be used, with an appropriate grade given on the basis of histological features (pleomorphism, mitotic activity, tumour necrosis, microvascular proliferation). Gliomas may occur at any age and at any anatomical site within the nervous system, and are the most common group of solid tumours encountered in children.

Tumours may present to the pathologist through a variety of mechanisms:

- They can grow to a large size producing only rather non-specific symptoms such as headache. When slow growing and large, the tumour acts as a mass lesion. Should the lesion then enlarge rapidly, e.g. after tumour haemorrhage, the brain is unable to compensate for the rapidly expanding mass lesion, resulting in brain herniation and death (Poon and Solis 1985) (Figure 13.15). Although intratumoral

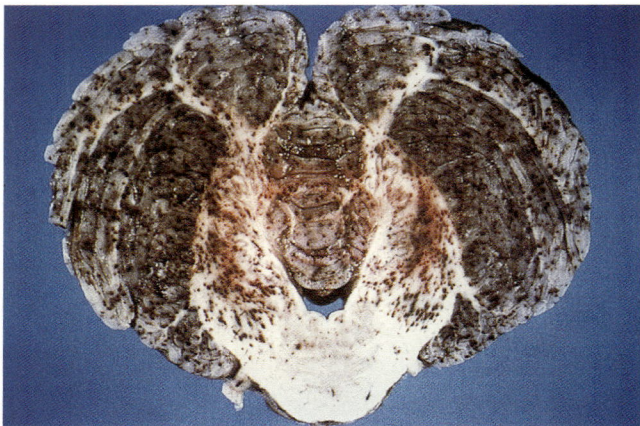

Figure 13.14 Section through the cerebellum and brain stem demonstrating multiple petechial haemorrhages secondary to malarial infection.

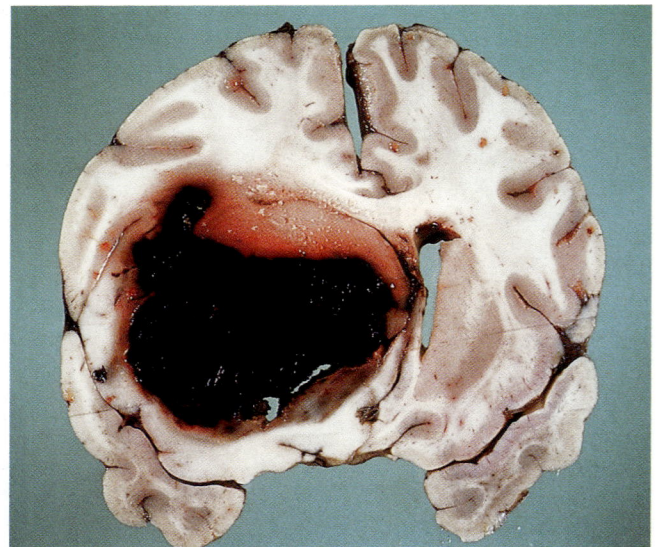

Figure 13.15 Acute haemorrhage into a glioblastoma, resulting in a rapid rise in intracranial pressure. There is compression of the ventricular system and an obvious subfalcine hernia.

haemorrhage may occur in any tumour, primary or secondary, it is particularly associated with oligodendrogliomas, glioblastomas and metastatic melanomas.

- As discussed below, seizures are often associated with malignancy, and death may follow a seizure.
- Mild traumatic head injury in a patient with an underlying malignancy may result in intratumoral intracerebral haemorrhage and death after brain herniation (personal observation).

Macroscopically, gliomas result in diffuse expansion of both white and grey matter as they diffusely infiltrate brain tissue. They are often rather gelatinous and may have cystic areas. Oligodendrogliomas frequently show calcification, although this may be seen in other gliomas. Gliomas of the brain stem or spinal cord can be identified as an area of diffuse expansion with, on sectioning, loss of the normal anatomical structure. High-grade gliomas, such as glioblastoma, contain areas of necrosis that macroscopically appear yellow and granular. A specific tumour of which the pathologist should be aware is the colloid cyst of the third ventricle. This benign cystic lesion can be associated with sudden death, particularly in young to middle-aged individuals (Byard and Moore 1993, Filkins *et al.* 1996). This lesion lies within the third ventricle where it can obstruct the foramen of Monro, resulting in acute hydrocephalus with dilatation of the lateral ventricles.

EPILEPSY

The clinical condition of epilepsy is best, albeit rather simplistically, defined as a continuing tendency to epileptic seizures. An important point in this definition is that a single seizure does not automatically equate with a diagnosis of epilepsy. Status epilepticus is a defined clinical entity; for therapeutic purposes it is defined as 'continuous seizures lasting at least 5 minutes or two or more discrete seizures between which there is incomplete recovery of consciousness' (Lowenstein and Alldredge 1998). Certification of death as status epilepticus therefore requires a clear clinical history and should not be applied to a death secondary to a suspected seizure (Langan *et al.* 2002).

There is an increased risk of premature death associated with epilepsy, and a significant excess of accident-related deaths (Hirsch and Martin 1971, Kenneback *et al.* 1997). Postmortem examination in known individuals with epilepsy or in an individual with witnessed seizures may show some of the features associated with seizures, although they are not specific, such as bruising of the tongue and petechial haemorrhages in the mucous membranes. Postmortem examination of deaths in epilepsy should include toxicological examination of blood and urine for levels of antiepileptic drugs, ethanol and recreational drugs, and retention of the brain for detailed neuropathological examination after fixation (Black and Graham 2001, see also Chapter 2).

In many cases of known epilepsy (between 50 and 70 per cent) no macroscopic or microscopic lesions are identified to account for the epilepsy (idiopathic). Epilepsy may be associated with malformations, tumours, metabolic disorders, ischaemia, both recent and long-standing head trauma, and the sequelae of infections. A number of histological features are described that are considered to be a consequence of epilepsy rather than the cause. These include neuronal loss and gliosis in sectors CA1 and CA4 of the hippocampus (mesial temporal sclerosis), and subpial (Chaslin's) gliosis focally or diffusely throughout the cortex (Honavar and Medrum 2002). Detailed descriptions of the neuropathology associated with epilepsy may be found in more detailed texts (Scaravilli 1998).

Sudden unexpected death in epilepsy

This is defined as 'sudden unexpected witnessed or non-witnessed non-traumatic and non-drowning death in patients with epilepsy with or without evidence of a seizure and excluding documented status epilepticus where necropsy does not reveal a toxicological or anatomical cause for death' (Nashef 1997). Sudden unexpected death in epilepsy (SUDEP) is associated with young age (20–40 years) and poor seizure control. It is seen more frequently in males and in epilepsy secondary to head injury and alcoholism. In about 30 per cent of cases there are witnessed seizures before death, although in almost 50 per cent the circumstances suggest death during sleep (Black and Graham 2002). Postmortem examination may show non-specific seizure-associated pathology and neuropathology secondary to epilepsy.

Death in children with epilepsy

This is uncommon, unless there is an associated severe neurological disorder. SUDEP was seen in only a single case in a population-based cohort study of 692 children diagnosed with epilepsy over a 9-year period, with a 15-year follow-up (Camfield *et al.* 2002).

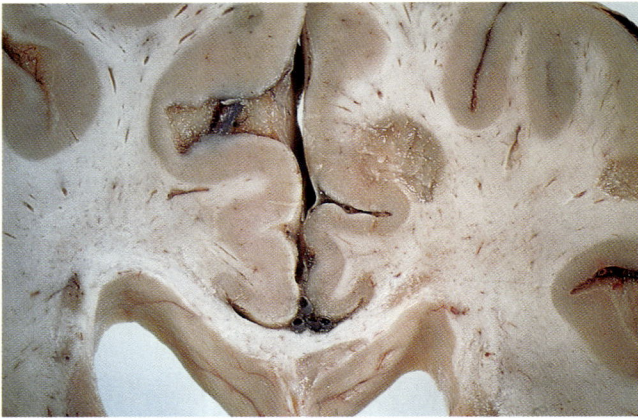

Figure 13.16 An acute plaque of demyelination with a granular appearance. Older plaques are seen in the periventricular area.

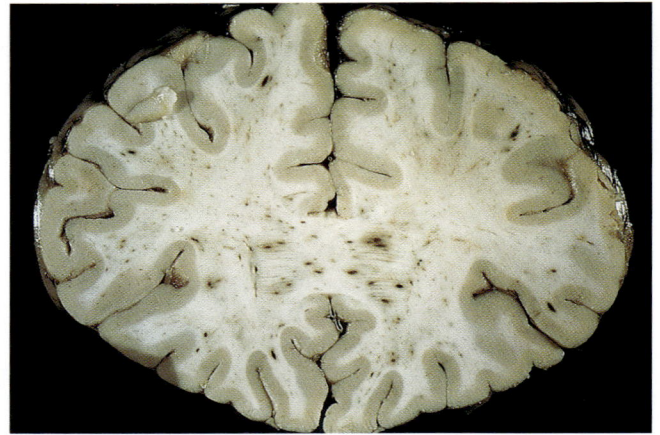

Figure 13.17 Petechial haemorrhages within the cerebrum from a case of acute haemorrhagic leukoencephalitis. Ball-and ring-type haemorrhages were seen microscopically.

DEMYELINATING DISORDERS

Multiple sclerosis

This is the most common form of demyelinating disease. It is a relapsing–remitting disorder in which plaques of demyelination can be widely distributed throughout the CNS. In general the forensic pathologist will encounter multiple sclerosis (MS) as an incidental finding rather than a direct cause of death, although death may occur if there is acute demyelination in the brain stem. In chronic MS the plaques have a predilection for subependymal and subpial sites, and for the cortical grey–white matter junction. Chronic plaques are sunken and grey and generally sharply defined. Acute plaques have a rather more granular appearance (Figure 13.16). The optic nerves should be examined if possible.

In some rare situations demyelinating disorders can present as an acute syndrome with rapid mortality (within days). These include the following.

Acute haemorrhagic leukoencephalitis (Hurst's disease)

This rare disorder is considered to be a hyperacute allergic reaction, seen at all ages and usually fatal within 2–3 days of the onset of symptoms (Hart and Earle 1975). The illness usually follows an upper respiratory tract infection and is characterized by fever, impaired consciousness and focal neurological signs. Macroscopically, the brain is swollen and congested, and tentorial and tonsillar hernias are usually apparent. Petechial haemorrhages are widely distributed

throughout the brain (Figure 13.17). Microscopically, these are ball- or ring-type haemorrhages with central necrotic small vessels (Gosztonyi 1978). Foci of perivascular demyelination are seen if the individual had survived longer than about 4 days from the onset of symptoms. A similar condition is perivenous encephalomyelitis (acute disseminated encephalomyelitis [ADEM]), which may follow measles infection or smallpox vaccination (DeVries 1960).

Central pontine myelinolysis

This uncommon disorder is seen in individuals with a variety of underlying disorders (Lampl and Yazdi 2002), although it is most commonly seen in people with chronic alcohol problems. This is covered in more detail in Chapter 14, p. 169.

DISORDERS ASSOCIATED WITH DEMENTIA

Cognitive decline is associated with ageing but can be an indication of a pathological condition termed 'dementia'. Dementia is not a specific disorder but rather a generic term for cognitive and behavioural disturbances produced by a variety of pathological processes. The pathologist may encounter dementia as a cause of abnormal behaviour, which may underlie self-harm or a road traffic death, or may be required to comment on neurodegenerative changes in assault victims who may have been displaying unusual aggressive behaviour. Allegations of maltreatment may also occur because the victims not infrequently become cachectic. Dementia is caused not only by neurodegenerative pathology, but also by neoplasia, infections and metabolic cerebral dysfunction induced by disorders such as chronic renal failure or endocrine disorders such as hypoparathyroidism.

Macroscopic examination should assess the presence or absence of cerebral atrophy and brain weight. The adult brain should weigh between 1200 and 1600 g, the weight usually being slightly greater in males. It must be remembered that, when interpreting brain weight, it is important to consider the stature of the individual; a person of small stature will have a small brain that may be outwith these guide weights. Beyond the age of about 50 years, there is a gradual decrease in brain weight of about 2 per cent per year (Esiri *et al.* 1997). If atrophy is present, the location and severity should be noted (frontotemporal distribution of Alzheimer's disease and the knife-edge frontal atrophy of Pick's disease being classic examples). In cases of suspected Creutzfeldt–Jakob disease (CJD) frozen tissue should be retained to assist with molecular diagnosis. The brain should be sectioned after adequate fixation. Histological sampling is a necessity to achieve a diagnosis. Guidelines for sampling the brain in cases of probable dementia have been published (Lowe 1998), and recommended blocks and special stains are outlined in Chapter 2, Appendix 6.

Assessment of neurodegenerative changes histologically has been detailed in a user-friendly flow chart that will allow accurate diagnosis in most cases (Lowe 1998). Some immunohistochemical assessment is required to assess the extent of neurodegenerative changes. In the absence of a good clinical history of cognitive and/or behavioural disturbances, the pathologist should not make a diagnosis of dementia. The pathological changes in such case should be described only with the comment being made that such changes are associated with specific neurodegenerative conditions.

Alzheimer's disease

This accounts for the majority of cases of dementia. It is most common after the age of 65 years, the incidence increasing with age, although early onset forms are well recognized. Macroscopically, the brain will show a frontotemporal pattern of atrophy, the changes being most severe in the medial aspects of the temporal lobe (Figure 13.18). Microscopically, the characteristic features are the abundance of neurofibrillary tangles and neuropil threads (tau immunoreactive) and β-amyloid deposition (Aβ protein immunoreactive). These features can be associated with conditions other than Alzheimer's disease and it is the distribution and quantity that determine the diagnosis. The assessment is based on the published CERAD

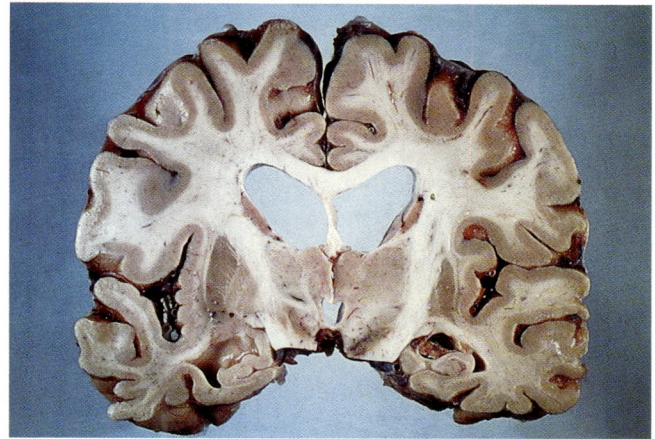

Figure 13.18 Bilateral temporal lobe atrophy with associated enlargement of the ventricular system in a case of Alzheimer's disease.

(Consortium to Establish a Registry for Alzheimer's Disease) criteria, which will allow a diagnosis of possible or definite Alzheimer's disease. Cerebral amyloid angiopathy (CAA) is strongly correlated with Alzheimer's disease, and the affected vessels are usually seen in the subarachnoid space and superficial cortex. CAA is associated with lobar intracerebral haemorrhage (see Chapter 7, p. 90) (Vinters and Gilbert 1983). Vessels with amyloid deposition can be detected microscopically using a haematoxylin and eosin (H&E) stain, but immunohistochemistry for Aβ protein is more sensitive.

Vascular dementia

The degree to which vascular pathology contributes to dementia is controversial and poorly understood (Vinters *et al.* 2000). Dementia may develop as the result of a single large infarct or of multiple small infarcts (multi-infarct dementia). The location of the infarcts is considered to be important, with lesions in the left hemisphere being more likely to cause dementia. Unsurprisingly infarcts involving the limbic system will also result in cognitive impairment. The role of subcortical infarcts (usually lacunar infarcts) in producing cognitive impairment is unclear, although a number of studies have identified these lesions as the most commonly seen in cases of vascular dementia (Esiri 2000). At postmortem examination of the brain the small lacunar infarcts are seen predominantly in relation to the basal ganglia, thalamus and adjacent white matter. Histological examination of such cases should be carried out to exclude the possibility of coexistent pathology such as Alzheimer's disease.

Dementia with Lewy bodies

This disorder has been increasing recognized in recent years and is associated with parkinsonism. Dementia with Lewy bodies (DLB) is often associated with Alzheimer-type pathology, but appears to have a distinct clinical progression that differs from Alzheimer's disease. The diagnosis is made by demonstrating Lewy bodies within the substantia nigra and within the hippocampus and neocortical regions. DLB is of particular importance in that neuroleptic drugs, used to treat agitation and hallucinations, may induce rapid clinical deterioration that may be life threatening (Allen and Burns 1998).

IATROGENIC NEUROPATHOLOGY

The pathologist may be asked to examine cases where an individual has died as a result of medical intervention in which the most significant pathology will be in the nervous system. This section considers neuropathology that can develop as a result of invasive intervention and medical intervention.

Invasive intervention

NEOPLASM SURGERY
The surgical procedure may be either a burr-hole biopsy for diagnosis or a craniotomy for debulking surgery. Both are associated with the risk of raised intracranial pressure through either haemorrhage or oedema, the risk of mortality being greater with burr-hole biopsies where the surgeon is unable rapidly to control the bleeding and the intracranial pressure. In burr-hole biopsies the risk is considered to be greater with free-hand needle biopsy rather than using stereotactic techniques, particularly for deeply seated lesions, although free-hand biopsy is now rarely done for such lesions (Lee *et al.* 1991).

SPINAL CORD INJURY
Spinal cord infarction has been reported after surgical treatment of abdominal aortic aneurysms in which the artery of Adamkiewicz is damaged (Hamano *et al.* 2000).

AIR EMBOLI
Air emboli may be introduced into the vascular system during spinal surgery, particularly when the sitting position is used for cervical spinal surgery (McCarthy *et al.* 1990). This is of greatest concern if there is a right-to-left cardiac shunt (such as a patent foramen ovale) and, if

untreated, this may result in cardiorespiratory arrest (Pham Dang *et al.* 2002). Air emboli are also associated with the insertion of central venous catheters.

RADIOLOGICAL COILING/EMBOLIZATION
Coiling of cerebral aneurysms is now frequently undertaken and is associated with a low rate of morbidity and mortality, but fatalities have been described after coiling. The coil may pass through the wall of the aneurysm resulting in subarachnoid haemorrhage or, alternatively, cause vasospasm or thrombosis of the stem vessel that results in massive infarction and subsequent death. Complications are less frequent than those associated with surgical clipping of aneurysms (Hohlreider *et al.* 2002). At postmortem examination careful assessment of the coil and its relationship to the aneurysm is required. Embolization is used in the treatment of arteriovenous malformations and can be used in some tumour types to reduce the size and vascularity of a tumour (particularly meningiomas) before surgery. Embolic material may not be confined to the target site, resulting in embolic infarcts and rarely death (Gruber *et al.* 2000). At postmortem examination the embolic material is easily identified within vessels.

Medical intervention

ANTICOAGULANT, ANTI-PLATELET AND THROMBOLYTIC DRUGS
These drugs have a risk of intracerebral haemorrhage associated with them, as well as haemorrhage at other sites. The risk of intracerebral haemorrhage is dependent on the specific drug and the underlying medical disorders. Underlying CAA may predispose to intracerebral haemorrhage with anticoagulant therapy (McCarron *et al.* 1999). The intracerebral haemorrhages may be multiple small lesions throughout the brain and spinal cord, or may be a single large lesion producing significant mass effect.

MALIGNANT HYPERTHERMIA
This is a rare disorder of skeletal muscle in which there is skeletal muscle rigidity in susceptible individuals after a certain trigger such as an inhaled anaesthetic (halothane) or skeletal muscle relaxant (suxamethonium). The condition is characterized by muscle rigidity, pyrexia, tachycardia and metabolic acidosis. If appropriate therapy is not initiated the patient can die within minutes. At postmortem examination muscle biopsies should be taken including frozen tissue to assess the genetic mutations associated with malignant hyperthermia (mutations in either the *RYR1* or the *CACNA1S* gene). Histological

sections of skeletal muscle show extensive acute fibre necrosis (MacLennan and Loke 2002).

SPINAL MANIPULATION

Increasingly, chronic low back and cervical pain is managed conservatively, with a significant proportion of patients visiting chiropractors for spinal manipulation. Although rare, spinal manipulation does have an associated mortality related to vertebrobasilar artery territory infarction as a result of vertebral artery dissection (Stevinson *et al.* 2001, Haldeman *et al.* 2002). Cervical manipulation carries the greatest risk. The damaged vessel should be identified within the vertebral canal and, when removed, serially sectioned to confirm the presence of intimal dissection and thrombosis (see Chapter 6, p. 77).

COMPLICATIONS OF PREVIOUS IRRADIATION

In the acute situation therapeutic cranial irradiation may be associated with oedema, although this is usually well treated with steroids. Delayed radiation injury may take the form of focal radionecrosis or diffuse radiation leukoencephalopathy. Radionecrosis requires at least 6–12 months to develop and can be recognized histologically by the presence of hyaline vessels, some of which may show fibrinoid necrosis. Radiation leukoencephalopathy is rather poorly defined but has been reported in cases 12 months or more after therapeutic cranial irradiation. Clinically there is neurological and intellectual decline; histologically the brain may show diffuse white matter vacuolation or focal white matter demyelination and necrosis (Ironside *et al.* 2002).

CONCLUSION

This chapter illustrates a range of non-traumatic neurological conditions that may be encountered in forensic practise. It will be clear that this chapter is not comprehensive and can act only as an introduction to the many potential causes of non-traumatic injury to the nervous system. When faced with a neurological condition as the cause of death, a successful postmortem examination requires consideration of all the information available combined with detailed examination of the nervous system.

REFERENCES

Adams, J.H. and Milled, D. 1973. Herpes simplex encephalitis: a clinical and pathological analysis of twenty-two cases. *Postgrad Med J*, **49**, 393–7.

Allen, H. and Burns, A. 1998. Current pharmacologic treatments for dementia. In Growdon, J.H. and Rossor, M.N. (eds), *Blue Books of Practical Neurology*, Vol 19. *The Dementias*. Boston: Butterworth-Heinemann, 335–58.

Ameri, A. and Bousser, M.G. 1992. Cerebral venous thrombosis. *Neurol Clin*, **10**, 87–111.

Auer, R.N. and Siesjö, B.K. 1988. Biological differences between ischemia, hypoglycemia, and epilepsy. *Ann Neurol*, **24**, 699–707.

Auer, R.N. and Sutherland, G.R. 2002. Hypoxia and related conditions. In Graham, D.I. and Lantos, P.L. (eds), *Greenfield's Neuropathology*, 7th edn, Vol 1. London: Arnold, 234–80.

Bell, J.E. 1998. The neuropathology of adult HIV infection. *Rev Neurol (Paris)*, **154**, 816–29.

Bell, J.E., Ironside, J.W., Donaldson, Y. and Simmonds, P. 1995. HIV and the brain: Contrasting patterns and viral load in homosexuals and drug abusers. *Neuropath Appl Neurobiol*, **21**, 150

Black, M. and Graham, D.I. 2001. Sudden unexplained death in adults. In: Love, S. (ed.), *Current Topics in Pathology 95*, *Neuropathology* Berlin: Springer, 125–48.

Black, M. and Graham, D.I. 2002. Sudden unexplained death in adults caused by intracranial pathology. *J Clin Pathol*, **55**, 44–50.

Budka, H., Wiley, C.A., Kleihues, P. *et al.* 1991. HIV-associated disease of the nervous system: review of nomenclature and proposal for neuropathology-based terminology. *Brain Pathol*, **1**, 143–52.

Byard, R.W. and Moore, L. 1993. Sudden and unexpected death in childhood due to a colloid cyst of the third ventricle. *J Forensic Sci*, **38**, 210–3.

Camfield, C.S., Camfield, P.R. and Veugelers, P.J. 2002. Death in children with epilepsy: a population-based study. *Lancet*, **359**, 1891–5.

Davis, L.E. and Kennedy, P.G.E. (eds) 2000. *Infectious Diseases of the Nervous System*. Oxford: Butterworth-Heinemann.

Day, N. and Turner, G. 2000. Cerebral malaria. In: Davis, L.E. and Kennedy, P.G.E. (eds). *Infectious Diseases of the Nervous System*. Oxford: Butterworth-Heinemann, 1–14.

Delattre, J.Y., Krol, G., Thaler, H.T. and Posner, J.B. 1988. Distribution of brain metastases. *Arch Neurol*, **45**, 741–4.

DeVries, E. 1960. *Postvaccinal Perivenous Encephalitis*. Amsterdam: Elsevier.

Esiri, M.M. 2000. Which vascular lesions are of importance in vascular dementia? *Ann NY Acad Sci*, **903**, 239–43.

Esiri, M., Hyman, B.T., Beyreuther, K. and Masters, C.L. 1997. Ageing and dementia. In: Graham, D.I., Lantos, P.L. (eds), *Greenfield's Neuropathology*, 7th edn, Vol 1. London: Arnold, 153–233.

Filkins, J.A., Cohle, S., Levy, B.K. and Graham, M. 1996. Unexpected deaths due to colloid cysts of the third ventricle. *J Forensic Sci*, **41**, 521–3.

Gosztonyi, G. 1978. Acute haemorrhagic leucoencephalitis (Hurst's disease). In Vinken, P.J., Bruyn, G.W. and

Klawans, H.L. (eds), *Handbook of Clinical Neurology*, Vol 34. *Infections of the Nervous System*. Amsterdam: North-Holland, 587–604.

Gray, F. and Alonso, J-M. 2002. Bacterial infections of the nervous system. In Graham, D.I. and Lantos, P.L. (eds), *Greenfield's Neuropathology*, 7th edn, Vol 2. London: Arnold, 151–94.

Gruber, A., Bavinzski, G., Killer, M. and Richling, B. 2000. Preoperative embolization of hypervascular skull base tumors. *Minim Invasive Neurosurg*, **43**, 62–71.

Haldeman, S., Kohlbeck, F.J. and McGregor, M. 2002. Stroke, cerebral artery dissection, and cervical spine manipulation therapy. *J Neurol*, **249**, 1098–104.

Hamano, T., Miyoshi, Y., Hirayama, M., Hiraki, S., Mutoh, T. and Kuriyama, M. 2000. Posterior thoracic spinal cord infarction: complication of thoracoabdominal aortic aneurysm. *Eur Neurol*, **44**, 59–60.

Hart, M.N. and Earle, K.M. 1975. Haemorrhagic and perivenous encephalitis: a clinical-pathological review of 38 cases. *J Neurol Neurosurg Psychiatry*, **38**, 585–91.

Hirsch, C.S. and Martin, D.L. 1971. Unexpected death in young epileptics. *Neurology*, **21**, 682–90.

Hohlrieder, M., Spiegel, M., Hinterhoelzl, J. *et al.* 2002. Cerebral vasospasm and ischaemic infarction in clipped and coiled intracranial aneurysm patients. *Eur J Neurol*, **9**, 389–99.

Honavar, M. and Meldrum, B.S. 2002. Epilepsy. In Graham, D.I. and Lantos, P.L. (eds), *Greenfield's Neuropathology*, 7th edn, Vol 1. London: Arnold, 899–942.

Ironside, J.W., Moss, T.H., Louis, D.N., Lowe, J.S. and Weller, R.O. (eds) 2002. *Diagnostic Pathology of Nervous System Tumours*. London: Churchill-Livingstone.

Kalimo, H., Kaste, M. and Haltia, M. 2002. Vascular diseases. In: Graham, D.I. and Lantos, P.L. (eds), *Greenfield's Neuropathology*, 7th edn, Vol 1. London: Arnold, 281–355.

Kamenar, E. and Burger, P.C. 1980. Cerebral fat embolism: a neuropathological study of a microembolic state. *Stroke*, **11**, 477–84.

Kenneback, G., Ericson, M., Tomson, T. and Bergfeldt, L. 1997. Changes in arrhythmia profile and heart rate variability during abrupt withdrawal of antiepileptic drugs. Implications for sudden death. *Seizure*, **6**, 369–75.

Kleihues, P. and Cavenee, W.K. (eds) 2001. *Pathology and Genetics of Tumours of the Nervous System*. Lyon: IARC Press.

Krauss, W.E. and McCormick, P.C. 1992. Infections of the dural spaces. *Neurosurg Clin N Am*, **3**, 421–33.

Lampl, C. and Yazdi, K. 2002. Central pontine myelinolysis. *Eur Neurol*, **47**, 3–10.

Langan, Y., Nashef, L. and Sander, J.W. 2002. Certification of deaths attributable to epilepsy. *J Neurol Neurosurg Psychiatry*, **73**, 751–2.

Lee, T., Kenny, B.G., Hitchock, E.R. *et al.* 1991. Supratentorial masses: stereotactic or freehand biopsy? *Br J Neurosurg*, **5**, 331–8.

Love, S. 2001. Autopsy approach to infections of the CNS. In Love, S. (ed.), *Current Topics in Pathology 95, Neuropathology*. Berlin: Springer, 1–50.

Love, S. and Wiley, C.A. 2002. Viral diseases In Graham, D.I. and Lantos, P.L. (eds) *Greenfield's Neuropathology*, 7th edn, Vol 2. London: Arnold, 1–106.

Lowe, J.S. 1998. Establishing a pathological diagnosis in degenerative dementias. *Brain Pathol*, **8**, 403–6.

Lowenstein, D.H. and Alldredge, B.K. 1998. Status epilepticus. *N Engl J Med*, **338**, 970–6.

McCarron, M.O., Nicoll, J.A., Ironside, J.W., Love, S., Alberts, M.J. and Bone, I. 1999. Cerebral amyloid angiopathy-related hemorrhage. Interaction of APOE epsilon2 with putative clinical risk factors. *Stroke*, **30**, 1643–6.

McCarthy, R.E., Lonstein, J.E., Mertz, J.D. and Kuslich, S.D. 1990. Air embolism in spinal surgery. *J Spinal Disord*, **3**, 1–5.

MacLennan, D.H. and Loke, J.C.P. 2002. Malignant hyperthermia and central core disease associated with defects in Ca²⁺ channels of the sarcotubular system. In Karpati G (ed.), *Structural and Molecular Basis of Skeletal Muscle Disorders*. Basel: ISN Neuropath Press, 99–102.

Miyamoto, O. and Auer, R.N. 2000. Hypoxia, hyperoxia, ischemia, and brain necrosis. *Neurology*, **54**, 362–71.

Nashef, L. 1997. Sudden unexpected death in epilepsy: terminology and definitions. *Epilepsia* **38**(suppl 11): S6–8.

Navia, B.A., Cho, E.S., Petito, C.K. and Price, R.W. 1986. The AIDS dementia complex: I.I. Neuropathology. *Ann Neurol*, **19**, 525–35.

Palmer, A.C., Calder, I.M. and Hughes, J.T. 1987. Spinal cord degeneration in divers. *Lancet*, **ii**, 1365–6.

Pham Dang, C., Pereon, Y., Champin, P., Delecrin, J. and Passuti, N. 2002. Paradoxical air embolism from patent foramen ovale in scoliosis surgery. *Spine*, **27**, E291–5.

Poon, T.P. and Solis, O.G. 1985. Sudden death due to massive intraventricular hemorrhage into an unsuspected ependymoma. *Surg Neurol*, **24**, 63–6.

Posner, J.B. 1995. *Neurologic Complications of Cancer*. Philadelphia: FA Davis.

Pulsinelli, W.A. 1997. Selective neuronal vulnerability and infarction in cerebrovascular disease. In Welch, K.M.A., Caplan, L.R., Reis, D.J., Siesjö, B.K. and Weir, B. (eds), *Primer on Cerebrovascular Disease*. San Diego: Academic Press, 104–7.

Scaravilli, F. 1998. *Neuropathology of Epilepsy*. Singapore: World Scientific Publishing.

Siesjö, B.K. 1992. Pathophysiology and treatment of focal cerebral ischemia. Part II: Mechanisms of damage and treatment. *J Neurosurg*, **77**, 337–54.

Stevinson, C., Honan, W., Cooke, B. and Ernst, E. 2001. Neurological complications of cervical spine manipulation. *J R Soc Med*, **94**, 107–10.

Strittmatter, C., Lang, W., Wiestler, O.D. and Kleihues, P. 1992. The changing pattern of human immunodeficiency virus-associated cerebral toxoplasmosis: a study of 46 postmortem cases. *Acta Neuropathol*, 83, 475–81.

Toro-Gonzalez, G., Navarro-Roman, L., Roman, G.C. *et al.* 1993. Acute ischemic stroke from fibrocartilaginous embolism to the middle cerebral artery. *Stroke*, **24**, 738–40.

Vinters, H.V. and Gilbert, J.J. 1983. Cerebral amyloid angiopathy: incidence and complications in the aging brain. I.I. The distribution of amyloid vascular changes. *Stroke*, **14**, 924–8.

Vinters, H.V., Ellis, W.G., Zarow, C. *et al.* 2000. Neuropathologic substrates of ischemic vascular dementia. *J Neuropathol Exp Neurol*, **59**, 931–45.

Walsh, T.J., Hier, D.B. and Caplan, L.R. 1985. Aspergillosis of the central nervous system: clinicopathological analysis of 17 patients. *Ann Neurol*, **18**, 574–82.

Wispelwey, B., Dacey, R.G.J. and Scheld, W.M. 1991. Brain abscesses. In Scheld, W.M., Whitley, R.J. and Durack, D.T. (eds), *Infections of the Central Nervous System*. New York: Raven Press, 457–86.

14

Alcohol, drugs and toxins

Colin Smith and Alexander R W Forrest

Alcohol and recreational drugs are widely used in western society and with their increased prevalence comes a greater understanding of the potential pathological consequences related to their use. In addition, the nervous system may be damaged by compounds with neurotoxic effects. This chapter covers the neurotoxic effects of alcohol, recreational drugs and neurotoxins, and the potential implications relating to forensic practice.

ALCOHOL

Alcohol produces neurological dysfunction both directly and secondary to vitamin deficiencies and metabolic dysfunction.

Thiamine deficiency

The neuropathological features associated with thiamine deficiency are most commonly seen in people with alcohol problems but may be seen in other rare causes of extreme malnutrition (Ihara *et al.* 1999). Thiamine deficiency underlies Wernicke–Korsakoff syndrome and cerebellar degeneration. Clinically, Wernicke's encephalopathy typically presents with ophthalmoplegia, ataxia and alterations in mental status, which may progress to coma and death in acute cases. Chronic cases may progress to Korsakoff's psychosis, a rare irreversible psychiatric condition seen in chronic alcoholism. Cerebellar degeneration typically results in an ataxic gait and clumsiness, particularly of the lower limbs. It must be remembered that chronic alcoholism is frequently associated with recurrent falls caused by intoxication and that the clinical symptoms associated with Wernicke's syndrome and chronic cerebellar degeneration are likely to compound the problem. Evidence of

both recent and old head injury is common in chronic alcoholism.

Wernicke's encephalopathy

MACROSCOPIC EXAMINATION

It is important to remember that microscopic examination is mandatory because the brain may appear macroscopically normal in up to 25 per cent of cases (Harper 1983). Areas of haemorrhagic necrosis are seen in the mamillary bodies (Figure 14.1) and in the hypothalamic region surrounding the third ventricle. Similar lesions may be seen in the periaqueductal region of the brain stem. In long-standing cases, often with recurrent episodes, the typical pathology is that of shrunken mamillary bodies with brown discoloration (Figure 14.2).

MICROSCOPIC EXAMINATION

In the acute phase the principal pathology is that of perivascular haemorrhages in the regions described above, often

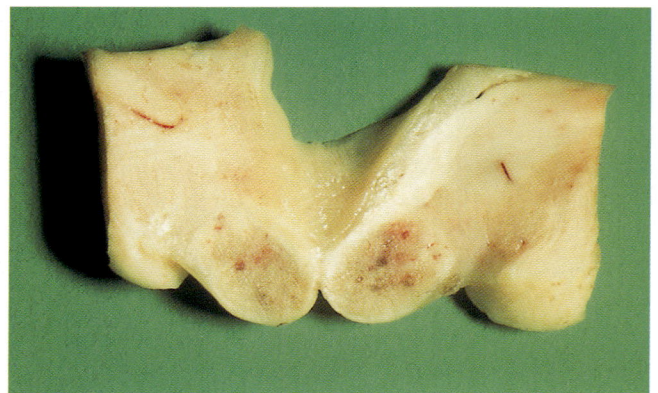

Figure 14.1   Acute Wernicke's encephalopathy: haemorrhages are seen within the mamillary bodies.

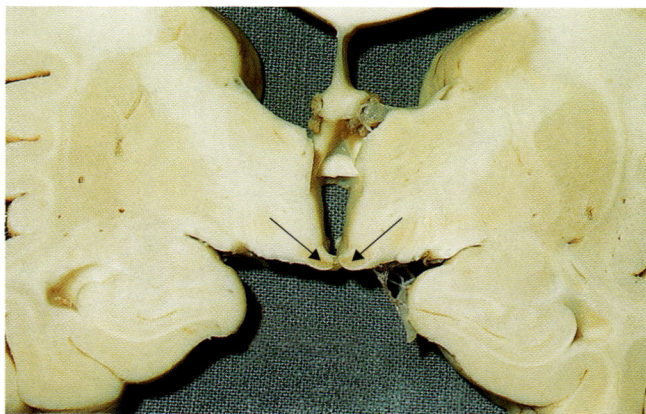

Figure 14.2 Chronic Wernicke's encephalopathy: the mamillary bodies are shrunken and show brown discoloration (arrows).

Figure 14.3 Histological appearances of Wernicke's encephalopathy (a) hematoxylin and eosin, (b) reticulin.

associated with microvascular proliferation (Figure 14.3). Unlike ischaemic necrosis the neuronal population is relatively well preserved in the mamillary bodies. In subacute and chronic cases gliosis is best demonstrated by immunohistochemistry for glial fibrillary acidic protein (GFAP). The increased density of capillaries can be demonstrated by a reticulin stain and Perls' stain is useful for highlighting haemosiderin indicative of previous haemorrhage. Necrotic lesions are frequently seen in the thalamus, although detailed examination of thalamic nuclei is outwith the normal neuropathological examination in forensic cases.

Alcohol and atrophy

Neuroimaging and neuropathological studies have shown that alcoholics have atrophic brains in comparison with controls (Fadda and Rossetti 1998, Harper *et al.* 2003). Atrophy in alcoholics can be accounted for predominantly by white matter loss. Pre-frontal white matter is the most significantly reduced. In uncomplicated alcoholics brain weights have been shown to be reduced to 94 per cent of the control weight. Alcoholics with Wernicke/Korsakoff syndrome had brain weights of 91 per cent of the control weight. An association between atrophy and the rate and amount of alcohol consumed over a lifetime has been demonstrated. There is some evidence that atrophic change may revert to the same rate as control populations following abstinence (Pfefferbaum *et al.* 1998). Studies examining the role of simultaneous alcohol and cocaine abuse have proved inconclusive as to whether atrophy is increased by simultaneous cocaine use and further study in this area is needed (Bjork *et al.* 2003). Neuropathologists may be asked to comment on the role of atrophy in alcoholics and the development of subdural haematoma formation.

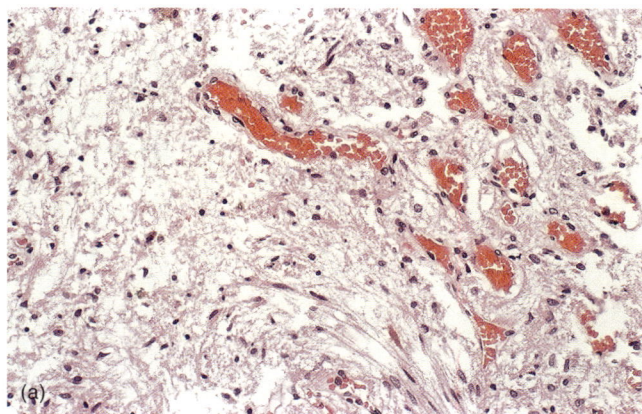

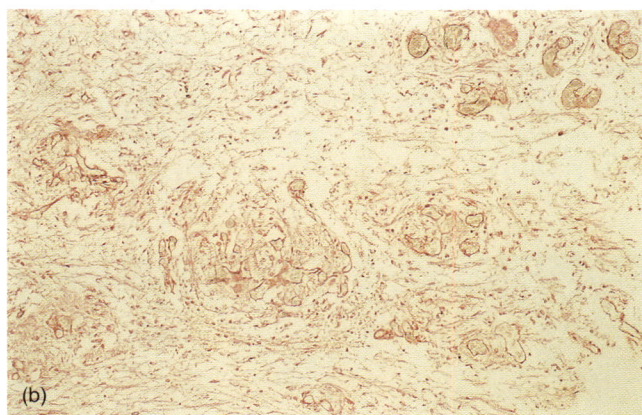

Cerebellar degeneration

MACROSCOPIC
Cerebellar atrophy, predominantly in the superior vermis, is a frequent finding in chronic alcoholism (Figure 14.4).

MICROSCOPIC
Microscopically there is obvious loss of Purkinje cells with associated Bergmann gliosis in the superior vermal region. Similar changes have been described in malnutrition without alcohol (Adams 1976).

Other metabolic disorders related to alcohol

CENTRAL PONTINE MYELINOLYSIS
This disorder is seen in individuals with a variety of underlying disorders (Lampl and Yazdi 2002), although it is most commonly seen in people with chronic alcoholism. The disorder is thought to result from electrolyte

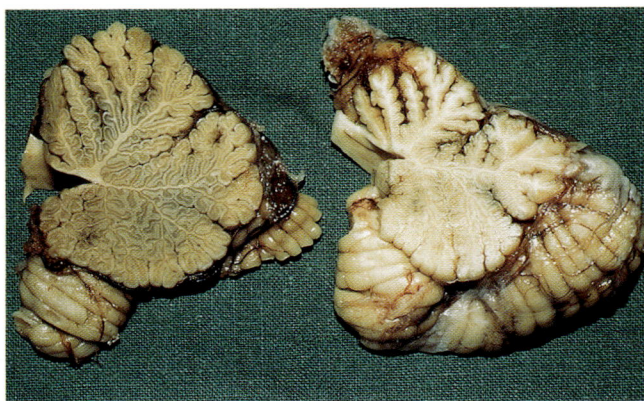

Figure 14.4 Marked superior vermal atrophy secondary to chronic alcoholism. The vermal section on the left is from an age-matched control and shows only very minor superior vermal atrophy. The vermal section on the right, from someone who is alcoholic, by contrast shows marked atrophy. Clinically this presented as an ataxic gait.

Figure 14.5 Sections through the pons from a case of central pontine myelinolysis. The area of myelin loss is central within the basis pontis and is seen as a shrunken grey area.

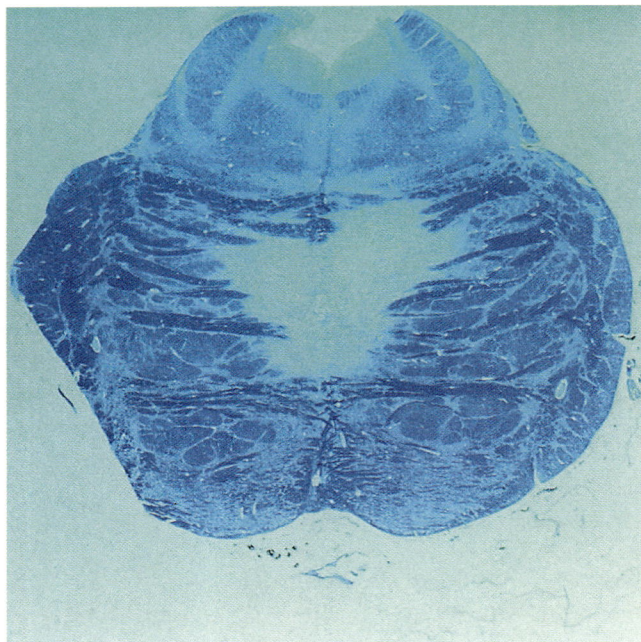

Figure 14.6 Histological section demonstrating the central area of myelin loss (area of pallor) (luxol fast blue/cresyl violet) in central pontine myelinolysis.

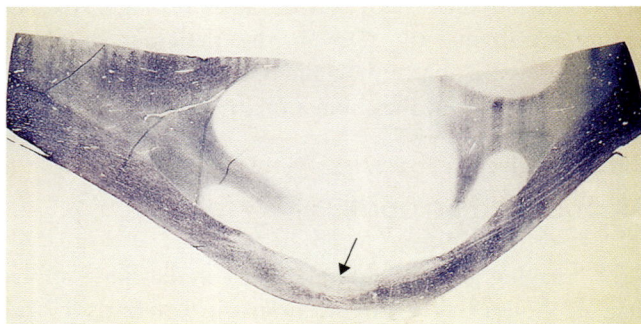

Figure 14.7 Histological section from a case of Marchiafava–Bignami disease. The corpus callosum is reduced in size and there is a central area of myelin pallor (arrow) (luxol fast blue/cresyl violet).

abnormalities, with profound hyponatraemia being seen in most cases. Macroscopically a pale region is identified in the centre of the basis pontis (Figure 14.5), and this corresponds to a region of pallor when examined microscopically with the luxol fast blue (LFB)/cresyl violet (CV) stain (Figure 14.6). As with any demyelinating disorder β-amyloid precursor protein (APP) immunoreactivity may be seen in relation to the area of pontine demyelination (Medana and Esiri 2003). It is now recognized that extrapontine lesions are seen in approximately 10 per cent of cases (Wright *et al.* 1979), particularly in relation to the thalamus and striatum.

MARCHIAFAVA–BIGNAMI DISEASE

This is a very rare condition particularly seen in people with alcoholism from Mediterranean regions. Clinically the acute form is associated with seizures and loss of consciousness and usually results in death. The chronic form is unlikely to come to the pathologist's attention. The lesion is characterized by an area of demyelination in the corpus callosum, particularly anteriorly in the corpus callosum. Microscopically the lesions are highlighted by a myelin stain such as LFB, and macrophages are abundant (Figure 14.7).

METHANOL

Methanol is found in a number of common products (solvents, toiletries) that can be drunk to produce intoxication and can be produced in the home brewing of spirits.

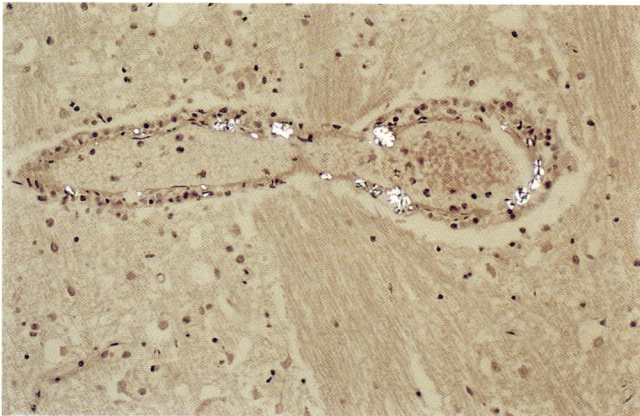

Figure 14.8 Oxalate crystals in a case of ethylene glycol poisoning.

Toxicity in the acute phase can lead to blindness and cardiorespiratory collapse. At postmortem examination the brain is swollen and petechial haemorrhages may be seen. The haemorrhages are particularly pronounced in relation to the third and fourth ventricles and can also be subarachnoid. If the patient has survived for a period of time before death, bilateral necrosis of the putamen may be seen and there is necrosis of the cerebellar cortex (McLean *et al.* 1980). In the eyes there is loss of retinal ganglion cells and optic nerve axons (Naeser 1988).

ETHYLENE GLYCOL

Ethylene glycol is considered under alcohol because it is usually consumed in anti-freeze by people who are alcoholics. At postmortem examination there is non-specific cerebral oedema and petechial haemorrhages. Oxalate crystals are seen in the brain and meninges in vessels and in perivascular spaces (Figure 14.8).

RECREATIONAL DRUGS

Opiates

Opiates are commonly abused. Heroin (diamorphine) is widely misused, but other opiates and drugs with opiate action including methadone, pethidine, morphine, oxycodone and dihydrocodeine are also misused. Opiate abuse can result in a wide range of lesions within the nervous system. Some of these lesions may be directly caused by opiates whereas many others are the result of the impurities mixed with the drug. Most cases that come to the pathologist's attention involve intravenous administration of heroin. Many impurities are introduced into the bloodstream and can travel to the brain or spinal cord. Non-sterile

injection techniques can introduce infection that, via haematogenous spread, can lead to cerebral abscesses or meningitis. Human immunodeficiency virus (HIV) infection may result in central nervous system pathology (see Chapter 13, p. 159). Cerebral infarcts may be seen as a consequence of emboli of injected materials (Karch 2002).

The most common pathology seen in relation to a fatal opiate overdose is, however, that of ischaemic damage secondary to hypotension, as described in Chapter 13, p. 155. The brain is swollen and lesions are seen at the parieto-occipital watershed region and often within the globus pallidus (Andersen and Skullerud 1999, Büttner *et al.* 2000). Hypoxic brain damage may occur in heroin addicts who are resuscitated following the administration of reversal agents, such as naloxone, where cardio-respiratory arrest has been prolonged.

In 1982, spongiform leukoencephalopathy was reported amongst heroin smokers in The Netherlands (Wolters *et al.* 1982), and subsequently cases have been reported from other countries. The brains are oedematous and on microscopy there is damage to the white matter with vacuolation. There is an attenuation of myelin around cavities formed by colaescence of the vacuoles. Oligodendria are reduced. No inflammatory cells are present.

1-methyl-4-phenyl-1,2,3,6-tetrahydropyridine (MPTP) is being produced by clandestine laboratories atempting to produce 1-methyl-4-phenyl-4-propionpiperidine (MPPP), which is a meperidine analogue. Meperidine is a synthetic opiate. MPTP is converted by astrocytes into a toxic compound. When taken in sufficient quantities MPTP has been associated with parkinsonism (Davis *et al.* 1979, Langston *et al.* 1983). Degenerative changes can be seen in the substantia nigra without Lewy body formation. Cases have been reported in intravenous drug taking populations using what they believed was synthetic heroin and also in industrial chemists.

Cocaine

Cocaine abuse is common and cocaine use is now recognized as a common cause of cerebro–vascular pathology particularly in younger adults. Cerebral infarction is now well recognised as a complication of cocaine use. Many cases follow within three hours of cocaine use. In comparison with non-drug related cerebral infarction, where individuals are characteristically over 65 years of age, most individuals are in their mid-thirties (Karch 2002). Although vasculitis has been reported as a cause of cerebrovascular disease in cocaine users it appears to be a rare cause. Using MRI, cocaine has been shown to cause vasoconstriction. Ischaemic infarcts may be the result of vasoconstriction, increased platelet aggregation and activation (Sloan 1997).

Subarachnoid haemorrhage is common in cocaine users. The majority are associated with berry aneurysms. Intracerebral haemorrhage occurs most commonly in the basal ganglia and thalamus, but may also occur in the cerebellum and sub-corticol white matter. There is an association between hypertensive heart disease and intracerebral haemorrhage: heart weights in cocaine users dying of intracerebral haemorrhage being heavier than those with ruptured aneurysms. There is also a high incidence of underlying vascular anomalies and malformations (Büttner *et al.* 2003). A study by Fessler and colleagues (1997) revealed that of 33 patients with neurological symptoms following cocaine abuse nearly half (16 of 33) had subarachnoid haemorrhage, 6 had intracerebral haemorrhage and 7 had cerebral infarction.

Cocaine abuse during pregnancy has been associated with fetal vascular lesions (Kinney and Armstrong 2002). The actual incidence of vascular lesions varies between different published series, but there does appear to be an increased risk of fetal brain haemorrhages and cerebral infarcts.

Cocaine use has been associated with movement disorders including choreoathetosis, akathisia and parkinsonism and may cause grand mal seizures. Cocaine is associated with excited delerium and changes in neurotransmitters have been reported in these deaths. The brains of cocaine users have been shown to have altered dopamine receptor status. Consideration should be given to the retention of fresh frozen brain in cases of postulated excited delerium. Discussion with an appropriate laboratory may be necessary before commencement of the postmortem examination (see Chapter 2, p. 27).

Amphetamines

The amphetamines are a group of manufactured stimulants including amphetamine, methamphetamine, and the amphetamines derivatives (including 3,4-methylenedioxymethylamphetamine (MDMA), and related drugs). Amphetamine and methamphetamine are easy to synthesise illicitly and are widely abused as stimulants. A variety of amphetamine derivatives have been synthesised and many have nonstimulant mind altering properties that make them desirable as drugs of misuse. At present methamphetamine use is not common in the UK, though it is popular in the USA, particularly on the west coast. In the UK amphetamine and MDMA use are more popular. Like cocaine, amphetamines are particularly associated with effects on the cardiovascular and cerebrovascular

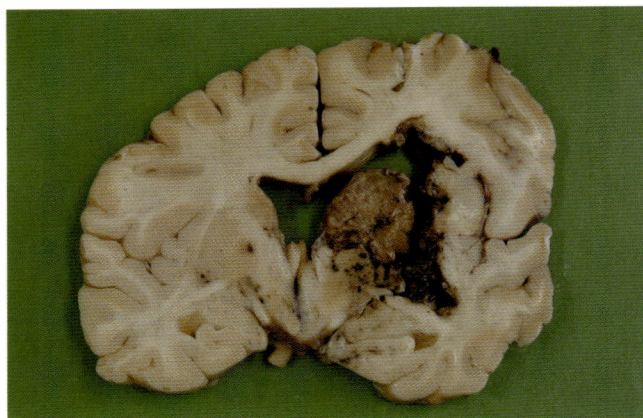

Figure 14.9   Intracerebral haemorrhage in a 33-year-old amphetamine user.

systems. Haemorrhagic and ischaemic strokes are associated with amphetamine use. Intracerebral haemorrhage is more common in the frontal lobes, and may also be seen in the basal ganglia (Karch 2002) (Figure 14.9). In the UK the most common amphetamine derivative used recreationally is MDMA, which goes under the street name of 'ecstasy'. It is important to note that, whilst at the time of writing a tablet described by the seller as 'ecstasy' is likely to contain MDMA, it is by no means certain that that will be the case. MDMA has been associated with deaths from heatstroke and water intoxication. The neuropathological features identified at postmortem examination in these cases include severe cerebral oedema in water intoxication, and microscopic perivascular haemorrhages in cases of heatstroke (Milroy *et al.* 1996). In addition, intracerebral haemorrhages have been described on imaging and in one case there was a fatal outcome in an individual with an underlying arteriovenous malformation (Harries and De Silva 1992).

NEUROTOXINS

Neurotoxins are discussed under the following headings: gases, heavy metals and organic solvents.

Gases

CARBON MONOXIDE

Carbon monoxide (CO) poisoning is still seen as a frequent form of suicide and can still be seen in relation to poorly maintained gas heaters. Non-combustion sources of carbon monoxide may be encountered and include the consumption of dichloromethane. At postmortem examination the brain is swollen and on sectioning has a characteristic pink-red colour caused by the accumulation of carboxyhaemoglobin. Where there has been survival for several days, infarcts are seen within the globus pallidus and substantia nigra. Histologically, neuronal loss is common in the hippocampus and Purkinje cell layer of the cerebellar cortex (Auer and Sutherland 2002). In long-term survivors a myelinopathy may develop (Grinkler's myelinopathy), the damage being seen throughout the cerebrum, although being most severe posteriorly and deeply (Neubuerger and Clarke 1945).

OTHER GASES

Carbon disulphide (CS$_2$) and methyl bromide have been documented to cause fatal intoxications in industrial settings (Osteowska 1971). CS$_2$ has been used in textile and rubber industries whereas methyl bromide was a component of fire extinguishers. Both can result in sudden loss of consciousness and coma. Postmortem studies have demonstrated neuronal loss and gliosis. Methyl bromide intoxication has been associated with petechial haemorrhages and cerebral oedema in the acute stages, and in one case with 30-day survival the lesions resembled Wernicke's encephalopathy.

Heavy metals

ARSENIC

Arsenic has been a classic poison in homicides, with such poisoners as Armstrong and the Seddons being notable British cases. Arsenic is available in a trivalent form and an organic pentavalent form. In acute deaths the findings may be minimal. Where poisoning is more chronic, symptoms relating to the gastrointestinal tract may result in a mistaken diagnosis of gastroenteritis. A peripheral neuropathy may develop with Wallerian degeneration seen in the distal nerves. Whilst homicidal arsenic poisoning is rare, it still occurs and a recent case seen by the Sheffield department involved the mistaken diagnosis of gastroeneritis. A peripheral neuropathy was ascribed to diabetes mellitus by the clinicians. Postmortem toxicology confirmed arsenic poisoning, with hair analysis confirming the repeated administration of arsenic.

ORGANIC ARSENIC

Organic arsenic has been used in the treatment of trypanosomiasis and can be associated with fatal intoxication. Acute haemorrhagic leukoencephalitis has been described with perivascular demyelination and 'ball and ring'-type haemorrhages particularly in the pons and midbrain (Adams et al. 1986).

MERCURY

Mercury intoxication has been described in two situations: after industrial pollution of the environment resulting in the element being introduced into the human food chain, and the use of mercury as a fungicide. Intoxication after industrial pollution was first described in Minimata in Japan in the mid-1950s (Shiraki 1979), and intoxication with fungicide was the cause of many deaths in Iraq in 1971 (Bakir et al. 1973). The neuropathological features associated with mercury intoxication include severe atrophy of the calcarine cortex and cerebellar cortex, with less severe cortical atrophy elsewhere in the cerebrum. Within the cerebellum the atrophy is most pronounced in the granule cell layer with preservation of the Purkinje cells. Axonal torpedoes can be seen in Purkinje cells and spiny dendritic processes have been described (Jacobs and LeQuesne 1992).

THALLIUM

Thallium is used in the glass and semiconductor industry and, rarely in the developed world, as a pesticide and has been described in accidental, suicidal and homicidal poisoning. The clinical presentation is of unexplained gastrointestinal disturbance followed by a peripheral neuropathy. Unexpected hair loss may occur. Motor weakness and cranial neuropathies can develop. Uniquely, thallium has been identified in cremated remains following homicidal poisoning, and may also be identified by radiology at postmortem, as it accumulates in the liver and is radiopaque. The neuropathological features are rather non-specific and include cerebral oedema with petechial haemorrhages. Chromatolysis of motor neurons is seen and sampling of the distal peripheral nervous system may show loss of myelinated axons (Cavanagh et al. 1974).

OTHER HEAVY METALS

Other heavy metals have been described to cause an encephalopathy in the past but are now no longer seen. These include aluminium (dialysis encephalopathy) and lead (used in paint and associated with childhood intoxication). Others are seen in only very specific situations (manganese intoxications have been described in manganese miners).

SAMPLE COLLECTION FOR TOXICOLOGICAL ANALYSES AT POSTMORTEM EXAMINATION

Interpreting the results of toxicological analyses of postmortem samples is difficult under the best of circumstances;

when the sample collection and documentation are sub-optimal, it can at best be well high impossible and at worst productive of positively misleading opinions.

The neuropathologist may not have a great deal of choice as to where samples are submitted for postmortem toxicological analysis. The Judicial Authority, on whose authority many such postmortem examinations are performed, may have fairly fixed views on the matter. Where there is discretion, the best laboratory may not be the local clinical laboratory or public analyst. Postmortem toxicology, in terms of both analytical techniques and interpretation, is different from both clinical toxicology and the determination of water in coal or the proportion of fat in sausages. Useful touchstones to determine whether or not a laboratory actually has some expertise and professional interest in postmortem toxicology are to establish whether or not they participate in a proficiency testing scheme appropriate for forensic toxicology and, in the UK, has staff registered with the Council for the Registration of Forensic Practitioners. In Europe, good examples of appropriate schemes are the Quartz scheme managed by the Laboratory of the Government Chemist in London (www.lgc.co.uk/pts_schemes_children.asp?id= 53_1_3_5) and the scheme provided by the Norwegian National Institute for Forensic Toxicology (Statens Rettstoksikologiske Institutt – www.nifu.no/instkat/enginst/ institutes/sri.html).

The next stage is to establish whether there are any samples available that were collected in life. These will be more straightforward both to analyse and in the interpretation of the results produced. Unfortunately, when a patient dies in the emergency admissions unit only point of care tests such as blood gases, electrolytes and glucose may be done between the time of admission and the time of death. The samples for such analyses are often destroyed in the process of analysis or disposed of immediately afterwards. However, it is well worth looking for samples that have been sent to the laboratory; although these too can be ephemeral, samples sent for immunology or cross-matching can be kept for weeks rather than days. They are always worth looking for in appropriate circumstances, particularly if they were collected before, e.g. intensive therapy unit (ITU) admission.

The collection of samples *post mortem* comes next. The blood samples should be collected before evisceration; the removal of viscera may move blood from regions where the concentration of particular drugs is high to areas where it is low, confounding the interpretation of the analytical results. Ideally, the blood should be collected from the femoral vein distal to a clamp or ligature and without milking the calf or thigh. The site of collection and the method of collection should always be specified on the

toxicology request form. Normally, 10–15 ml of blood in a sterile universal container with about 2–5 ml in fluoride oxalate is sufficient, although if a second postmortem examination is likely more should be collected. The main purpose of the fluoride oxalate sample is to preserve the sample for alcohol analysis (Blackmore 1968). Although a tube intended for clinical blood glucose analyses, e.g. a grey stoppered Vacutainer, is adequate for an aseptically collected blood sample collected from a living individual for a clinical blood ethanol determination, such tubes do not contain enough fluoride to inhibit microbial overgrowth, with consequent alcohol production, in postmortem samples. Fluoride added to a concentration of 1.5–2 per cent coupled with prompt storage of the sample at 4°C will usually prevent microbial overgrowth. The 'RTA' ('road traffic accident') vials used for forensic blood alcohol sample collection from the living in UK practice contain sufficient fluoride to preserve postmortem samples for alcohol analysis. They are not ideal for use in the postmortem room, or indeed anywhere else, because the rubber septum seal poses a very real risk of needle-stick injury to the person filling the vial.

The collection of a fluoride-preserved sample is also important in that it will help to prevent the breakdown of a number of drugs after collection, including cocaine, clonazepam and nitrazepam (Robertson and Drummer 1998). Of course, it will do nothing to prevent the breakdown of these drugs between death and sample collection. Where, for example, public holidays mean that there is a delay of several days between death and postmortem examination, there is a case for the collection of a blood sample by external puncture before the postmortem examination. The time and site of collection should be documented.

When the body is showing significant external signs of postmortem change, it may assist the interpretation of alcohol results if samples are taken from several sites in the body. In the presence of postmortem alcohol production, blood alcohol concentrations will usually vary significantly between different collection sites.

Where available urine should always be collected: 10–20 ml in a universal container and about 5 ml preserved with fluoride are ideal. Drug screening is most easily done on urine; however, when death occurs immediately after an intravenous drug overdose the relevant drug may not be present in detectable concentrations in urine. Always let the laboratory know the circumstances of the death insofar as they are known at the time of the postmortem examination.

Ideally the total stomach contents should be sent to the laboratory; if this is not possible a portion should be submitted together with a note of the total volume. When dealing with drugs, such as the tricyclic antidepressants, that show significant postmortem redistribution, the presence of a large amount in the stomach content may be the best evidence of an overdose. If an overdose of a modified release formulation of a drug is suspected, it is often worthwhile taking samples of small bowel content.

In most cases samples of blood, urine and stomach content are often all that is required for adequate investigation of a death; in special circumstances other samples may be needed. One interesting source of material for toxicological analysis is intracranial haemorrhage. As well as testing for carboxyhaemoglobin in fire deaths, other drugs may be measured. This can be particularly useful where a period of time has elapsed between collapse and death, where the drug may have cleared from peripheral blood and urine (Hirsch and Adelson 1973, McIntyre *et al.* 2003). Also, in most circumstances the laboratory will need more information than just the names of the deceased and the pathologist. In general, at least as much information as is available to the pathologist should be sent to the laboratory with the samples. All the samples should be labelled with the patient's name, the date and time of collection, and the nature and origin of the sample.

SPECIAL CASES AND SPECIAL SAMPLES

Sudden unexpected death in epilepsy

Many antiepileptic drugs are not detected on routine toxicological screening; they have to be specifically looked for. If the laboratory is not on notice to look for them they will not be detected. At postmortem examination the concentrations of antiepileptic drugs are often somewhat lower than the levels found in the same patient in life; this needs to be borne in mind when interpreting such results. It has been suggested that prolactin may be raised in postmortem blood samples after sudden unexpected death in epilepsy (SUDEP). The test is useless (Opeskin *et al.* 2000).

Metabolic deaths

Vitreous humour and cerebrospinal fluid (CSF) may both be usefully analysed for urea and creatinine. More reproducible results may be obtained using standard clinical chemistry instrumentation if the sample is ultrafiltered using a Centricon device to remove molecules with a molecular weight of more than 3000 Da. In both CSF and vitreous humour the creatinine concentration, when measured by standard clinical methods, but not when measured by highly specific methods, will tend to be lower than in

plasma collected from the same patient shortly before death. The low protein concentration in these fluids means that the chloride concentration is higher than in plasma as a result of Donnan equilibrium effects. Many ion-sensitive electrodes, now commonly used for the measurement of electrolytes in biofluids, are sensitive to products of putrefaction. These can make the measurement of potassium, in particular, unreliable in postmortem samples.

The blood glucose concentration falls rapidly after death; in general, if the vitreous humour glucose concentration is greater than about 1 mmol/l, it is likely that it was elevated during life except for circumstances such as rapid chilling of the body after death.

Often the first pointer to unsuspected diabetic ketoacidosis will be a large acetone peak in the gas chromatograph tracing during an assay of blood or urine for ethanol. In diabetic ketoacidosis, the blood, vitreous and urine glucose will be elevated in postmortem samples as will the β-hydroxybutyrate concentration in blood and vitreous humour. Glycated haemoglobin is reasonably stable in the early postmortem period and can be used to confirm presumptively high blood glucose concentrations in life.

Hypoglycaemia in life is much more difficult to confirm after death, although formal neuropathological examination of the brain may be very useful. The plasma and urine cortisol will both be elevated and can be measured in postmortem samples; this is a non-specific finding. If overdose of insulin is suspected, take the injection site, if it can be identified, and control tissue. Blood should be centrifuged as soon as possible and the serum frozen unless the sample can be taken directly to the laboratory in the same building. Red cell glutathione reductase will rapidly degrade the immunogenicity of insulin by reducing the S-S bonds that maintain its structure.

In a significant proportion of sudden unexpected deaths in people with alcoholism, high concentrations of ketones may be found in blood indicating alcohol-associated ketoacidosis (Pounder *et al.* 1998). The principal ketone body present is β-hydroxybutyrate. This is because the reduced NAD produced by alcohol metabolism drives the acetoacetate to β-hydroxybutyrate equilibrium towards β-hydroxybutyrate. It is important to remember that β-hydroxybutyrate does not react with the ketone block in urine testing sticks. It is much easier to measure β-hydroxybutyrate in the routine laboratory than the total ketone concentration in postmortem blood. The sample needs pre-treatment with perchloric acid to precipitate proteins followed by centrifugation; it can then be assayed by a routine clinical method for β-hydroxybutyrate; *post mortem* a β-hydroxybutyrate of up to at least 800 μmol/l can be regarded as normal. Alcoholic ketoacidosis is associated with levels of several thousand micromoles per litre.

Heavy metal poisoning

Inductively coupled plasma mass spectroscopy (ICP-MS) has made the screening of postmortem blood and urine samples for heavy metals relatively straightforward. Hair and nail samples may be taken to investigate the possibility of chronic exposure. Hair and nail samples may also be useful in many cases where chronic exposure to a drug is suspected. When collecting hair for toxicology, collect a hank of hair about the thickness of a pencil from the crown of the head. This can be obtained either by cutting close to the scalp or by plucking. Bind the bundle together with an elastic band and wrap, i.e. with the hair held as straight as possible in aluminium foil, unless contrary advice is received from the laboratory. One of the major issues in hair analysis is external contamination of the hair. It is best to obtain the hair before the body is opened to avoid contamination of the hair with body fluids. Not all laboratories offering to analyse hair samples can provide an equal quality of service and interpretation. Thus, care should be taken to ensure that the laboratory to which the sample is submitted is fit for the purpose.

Disasters involving intrathecal or epidural drug administration

If death occurs fairly promptly after the error, it is worth collecting CSF for analysis. To avoid the possibility of contamination of the sample by the presence of high concentrations of the drug in the intrathecal or epidural cannula, it is worth taking a sample by cisternal puncture. A sample of the drug in from the cannula and/or the infusion device and the original ampoules should be submitted for analysis, to exclude the possibility of manufacturing or dilution error.

TISSUE SAMPLES AND BILE

Many drugs are present in bile in high concentrations. When liver failure has supervened, e.g. after paracetamol overdose, any bile present in the gallbladder *post mortem* may have been secreted well before death. Consequently, it can be useful to submit bile in such circumstances. It can be obtained with needle and syringe from the gallbladder. If the gallbladder is absent, useful amounts may be obtained by aspiration from the common bile duct.

Muscle tissue is useful when blood is unavailable. For many, but not all, drugs, the concentration in muscle is similar to the concentration in blood. Muscle is particularly helpful for the toxicological investigation of embalmed bodies. About 50 g from the quadriceps or the psoas muscle is adequate.

It is even more difficult to interpret the results of liver analyses when compared with blood. Many drugs are present in the liver in high concentrations. Where other samples are available only in small amounts, it is worthwhile submitting about 50 g of liver tissue, collected well away from the stomach and major bile ducts. The laboratory can then do screening tests on the liver and quantitative analyses on other samples once the drugs of interest have been identified. Heart blood or even 'blood' from the paracolic gutter can also be used for screening, provided that their origin is clearly specified to the laboratory. Interpretations of quantitative analyses of such samples will be misleading.

Occasionally other samples may be required, e.g. lung tissue is sometimes submitted after deaths apparently associated with volatile substance misuse. Blood from the left side of the heart or pulmonary vein can reasonably be substituted. A glass screw-topped universal container with a PTFE (polytetrafluoroethylene) seal should be used for such samples. It should be filled to the brim, with minimal dead space. If there is a delay in analysis it should be stored at 4°C, not frozen.

INTERPRETING THE RESULTS

Toxicology results should never be interpreted in isolation; they always need to be interpreted in the context of the case with all the information available. The interpretation often requires the joint efforts of the pathologist, toxicologist and clinician (Drummer and Gerostamopoulos 2002, Leikin and Watson 2003, Drummer *et al*. 2004). The whole area is made much more difficult by the phenomenon of postmortem changes in blood drug concentrations which vary from site to site in the body (Pounder 1993). Tricyclic antidepressants can be present in concentrations in left ventricular blood that are orders of magnitude greater than those in plasma shortly before death. Morphine concentrations can vary by 40 per cent between left and right femoral vein samples. The therapeutic range in plasma in life can be a grossly misleading guide to the interpretation of the drug concentration in postmortem samples, particularly when the sample has not been obtained from the femoral vein in a standardized manner. Some help in the interpretation can be found when the drug is present in high concentration in stomach contents. Tissue measurements, particularly in muscle, may help in some, but not all, cases (Langford *et al*. 1998). The relative concentration of drug metabolites to the parent compound may help; in an acute overdose the metabolite is often present in much lower concentrations than the parent drug. Collections of postmortem drug concentrations are starting to appear and these can be very helpful (Druid and Holmgrem 1997). The most useful single textbook is that of Baselt (2004). The current edition of *Clark's Isolation and identification of Drugs* was published, with an on-line version in 2004 (Moffat, Ossleton and Widdop 2004).

Alcohol

The single most important neurotoxin is, of course, ethanol. The interpretation of its concentration in blood and urine is fraught with difficulty even in life; the possibility of postmortem production makes it even more difficult to interpret in postmortem samples. Prompt sample collection after death, adequate preservation of the sample after collection and the use of an appropriate analytical method do assist in the task of interpreting the postmortem results. Where these conditions do not apply, attempts at interpretation can be misleading (Mayes *et al.* 1992).

The acute intoxicant effect of alcohol varies significantly from person to person depending on a number of factors. These include age, physique, sex, race and tolerance to the effects of alcohol, reflecting the individual's drinking habits. The Mellanby effect, where the intoxicant effect of alcohol is greater when the concentration in blood is rising than when it is falling, also needs to be taken into account. Table 14.1, relating the effect of alcohol to the blood alcohol concentration, should be regarded as a very rough guide indeed (Dubowski 1989).

Coroners frequently ask for an estimate of the amount of alcohol that a deceased person has consumed. A very rough estimate can be made as follows (Widmark 1981):

Take the postmortem blood alcohol concentration – say 120 mg/100 ml.

For each hour between the time drinking started and death add 18 mg/100 ml. So if the consumption started 5 h before death add 90 mg/100 ml to make a total of 210 mg/100 ml in our example.

Convert the units to grams of alcohol per litre of blood: 210 mg/100 ml is the equivalent of 2.1 g/l.

Table 14.1 Acute intoxicant effects of alcohol

Blood alcohol concentration (mg/100 ml)	Behavioural effects
10–50	Subclinical
30–120	Euphoria
90–250	Excitement
180–300	Confusion
250–400	Stupor
350–500	Coma
>450	Death

Reproduced from Dubowski, K.M. 1989. *Stages of Acute Alcoholic Influence/Intoxication.* Oklahoma City: University of Oklahoma College of Medicine.

Estimate the total body water of the deceased; for men of average physique use a value for body water of 0.68 l/kg, for women use 0.55 l/kg. Thus an 80-kg man will have a body water of 80 × 0.68 l, i.e., 54.4 l.

The total amount of alcohol consumed is then obtained, in this example, by multiplying 2.1 g/l by 54.4 l. This gives 114.24 g.

As a pint of beer contains about 18 g of alcohol, this suggests that the deceased would have consumed about 6 pints of beer over the 5 h preceding his death.

Like all attempts to estimate ante-mortem drug consumption from postmortem drug concentrations, such calculations are open to criticism. If used in the coroner's court it would be prudent to quote an imprecision of at least ±50 per cent.

The interaction of alcohol with the compromised brain is of considerable forensic interest. Although modest alcohol use may protect against ischaemic stroke, heavier drinking is associated with an increased incidence of both haemorrhagic and ischaemic strokes (Hillbom and Kaste 1990, Berger *et al.* 1999).

Acute intoxication with alcohol is associated with a thrombophilic tendency (Numminen *et al.* 2000). This may contribute to the association between binge drinking and both thrombotic strokes and myocardial infarctions. Fluid shifts and the increase in blood pressure may contribute to the increased incidence of subarachnoid haemorrhage and binge drinking (Gray *et al.* 1999).

This 'J'-shaped effect, with low levels of alcohol having a protective effect and higher levels an adverse effect, is also seen in acute intoxication with alcohol associated with head injury, at least in animal models. Alcohol concentrations of around 100 mg/100 ml may have a neuroprotective effect after head injury in animal models, possibly as a result of the inhibition of the activity of N-methyl-D-aspartate (NMDA) by alcohol (Kelly *et al.* 1997, Janis *et al.* 1998). The stimulation by alcohol of corticosterone production by the adrenals may also have a protective effect (Gottesfeld *et al.* 2002). Higher alcohol concentrations, of around 220 mg/100 ml, are associated with depressed respiratory drive following experimental head injury. This may be associated with increased endogenous opiate production Zink *et al.* 2001). The practical effect of this in humans is moot; other variables such as the quality of immediate care and the maintenance of oxygenation and haemodynamic status may confound the putative protective effects of modest alcohol concentrations and the deleterious effects of higher amounts of alcohol. In humans, higher alcohol concentrations at the time of injury tend to be associated with a poorer early cognitive outcome. Whether this is an acute toxic effect of alcohol or a result of the interaction of the effects of an acute head injury and chronic brain damage associated with regular alcohol use, or a combination of both, is a matter for ongoing research (Bombardier and Thurber 1998).

Toxicological eschatology

The neuropathologist may be asked about the interpretation of blood drug levels before brain-stem death testing is carried out. Where patients are sedated with high doses of thiopental, used as a neuroprotective agent after head injury, the issue may arise as to when the drug has been cleared from the body such that brain-stem death tests may be performed. It has been suggested that, if brain-stem reflexes are absent when the thiopental concentration in blood has fallen to less than 40 mg/l, then this implies that there is a cause other than just the presence of thiopental (Grattan-Smith and Butt 1993). However, to assess potential barbiturate suppression of the brain stem, not only thiopental, but also its active metabolite pentobarbital should be measured. As the effects of all drugs vary from person to person, very great care needs to be taken before assuming that a low barbiturate concentration is having no functional effect on the patient. If there is any doubt, and in the author's (ARWF) view there is always doubt, then imaging techniques may be needed to confirm brain-stem death until any neurodepressant drugs in blood have fallen to very low, often virtually undetectable, concentrations (Schober *et al.* 1987). In this situation, as in every other, toxicology results cannot be considered in isolation but only in the clinical context of the case.

CONCLUSION

Postmortem toxicology is fraught with difficulties. To apply it successfully requires thought, teamwork, constant review of the literature and attention to detail.

REFERENCES

Adams, J.H., Haller, L., Boa, F.Y., Doua, F., Dago, A. and Konian, K. 1986. Human African trypanosomiasis (*Tb. gambiense*): a study of 16 fatal cases of sleeping sickness with some observations on acute reactive arsenical encephalopathy. *Neuropathol Appl Neurobiol*, **12**, 81–94.

Adams, R.D. 1976. *Nutritional Cerebellar Degeneration*. Amsterdam: Elsevier.

Andersen, S.N. and Skullerud, K. 1999. Hypoxic/ischaemic brain damage, especially pallidal lesions, in heroin addicts. *Forensic Sci Int*, **102**, 51–9.

Auer, R.N. and Sutherland, G.R. 2002. Hypoxia and related conditions. In Graham, D.I. and Lantos, P.L. (eds),

Greenfield's Neuropathology, 7th edn, Vol. 1. London: Arnold, 234–80.

Bakir, F., Damluji, S., Amin-Zaki, L., Murtadha, M., Khalidi, A. and Al-Rawi, N.Y. 1973. Methylmercury poisoning in Iraq. *Science*, **181**, 230–41.

Baselt, R.C. 2004. *The Disposition of Toxic Drugs and Chemicals in Man*. Foster City, CA: Chemical Toxicology Institute.

Berger, K., Ajani, U.A., Kase, C.S. *et al.* 1999. Light-to-moderate alcohol consumption and risk of stroke among US. male physicians. *N Engl J Med*, **341**, 1557–64.

Blackmore, D.J. 1968. The bacterial production of ethyl alcohol. *J Forensic Sci Soc*, **8**, 73–8.

Bombardier, C.H. and Thurber, C.A. 1998. Blood alcohol level and early cognitive status after traumatic brain injury. *Brain Injury*, **12**, 725–34.

Büttner, A., Mall, G., Penning, R. and Weis, S. 2000. The neuropathology of heroin abuse. *Forensic Sci Int*, **113**, 435–42.

Büttner, A., Mall, G., Penning, R., Sachs, H. and Weis, S. 2003. The neuropathology of cocaine abuse. *Leg Med*, **5**, S240–2.

Bjork, J., Steven, G. and Hommer, D. 2003. Cross-sectional volumetric analysis of brain atrophy in alcohol dependence: effects of drinking history and comorbid substance use disorder. *Am J Psychiatry*, **160**, 2038–45

Cavanagh, J.B., Fuller, N.H., Johnson, H.R. and Rudge, P. 1974. The effects of thallium salts, with particular reference to the nervous system changes. A report of three cases. *Q J Med*, **43**, 293–319.

Davis, G.D., Williams, A.C., Markey, S.P., Ebert, M.H., Caine, F.D., Reichert, C.M. and Kopin, I.J. 1979. Chronic parkinsonism secondary to intravenous injection of meperidine analogues. *Psych Res*, **1**, 249–54.

Druid, H. and Holmgren, P. 1997. A compilation of fatal and control concentrations of drugs in postmortem femoral blood. *J Forensic Sci*, **42**, 79–87.

Drummer, O.H. and Gerostamoulos, J. 2002. Postmortem drug analysis: analytical and toxicological aspects. *Therap Drug Monit*, **24**, 199–209.

Drummer, O., Forrest, A.R., Goldberger, B., Karch, S. B. and the International Toxicology Advisory Group. 2004. Forensic science in the dock. *BMJ*, **329**, 636–7.

Dubowski, K.M. 1989. *Stages of Acute Alcoholic Influence/ Intoxication*. Oklahoma City: University of Oklahoma College of Medicine.

Fadda, F. and Rossetti, Z.L. 1998. Chronic ethanol consumption: from neuroadaptation to neurodegeneration. *Prog Neurobiol*, **56**, 385–431.

Fessler, R.D, Esshaki, C.M., Stankewitz, R.C., Johnson, R.R. and Diaz, F.G. 1997. The neurovascular complications of cocaine. *Surg Neurol*, **49**, 339–45.

Gottesfeld, Z., Moore, A.N. and Dash, P.K. 2002. Acute ethanol intake attenuates inflammatory cytokines after brain injury in rats: a possible role for corticosterone. *J Neurotrauma*, **19**, 317–26.

Grattan-Smith, P.J. and Butt, W. 1993. Suppression of brain-stem reflexes in barbiturate coma. *Arch Dis Child*, **69**, 151–2.

Gray, J.T., Puetz, S.M., Jackson, S.L. and Green, M.A. 1999. Traumatic subarachnoid haemorrhage: a 10-year case study and review. *Forensic Sci Int*, **105**, 13–23.

Harper, C. 1983. The incidence of Wernicke's encephalopathy in Australia – a neuropathological study of 131 cases. *J Neurol Neurosurg Psychiatry*, **46**, 593–8.

Harper, C., Dixon, G., Sheedy, D. and Garrick, T. 2003. Neuropathological alterations in alcoholic brains. *Prog Neuropsychopharmacol Biol Psychiatry*, **27**, 951–61.

Harries, D.P. and De Silva, R. 1992. 'Ecstacy' and intracerebral haemorrhage. *Scot Med J*, **37**, 150–2.

Hillbom, M. and Kaste, M. 1990. Alcohol abuse and brain infarction. *Ann Med*, **22**, 347–52.

Hirsch, C.S. and Adelson, L. 1973. Ethanol in sequestered hematomas. *Am J Clin Pathol*, **59**, 429–33.

Ihara, M., Ito, T., Yanagihara, C. and Nishimura, Y. 1999. Wernicke's encephalopathy associated with hemodialysis: report of two cases and review of the literature. *Clin Neurol Neurosurg*, **101**, 118–21.

Jacobs, J.M. and LeQuesne, P.M. 1992. Toxic disorders In Adams, J.H. and Duchen, L.W. (eds), *Greenfield's Neuropathology*, 5th edn. London: Arnold, 234–80.

Janis, L.S., Hoane, M.R., Conde, D., Fulop, Z. and Stein, D.G. 1998. Acute ethanol administration reduces the cognitive deficits associated with traumatic brain injury in rats. *J Neurotrauma*, **15**, 105–15.

Karch, S.B. 2002. *Karch's Pathology of Drug Abuse*. London: CRC Press.

Kelly, D.F., Lee, S.M., Pinanong, P.A. and Hovda, D.A. 1997. Paradoxical effects of acute ethanolism in experimental brain injury. *J Neurosurg*, **86**, 876–82.

Kinney, H.C. and Armstrong, D.D. 2002. Perinatal neuropathology. In Graham, D.I. and Lantos, P.L. (eds), *Greenfield's Neuropathology*, 7th edn, Vol. 2. London: Arnold, 151–94.

Langston, W., Ballard, P., Tetrud, J.W. and Irwin, I. 1983. Chronic parkinsonism in humans due to a product of meperidine-analog synthesis. *Science*, **219**, 979–80.

Lampl, C. and Yazdi, K. 2002. Central pontine myelinolysis. *Eur Neurol*, **47**, 3–10.

Langford, A.M., Taylor, K.K. and Pounder, D.J. 1998. Drug concentration in selected skeletal muscles. *J Forensic Sci*, **43**, 22–7.

Leikin, J.B. and Watson, W.A. 2003. Postmortem toxicology: what the dead can and cannot tell us. *J Clin Toxicol*, **41**, 47–56.

McIntyre, I.M., Hamm, C.E., Sherrard, J.L., Gary, R.D., Riley, A.C. and Lucas, J.R. 2003. The analysis of an intracerebral hematoma for drugs of abuse. *J Forensic Sci*, **48**, 680–2.

McLean, D.R., Jacobs, H. and Mielke, B.W. 1980. Methanol poisoning: a clinical and pathological study. *Ann Neurol*, **8**, 161–7.

Mayes, R., Levine, B., Smith, M.L., Wagner, G.N. and Froede, R. 1992. Toxicologic findings in the USS Iowa disaster. *J Forensic Sci*, **37**, 1352–7.

Medana, I.M. and Esiri, M.M. 2003. Axonal damage: a key predictor of outcome in human CNS diseases. *Brain*, **126**, 515–30.

Milroy, C.M., Clark, J.C. and Forrest, A.R.W. 1996. Pathology of deaths associated with 'ecstacy' and 'eve' misuse. *J Clin Pathol*, **49**, 149–53.

Moffat, A.C., Ossleton, M.D. and Widdop, B. (eds) 2004. *B.W. Clark's Analysis of Drugs and Poisons*. 3rd Edition. London: Pharmaceutical Press.

Naeser, P. 1988. Optic nerve involvement in a case of methanol poisoning. *Br J Ophthalmol*, **72**, 778–81.

Neubuerger, K.T. and Clarke, E.R. 1945. Subacute carbon monoxide poisoning with cerebral myelinopathy and multiple myocardial necroses. *Rocky Mountain Med J*, **42**, 29–34.

Numminen, H., Syrjala, M., Benthin, G., Kaste, M. and Hillbom, M. 2000. The effect of acute ingestion of a large dose of alcohol on the hemostatic system and its circadian variation. *Stroke*, **31**, 1269–73.

Opeskin, K., Clarke, I. and Berkovic, S.F. 2000. Prolactin levels in sudden unexpected death in epilepsy. *Epilepsia*, **41**, 48–51.

Osetowska, E. 1971. Gases. In Minckler, J. (ed.), *Pathology of the Nervous System*, Vol. 2. New York: McGraw-Hill, 1638–44.

Pounder, D.J. 1993. The nightmare of postmortem drug changes. *Leg Med*, 163–91.

Pounder, D.J., Stevenson, R.J. and Taylor, K.K. 1998. Alcoholic ketoacidosis at autopsy. *J Forensic Sci*, **43**, 812–6.

Pfefferbaum, A., Sullivan, E.V., Rosembloom, M.J., Mathalon, D.H. and Kelvin, L. 1998. A controlled study of cortical grey matter and ventricular changes in alcoholic men over a 5-year interval. *Arch Gen Psychiatry*, **55**, 905–12.

Robertson, M.D. and Drummer, O.H. 1998. Stability of nitro-benzodiazepines in postmortem blood. *J Forensic Sci*, **43**, 5–8.

Schober, O., Galaske, R. and Heyer, R. 1987. Determination of brain death with ^{123}I-IMP and ^{99m}Tc-HM-PAO. *Neurosurg Rev*, **10**, 19–22.

Shiraki, H. 1979. Neuropathological aspects of organic mercury intoxication, including Minimata disease. In Vinken, P.J. and Brutn, G.W. (eds), *Handbook of Clinical Neurology*, Vol. 36. Amsterdam: North-Holland, 83–145.

Sloan, M.A. 1997. Toxicity/Substance abuse. In Welch, K.M.A., Caplan, L.R., Reis, D.J., Siesjö, B.K. and Weir, B. (eds), *Primer on Cerebrovascular Disease*. San Diego: Academic Press, 104–7.

Widmark, E.M.P. 1981. *Principles and Applications of Medico-Legal Alcohol Determination*. Davis, CA: Biomedical Publications.

Wolters, E.C., van Wijngaarden, G.K., Stam, F.C., Rengelink, H., Lousberg, R.J., Schipper, M.E. and Verbeeten, B. 1982. Leucoencephalopathy after inhaling 'heroin' pyrolysate. *Lancet*, **2**, 1233–7.

Wright, D.G., Laureno, R. and Victor, M. 1979. Pontine and extrapontine myelinolysis. *Brain*, **102**, 361–85.

Zink, B.J., Schultz, C.H., Stern, S.A. *et al.* 2001. Effects of ethanol and naltrexone in a model of traumatic brain injury with hemorrhagic shock. *Alcoholism: Clin Exp Res*, **25**, 916–23.

15

Clinical aspects of head injury

Damianos E Sakas

EPIDEMIOLOGY

Head injury is one of the major causes of morbidity and mortality in the modern world. Its significance had been recognized in antiquity. The first reference to head injuries was found in an Egyptian papyrus that was written 4500 years ago. In modern times, in the USA, trauma is the main cause of death in people younger than 40 years and brain injuries are responsible for 50 per cent of these deaths. Nearly 100 000 head trauma victims each year suffer lifelong debilitating loss of function, 5000 develop post-traumatic epilepsy and 2000 live in a persistent vegetative state (Goldstein 1990). In Australia, 2.5 deaths occur from head injury per 10 000 population, with road traffic accidents being responsible for about 65 per cent of the fatal head injuries; of those, half die before reaching hospital (Kaye 1991). Apart from the road traffic accidents, the other causes of head injuries are falls, assaults, and injuries at work, at home and during sports. The relative frequency of each cause varies among different age groups.

CLASSIFICATIONS OF HEAD INJURIES

Head injuries are a complex clinical problem with a wide spectrum from the mild concussion to the severe brain injury resulting in death. A thorough understanding of the head injuries should be based on the identification of the pathophysiological processes that developed as a result of the injury. This can be achieved by classifying the injuries according to the following specific criteria:

- Type of underlying tissue damage
- Mechanism of the injury

- Severity of the injury
- Extent of the tissue damage (focal or generalized).

The following classifications of head injuries in various types are very useful in clinical management and also in forensic practice. Depending on the *type* of damage they are classified as:

- Scalp injury
- Skull fracture
- Extradural haematoma
- Subdural haematoma
- Intracerebral haematoma
- Brain contusion
- Cerebral oedema
- Cranial nerve damage.

Depending on the causative mechanism, they are classified as:

- Resulting from linear acceleration–deceleration. The lesions are caused more commonly near the cortical surface when the brain parenchyma is rubbed against sharp parts of the inner skull.
- Resulting from rotational or angular acceleration. In these, the damage results from the relative difference in acceleration–deceleration because of the different specific gravities of the grey and white matter; this can cause shearing injury of nerve fibres.

Depending on the *severity* of the damage, they are classified as:

- Mild: the level of consciousness is affected mildly and the injured person has drowsiness or confusion. The Glasgow Coma Scale (GCS) score is 13–15.

- Moderate: the level of consciousness has been affected more seriously. The GCS score is 9–12.
- Severe: the injured person is comatose and the GCS score is 3–8.

Depending on whether they occur in a limited area or are widespread, they are classified as:

- Focal, such as haematomas or contusions
- Generalized, such as brain oedema or multiple contusions
- Mixed (focal and generalized).

CLINICAL FEATURES OF SIGNIFICANCE RELATING TO TIMING AND MECHANISM OF INJURY

Brain damage occurs both at impact and as a result of the development of secondary complications. It is useful to distinguish the post-traumatic lesions into two broad categories: primary and secondary injuries.

Primary injuries

A blow to the head can cause primary damage of several types.

PRIMARY TRAUMATIC AXONAL INJURY

This occurs as result of angular or rotational acceleration with or without impact (Gennarelli *et al.* 1982). Depending on severity it may produce minor concussion through to immediate coma. The pathology is covered in detail in Chapter 8.

CORTICAL CONTUSIONS AND LACERATIONS

These may occur under or opposite (contre-coup) the site of impact. They can be multiple and bilateral. The pathology is covered in Chapter 7. Multiple contusions do not depress the conscious level unless one of them acts in a space-occupying manner. A sizeable space-occupying contusion may cause either focal signs such as a hemiparesis, dysphasia and/or generalized signs usually as a result of raised intracranial pressure (ICP) such as headache, vomiting, drowsiness or confusion.

SCALP INJURY

Although the scalp injury may not be related to the severity of the underlying brain injury, one should always remember that a large scalp laceration may result in considerable blood loss and may cause death, particularly in association with alcohol intoxication.

SKULL FRACTURE

Any fracture suggests that the blow to the head was not mild and that the patient has a substantially increased risk that he or she could develop a haematoma later.

Secondary injuries

A blow to the head can cause secondary injury of the following types.

EXTRADURAL HAEMATOMA

The extradural haemorrhage (EDH) complicates 2–6 per cent of head-injured patients. Road traffic accidents and falls are responsible for 75–90 per cent of the cases. It is more common in the first two decades of life. Around 3 per cent occur in patients less than 2 years old and another 3 per cent in patients over 60 years old. In most adults, an EDH is created after a dural arterial branch has been torn by either a skull fracture or shearing of the surface vasculature or by calvarial deformation, resulting in direct laceration of the subjacent meningeal vessels. Several variables such as skull/scalp thickness, pre-existing cerebral compliance and the gradient of tissue deformation can contribute to the severity of tissue damage. The most life-threatening vascular tear is that of the middle meningeal artery or one of its branches that can be caused by a fracture of the temporal bone. Such an arterial rupture requires substantial impact such as in a road traffic accident or fall (Figure 15.1).

After the blow that caused the fracture and the arterial tear, the patient may become unconscious for a brief period because of the impact of the injury. However, he or she may regain consciousness temporarily. This is called the 'lucid interval' and its duration may vary from 30 min to several hours depending on the rapidity by which the blood is pouring out of the artery and is creating a sufficiently large clot to cause brain compression shift, functional compromise, depression of the conscious level and alteration of the neurological picture (Lobato *et al.* 1991). As the amount of the blood clot between the bone and the dura gradually increases, the patient will get progressively more drowsy and confused and may develop hemiparesis and fall into a coma. This may be followed by papillary dilatation, ipsilateral to the haematoma, and abnormal motor responses, i.e. flexion or extension. These signs will eventually become bilateral. The fully blown clinical picture of the EDH is caused by compression of the brain stem by the herniating uncus of the temporal lobe. This picture involves coma, bilateral extension motor responses and fixed dilated pupils (Figure 15.2).

An EDH may develop in patients who have suffered either a trivial or a severe primary brain injury. A deteriorating conscious state is the most important neurological

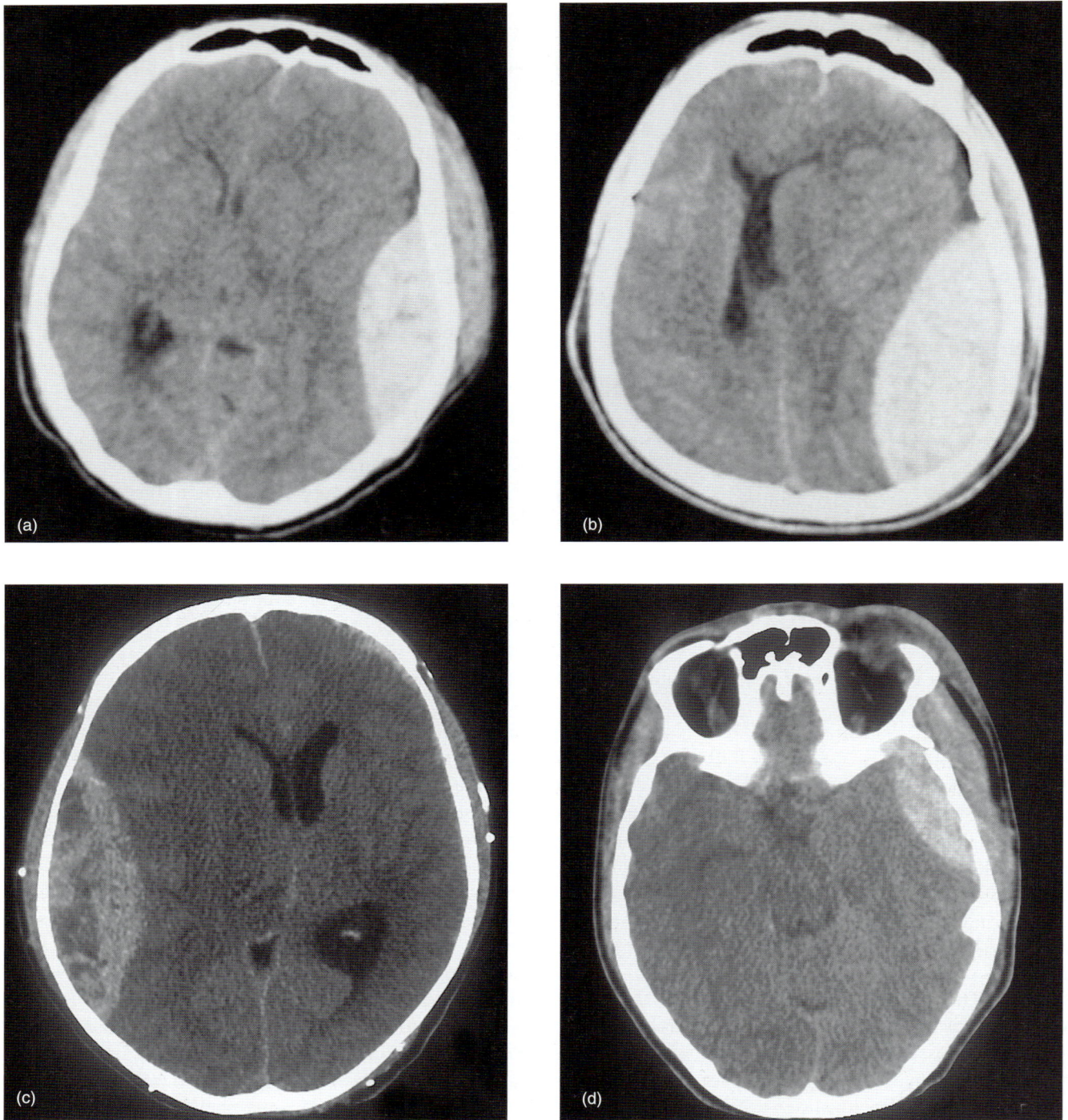

Figure 15.1   Acute extradural haematoma (EDH). (a) Quite large EDH in the left temporoparietal region. (b) A higher computed tomography (CT) image showing the large haematoma, midline shift and contralateral ventricular dilatation. (c) Sizeable EDH in the right temporoparietal region. (d) Small EDH in the left temporal basal region.

sign, especially after a 'lucid' interval. In a patient with a head injury it is mandatory that the drowsiness not be misinterpreted as a wish for rest or sleep.

It is important to remember that an EDH can cause a more gradual clinical picture whenever the fracture does not affect a major arterial branch but rather smaller vessels.

This can happen when the fracture extends through one of the following regions.

Venous sinuses

The superior sagittal sinus on the upper circumference of the head and the transverse sinus and sigmoid sinus in the

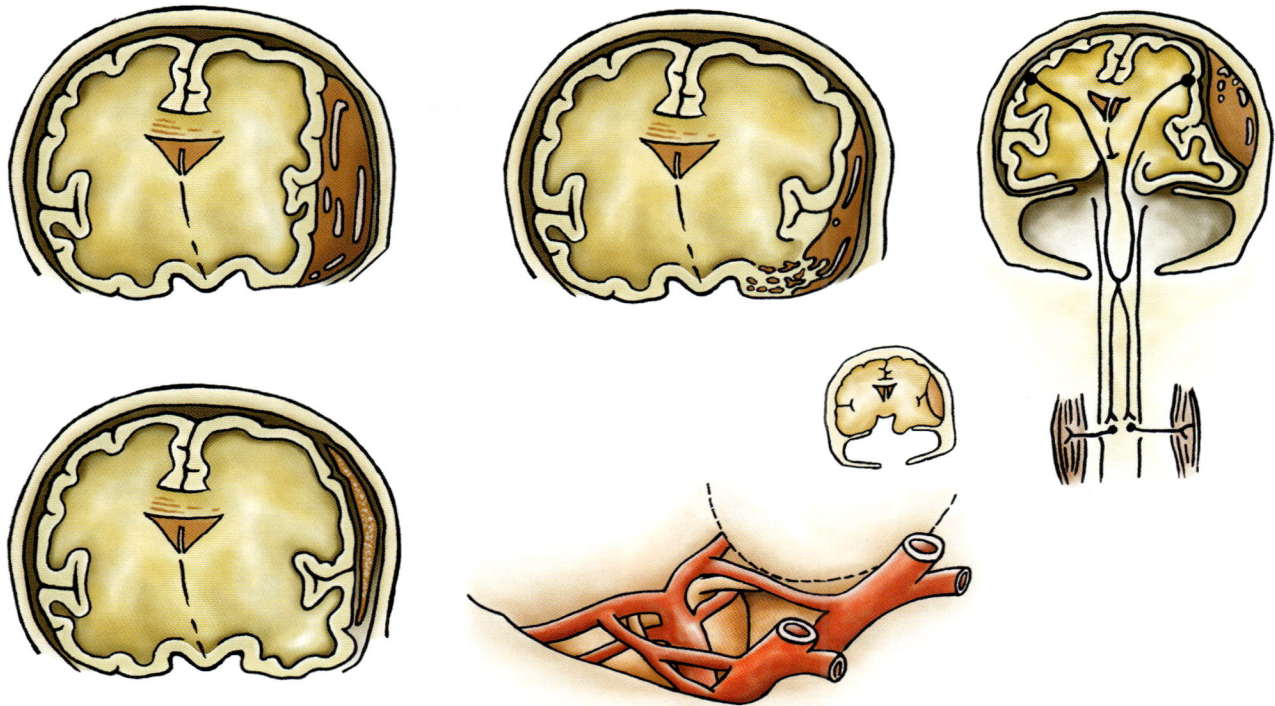

Figure 15.2 Progressive expansion of an intracranial haematoma, displacement of the cerebral hemisphere and cerebral herniation. (Modified from Lidsay, K. and Bone, I. 1997. *Neurology and Neurosurgery Illustrated*, 3rd edn. Edinburgh: Churchill Livingstone.)

posterior fossa can be torn by a fracture. In the last two, this may be more life threatening because an EDH in the posterior fossa (infratentorial compartment) can cause shift of the cerebellum, obliteration of the fourth ventricle, acute hydrocephalus and a relatively rapid deterioration compared with the deterioration that can result from a venous sinus rupture in the supratentorial compartment. In any case, it is unlikely that a posterior fossa haematoma caused by sinus rupture could expand with the same rapidity as a haematoma resulting from an arterial rupture.

Veins of the diploë
In this case, the haematoma is even more gradual in its development.

Small arterial branches
In certain very unusual cases, a haematoma caused by a rupture of a small artery or vein could remain contained within a certain size because of temporary tamponade of the source of bleeding. Such a haematoma may go unnoticed for several days without any focal symptoms and cause symptoms 1–2 weeks later (Borovich *et al.* 1985).

ACUTE SUBDURAL HAEMATOMA
The acute subdural haematoma (SDH) should be suspected in a patient with a severe head injury whose neurological state is either failing to improve or deteriorating. A cerebral haematoma can result from direct laceration of the brain by calvarial deformation or, more commonly, from shearing strains and pressure gradients generated during impact. The formation of pressure gradients within the brain derives from the fact that the brain possesses inertia and therefore lags slightly behind the skull during head motion (Gean 1995). It can also be created after a less severe trauma that has caused a rupture of a bridging vein or focal tear of a cortical artery (Figure 15.3). Rarely, a spontaneous acute SDH can develop because of a ruptured aneurysm or bleeding diathesis (see Chapter 5, p. 65).

CHRONIC SUBDURAL HAEMATOMA
The patients who suffer a chronic SDH can be divided into two groups according to the pathophysiological mechanism. In the first group, the cause is atrophy and shrinkage of the brain because of ageing. The brain becomes gradually more mobile; this increases the space traversed by a bridging vein between the cortex and the dura. Any movement of the brain, even after a relatively trivial injury, may result in tearing of the bridging vein. In this group, most patients are over 50 years of age. In a considerable

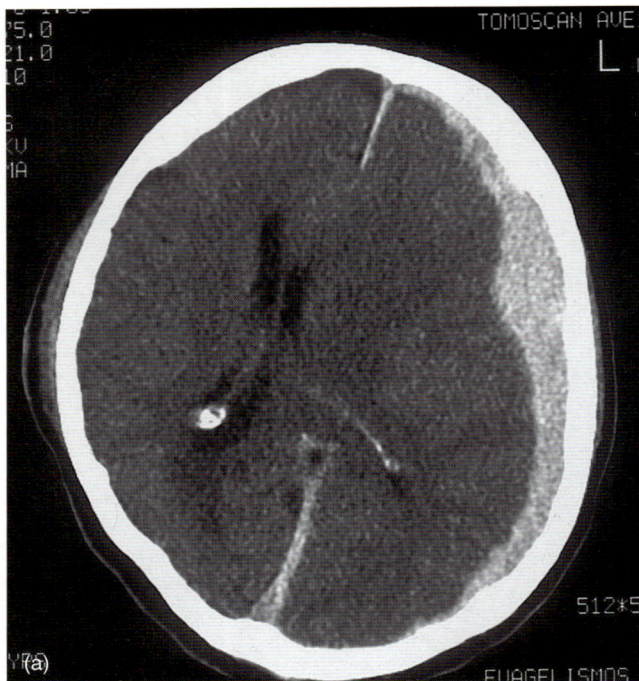

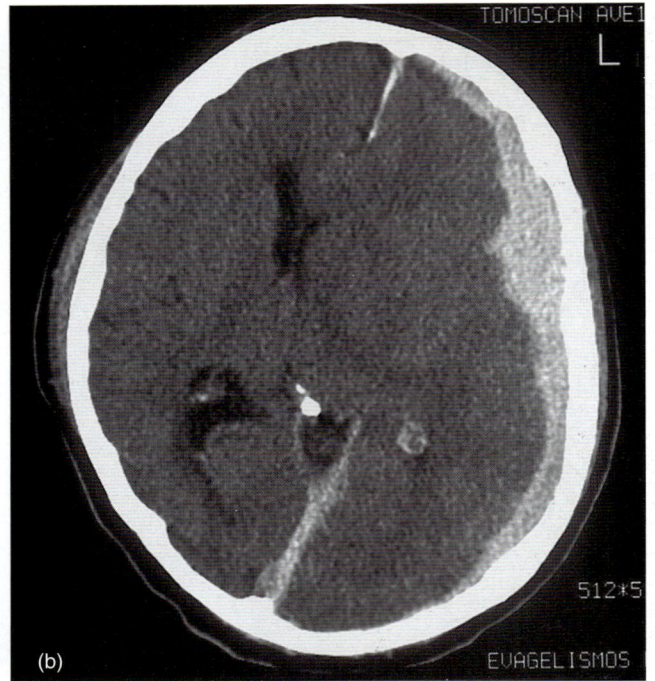

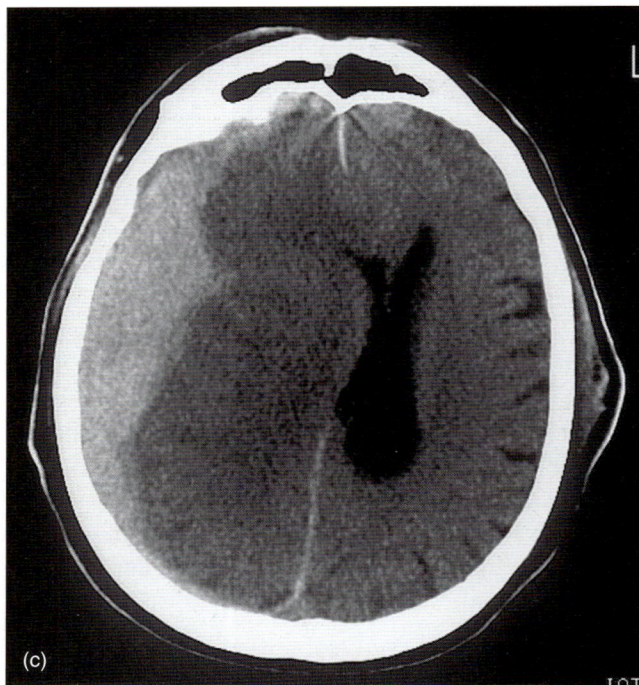

Figure 15.3 Acute subdural haematoma (SDH). (a, b) Acute SDH in the left frontotemporoparietal region causing obliteration of the ipsilateral lateral ventricle (CT). (c) Massive acute SDH covering most of the right hemisphere causing pronounced midline shift and dilatation of the contralateral ventricle due to compression of the third ventricle – an ominous prognostic indicator.

proportion of such patients there is no definite history of preceding head trauma. The second group involves patients who have suffered a significant head injury. This trauma causes a very small tear of a bridging vein that will result, after several days or weeks, in the formation of a blood clot. A similar mechanism may lead to the formation of a subacute SDH (within a week after trauma) (Figure 15.4).

INTRACEREBRAL HAEMORRHAGE

Intracerebral haematomas occur because of a contusion or laceration of brain tissue after a severe head injury, a penetrating injury or a depressed skull fracture. Intracerebral haematomas can be associated with SDHs. In some patients, a haemorrhage can evolve over the next few hours or days after trauma (see Chapter 7, p. 84). The brain shrinks as it grows older and its vasculature becomes less flexible

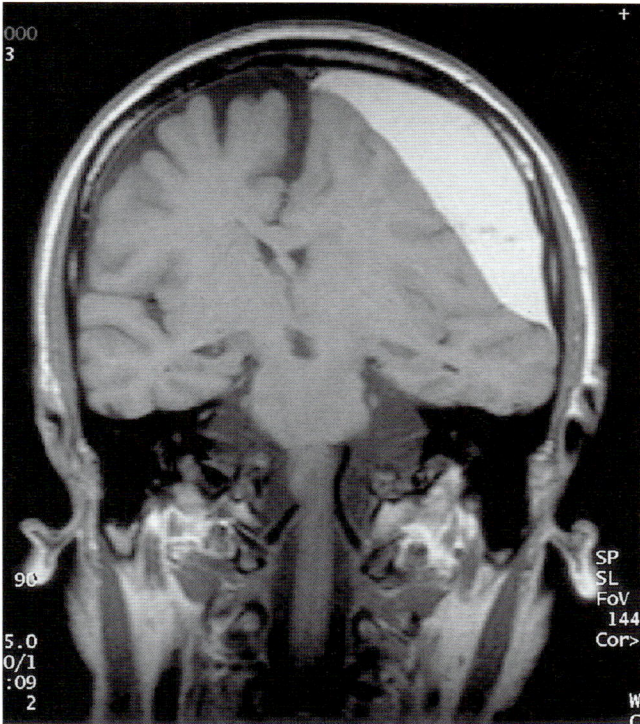

Figure 15.4 Chronic subdural haematoma causing displacement of the left cerebral hemisphere and midline shift (magnetic resonance imaging, T1, coronal view).

because of atherosclerosis. The decrease in cerebral plasticity combined with increase in vascular tethering is the main reason for the development of focal mass lesions more frequently in elderly than in younger patients.

The focal neurological signs will depend on the location of the haematoma. A large temporal haematoma is likely to manifest itself by a progressive contralateral hemiparesis and an ipsilateral dilated pupil. Further expansion of the haematoma will cause bilateral spastic limbs, a decerebrate posture and bilaterally dilated pupils. Occasionally, the hemiparesis can initially be ipsilateral, because of compression of the contralateral crus cerebri on the tentorial edge; it is rare, however, for the contralateral pupil to dilate first. The change in vital signs involves bradycardia and a rise of the systemic blood pressure in response to a rise in ICP (Cushing's reflex). This is followed by disturbances in respiration and ultimately Cheyne–Stokes breathing (Figure 15.5).

Haematomas at other than the temporal location can present with variations of the classic clinical picture. Frontal haematomas show first a deterioration of consciousness and pupil abnormalities and lateralizing signs occur late in their progression. Haematomas of the posterior cranial fossa may affect the vital signs first and cause bradycardia and a rise in systemic blood pressure before

any depression of the conscious level. The mechanism is not clear but the pressure on the brain stem by the adjacent clot may initially disturb the neuronal networks responsible for heart rate and blood pressure regulation.

CEREBRAL SWELLING

Cerebral swelling after trauma can be either focal around a haematoma or contusion or generalized throughout the cerebrum and/or cerebellum. The pathological processes are poorly understood but involve disturbance of not only vasomotor tone (vasocongestive swelling) but also all the known mechanisms of oedema, as outlined in Chapter 9. Generalized swelling will manifest itself by conscious state alterations, i.e. confusion, drowsiness or even coma. Focal swelling will cause focal signs such as hemiparesis.

HYDROCEPHALUS

In the early stages after a head injury, hydrocephalus can be caused by the following mechanisms: (1) obstruction of the fourth ventricle by blood, (2) swelling in the posterior fossa or (3) obstruction of the absorption of cerebrospinal fluid (CSF) as a result of a traumatic subarachnoid haemorrhage. This last type of hydrocephalus is more common many weeks after the head injury. The clinical picture of hydrocephalus resembles that of ICP with headache, vomiting and depression of conscious level.

CEREBRAL HERNIATION

Both types of lesions, i.e. haematomas or cerebral swelling, can cause cerebral herniation. The lower medial surface of the hemisphere may be displaced under the falx and the uncus of the temporal lobe may herniate through the tentorium, causing pressure on the third nerve and midbrain. The cardinal clinical features of tentorial herniation include contralateral hemiparesis and ipsilateral papillary dilatation, with loss of responsivity to light. In other cases, there may be a caudal displacement of the brain stem and/or cerebellum into the foramen magnum (see Chapter 9, p. 111). This will cause brain-stem disturbance with changes in blood pressure and heart rate progressing to cardiorespiratory arrest.

CSF RHINORRHOEA

Most dural fistulas result from anterior skull base fractures. In other cases, a fracture may not be found and the leak could result from an arachnoid tear of the olfactory rootlets in the area of cribriform plate, where there is no dural investment. In cases of CSF fistula, the major concern is the risk of meningitis (see later).

CERVICAL VASCULAR INJURY

Traumatic vascular lesions can be primary or secondary. A well-described secondary vascular lesion is post-traumatic

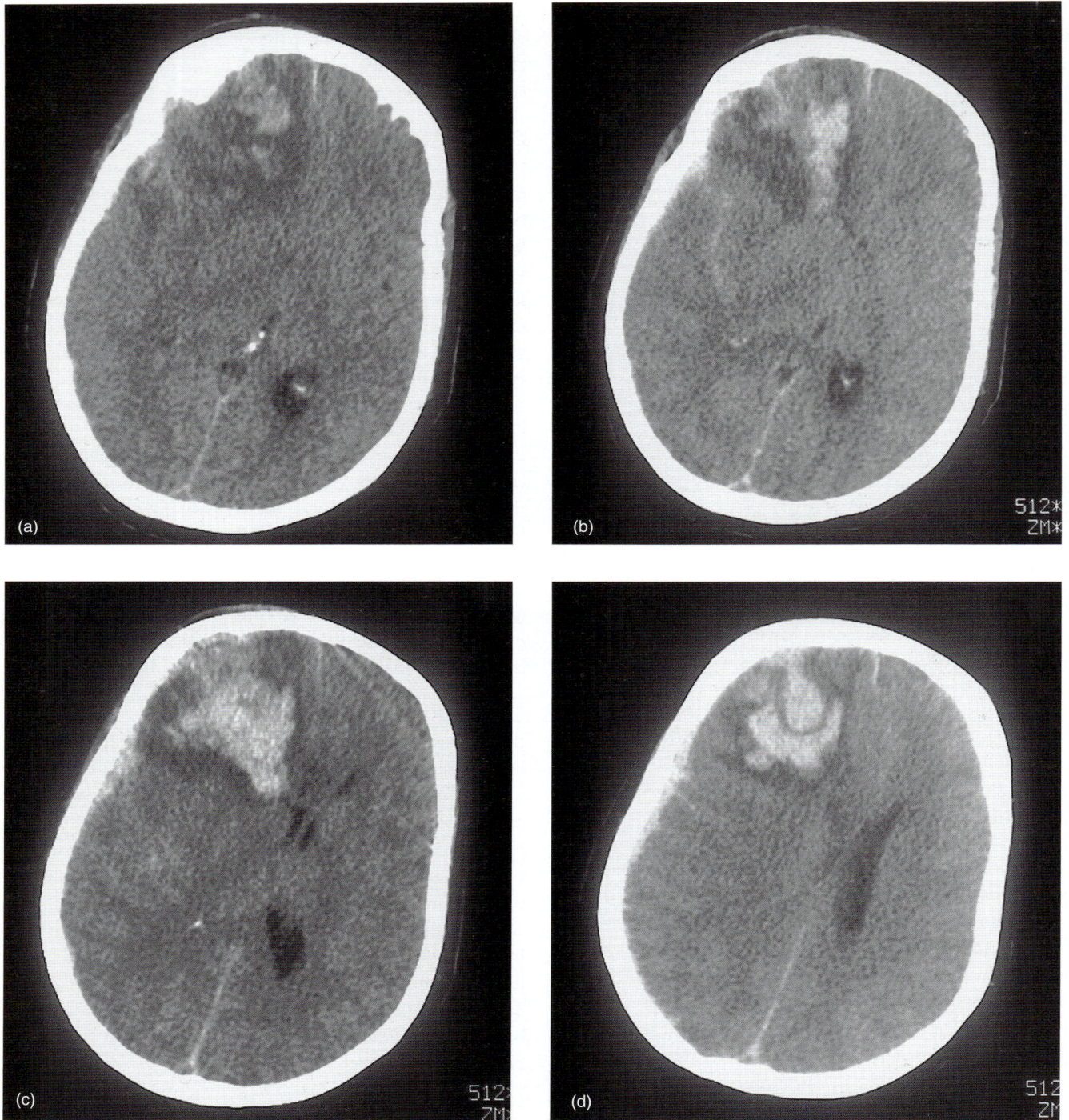

Figure 15.5   Intracerebral haematoma and contusion ('burst lobe') in the right frontal lobe (CT).

infarction. Cerebral infarction in a post-traumatic setting can be caused by many factors, including hypoxia–ischaemia, arterial embolic disease (e.g. dissection, fat embolism), increased ICP, venous thrombosis, and vascular compression resulting from brain herniation or an overlying extra-axial collection. The pathology of arterial dissection is covered in Chapter 6, p. 77.

CONCUSSION

The term 'concussion' describes the brief temporary post-traumatic alteration of consciousness. In addition to the alterations of consciousness, the clinical picture may include confusion or amnesia, delay in verbal responses, disorientation, incoherent or incomprehensible speech, difficulty of stance and gait, inappropriate emotional responses, loss of memory, etc. The term is derived from the Latin 'concutere' which means to shake. A blow to the head is more likely to cause a concussion when the head is free to move. The alteration of consciousness rarely lasts more than a day. The loss of consciousness may be associated with the retrograde amnesia that may last several days but this gradually decreases.

Concussion is normally not associated with computed tomography (CT) or magnetic resonance imaging (MRI) findings. A reliable assessment of the relative severity of the head injury is the duration of post-traumatic amnesia. If the amnesia after the head injury lasts more than 1 day, the concussion is regarded as being severe. Following a concussion, brain autoregulation is impaired and a second blow may have severe complications, including malignant brain oedema which would be less likely if the concussion had not occurred. This has been described by many as the second impact syndrome (SIS) and has been attributed to vasocongestive brain swelling. When it is severe, it can lead to brain herniation and death. There is some evidence that children and adolescents are at higher risk than adults (McCrory and Berkovic 1998, McCrory 2001).

COMA, PERSISTENT VEGETATIVE STATE AND 'LOCKED-IN' SYNDROME

Coma

The evaluation of the conscious state, its alterations and their time sequence are very significant with respect to assessing the effects of brain injury. The earliest clinical evidence of deteriorating conscious level is drowsiness, a condition in which the patient may be easily arousable and oriented in time, place and person. As the level of consciousness gets more depressed, the patient becomes more drowsy and confused.

Consciousness is a quality unique to humans because it includes the process of being conscious of one's own person, relations, memories, motives, etc. The term 'consciousness', therefore, has a wider meaning and it is not ideal for what is meant by it in clinical practice.

Consciousness can be defined as having two components: awareness (or arousal) and content. Awareness is a capability present in all mammals and, hence, the term 'awareness' may be more appropriate. Awareness or arousal can be divided into basal and goal directed types. Basal awareness depends on the normal function of the cerebral hemispheres and the ascending reticular formation of the brain stem. Goal-directed awareness depends on the normal function of the limbic system.

The impairment of arousal has a wide spectrum, with disorientation at one end and coma at the other. Between them there are several poorly defined states such as stupor, semi-stupor, obtundation, etc. The inaccuracy of those qualitative terms was overcome in clinical practice by utilizing a numerical scale for assessing the states of consciousness. This is the Glasgow Coma Scale (GCS), which allows evaluation of three different types of responses (Table 15.1); each patient receives a total score that is a sum of the scores of the patient in each category of response (Teasdale and Jennett 1974).

The best score that a patient can have is 15 (6 + 5 + 4) and the worst 3 (1 + 1 + 1). On the GCS, a patient is considered comatose when he or she has no speech and no eye opening, and the best motor response is that of localizing to pain.

Many pathological processes can depress the conscious level. Irrespective of the nature of each process, the depression of the conscious state can occur with one of the following three mechanisms:

- Diffuse hemispheric damage: this can result from either ischaemia or metabolic changes, whether they are the effect of medications or secondary to rises of ICP.

Table 15.1 Glasgow Coma Scale

Eye opening	Motor response	Verbal response	Score
–	Obeying	–	6
–	Localizing to pain	Oriented	5
Spontaneously	Withdrawing to pain	Confused	4
To speech	Flexing	Inappropriate	3
To pain	Extending	Incomprehensible	2
No response	No response	No response	1

Reproduced from Teasdale, G. and Jennett, B. 1974. Assessment of coma and impaired consciousness. A practical scale. *Lancet*, **ii**, 81–4.

- Brain-stem damage: this can also be the result of ischaemia, haemorrhage, medications, compression by a mass or a rise of the ICP, particularly when it is associated with uncal herniation.
- Bilateral thalamic damage: this is rare and can occur as the result of extensive bilateral haemorrhage.

COMA-RELATED POSTURES

Depending on the severity and location of the traumatic brain lesions, the body can adopt various specific postures (Figure 15.6). The following are the most recognized of these:

- Decorticate posture: this can be produced by lesions that disrupt the corticospinal pathways (above the midbrain) and, hence, terminate the inhibition of the cortex and basal ganglia on the brain stem and the spinal neuronal networks.
- Decerebrate posture: this can be produced by lesions at the level of the colliculi, which disrupt the vestibulospinal tract and the reticular formation at the pons.

Persistent vegetative state

In the persistent vegetative state (PVS), which usually develops after a severe brain injury and prolonged period of coma, the patient maintains the brain-stem-related functions but shows no signs of cortical function. Activities such as eye opening, eye movement to auditory stimuli, the cycle of sleep and arousal, and swallowing can frequently be mistaken for neurological recovery but these are brain-stem controlled and do not depend on cortical function. Patients in PVS can survive for many years, but after 6 months PVS is generally considered to be irreversible.

The neuropathology of the vegetative state and those with severe disability have been well reviewed by Adams and others (Adams *et al.* 2000, Jennett *et al.* 2001). In severely disabled individuals, some showed pathology similar to the vegetative state but others showed lesser degrees of axonal injury, thalamic damage and hypoxic–ischaemic injury, as well as increased incidence of contusions and intracranial haematoma. Major findings included traumatic axonal injury, neocortical ischaemic damage and previous intracranial haematoma. Damage to subcortical white matter or the relay nuclei of the thalamus was present. These lesions rendered any intact cortex unable to function (Adams *et al.* 2000).

Locked-in syndrome

The locked-in syndrome (LIS) results from a severe lesion of the corticospinal pathways at the anterior pons and follows a prolonged coma caused by severe brain injury. All the voluntary motor activity ceases and the patient's head, arms and legs remain immobile. The patient maintains eye movements because the third nerve nuclei are located higher in the upper brain stem. The patient in LIS understands the stimuli of the environment, can answer questions, opening and closing the eyes, and direct his or her eyes towards a chosen target.

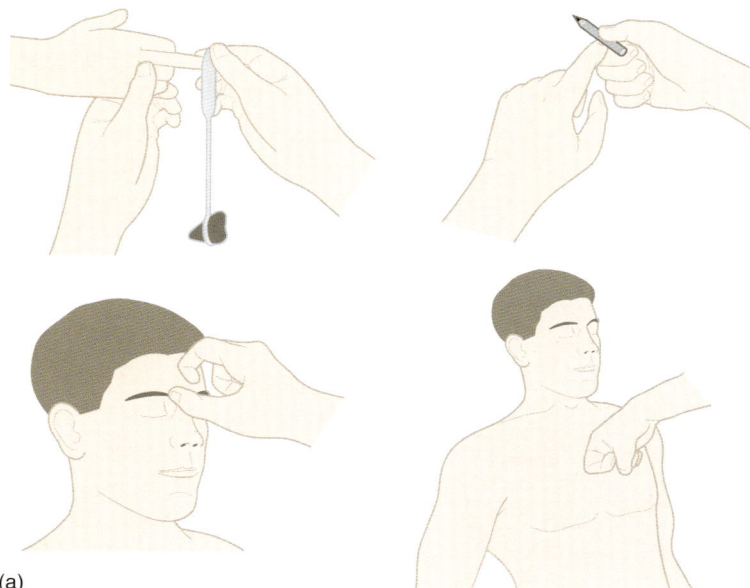

Figure 15.6   (a) Abnormal postures of patients in post-traumatic coma. Pressure with a hard object at the nail bed, supraorbital rim or chest to cause pain and elicit the motor response. Modified and redrawn from Plum, F. and Posner, J.B. 1982. Diagnosis of Stupor and Coma. 3rd edn. © 1966, 1972, 1980 by Oxford University Press Inc. Used by permission of Oxford University Press Inc. (Continued over)

(a)

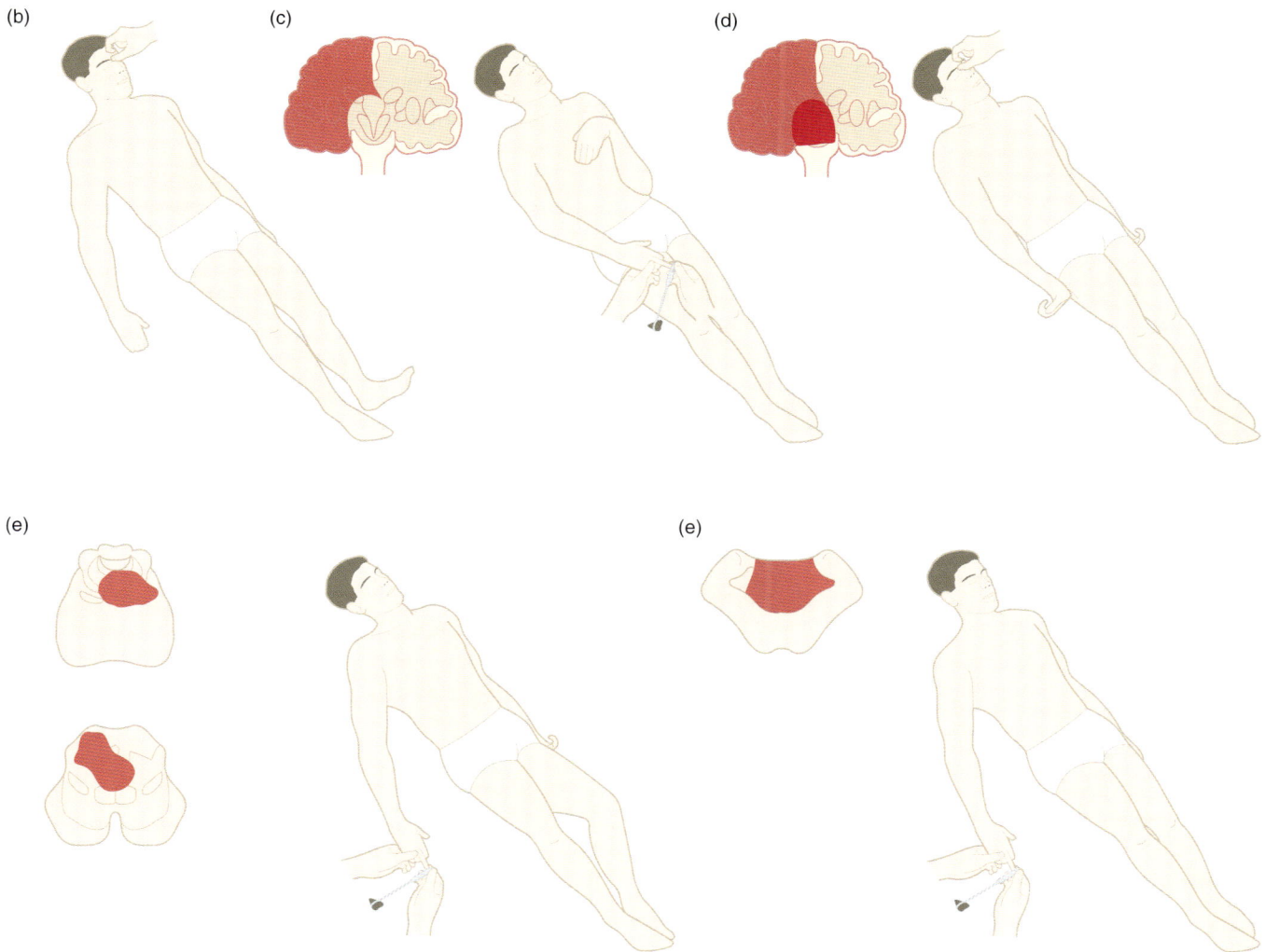

Figure 15.6 (*continued*)   (b) Cortical dysfunction: localizing to pain. This posture indicates the integrity of the spinothalamic tract. (c) Decortication: injury in the cortex and subcortical white matter. (d) Decerebration: injury at the mesencephalon and diencephalon. (e) Pontine injury. (f) Medulla injury. (continued over)

BRAIN DEATH AND ITS DIAGNOSIS

Definition of brain death

The definition 'brain death' describes the patient in full technological support of cardiorespiratory function, in whom irreversible cessation of cerebral function has been confirmed on the basis of twice-documented absence of brain-stem reflexes. It is mandatory that, before the appropriate tests of brain-stem reflexes are done, certain preconditions have been met. Otherwise, the tests should be postponed until these preconditions are satisfied. The conditions that must be checked for possible contribution to the patient's brain death state include metabolic and endocrinological abnormalities, pharmaceutical neuromuscular blockade, or sedative or depressive effects and

that the body temperature is above 35°C. In addition, the condition and pathophysiological mechanism leading to the cessation of neurological function should be considered compatible with the presumed brain death. The tests or criteria for documenting the absence of brain-stem function are:

- No pupillary reactivity to light.
- No corneal reflex, i.e. no orbicularis oculi contraction to corneal stimulation.
- No oculocephalic reflex, i.e. during sidewise head rotation, the eyes maintain original position (doll's eye) and do not follow.
- No vestibulo-ocular reflex (caloric testing), i.e. the slow injection of 50 ml iced water into the external acoustic meatus does not elicit any eye movements.

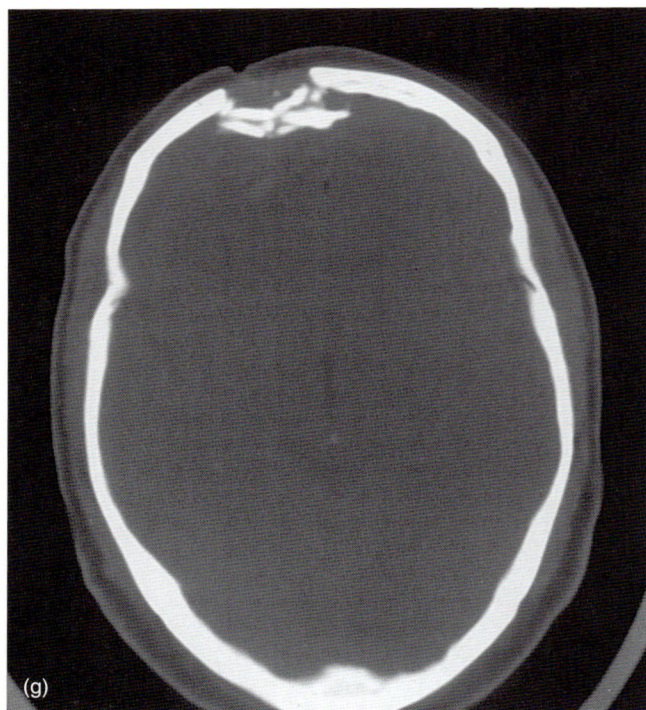

Figure 15.6 (*continued*)   (g) Compound depressed fracture in the right frontal region (CT).

- No gag or cough reflex on bronchial stimulation, i.e. the bronchial stimulation with the suction tube does not elicit any cough response.
- No contraction of facial muscles to painful stimuli.
- No respiratory movements on patient's disconnection from the ventilator after the arterial CO_2 pressure (Pa_{CO_2}) has been allowed to rise above 8 kPa (60 mmHg) (Kaye 1991).

In cases of doubt or difficulty in obtaining a clear-cut response in certain tests, other clinical, radiological or biochemical tests may provide additional supplementary confirmation of brain-stem death (Greenberg 2001). These tests include:

- Angiography and/or transcranial Doppler that can document absence of cerebral circulation.
- Electroencephalogram that can confirm electrical silence of the brain.
- High S-100b (calcium-binding protein released from traumatized astrocytes): protein serum levels that are compatible with no cerebral recovery (Dimopoulou *et al.* 2003).
- No heart rate changes on atropine injection because of the cessation of vagal tone.

The pathology of this is covered in Chapter 9, p. 111.

Issues of particular medicolegal importance

In cases in whom brain death has a 'metabolic' cause such as intoxication or hypoxia after cardiac arrest, etc., it is safer to postpone concluding that the patient is brain dead as compared with patients with obvious catastrophic cerebral lesions such as large brain-stem haemorrhage or multiple gunshot wounds to the head.

It is also important to check for possible orbital, spinal or limb fractures. Their presence may affect the pupillary reflex to light or the motor responses. The patient must be checked for the possibility of cervical fracture in particular before performing the oculocephalic reflex test.

SURGICAL MANAGEMENT: A BRIEF OUTLINE

The surgical management in most cases of acute traumatic brain lesions requires a craniotomy. Specific issues that are worthwhile mentioning with respect to the surgical management of the various traumatic lesions are described below.

Skull fracture

Compound or depressed fractures must be treated surgically. In any clinical or forensic examination, the hair should be shaved widely around the wound, which should be meticulously cleaned, débrided and inspected. In compound fractures, after cleaning and shaving of the scalp around the fracture, the wound is débrided, the foreign materials removed and the area washed with antiseptic solutions. The depressed fragments are elevated, the dura inspected and repaired, bone fragments placed in the bone defect and the wound closed in layers. If the fracture is depressed but not compound, a surgical separator is inserted underneath the bone extradurally through an adjacent burr hole to elevate the depressed bone. Bone fragments are placed to fill the defect.

Extradural haematoma

The craniotomy for evacuation of an EDH begins with an inversed question mark scalp incision, which starts at the zygomatic arch just in front of the tragus to spare the branch of the facial nerve to the frontalis muscle and the anterior branch of the superficial temporal artery. The incision is extended superiorly and posteriorly in front and above the pinna and then curves anteriorly along the superior temporal line of the skull to end just behind the frontal hairline. The bone flap is cut with a bone cutter and the clot removed with forceps and suction, and the bleeding artery coagulated. If an SDH is suspected as well at the same site, a small incision should be made on the dura to check for this. Finally, the dura is fixed with sutures all around at the edges of the craniotomy, the bone flap is replaced, and the muscle and scalp layers are sutured.

Acute subdural haematoma

The evacuation of the SDH can be achieved after a craniotomy overlying the blood collection. After the bone flap has been reversed one should assess the degree of dural tension. This must also be assessed on CT before surgery. If the midline shift on the scan is disproportionate to the size of the haematoma, it indicates that there is massive brain swelling associated with high ICP and that the brain could swell massively after a wide opening of the dura, making it impossible to close the dura afterwards. If such a situation is suspected, the surgeon should do multiple small incisions on the dura and try to evacuate the clot through them. On the other hand, if the swelling is not massive the surgeon could incise the dura, usually in a cruciate manner, and remove the clot with gentle suction.

Intracerebral haematoma, contusion, burst lobe

In these cases, a craniotomy is required at the appropriate site. After the bone flap has been reserved and the dura opened, the surgeon proceeds with a cortical incision at a site as near to the lesion as possible. After the haematoma has been visualized, gentle suction is used to evacuate the haematoma and all the contused, non-viable brain tissue. Haemostasis follows. The dura is sutured and secured at the edges of the bone, the bone flap replaced in its original position, and the muscle and scalp layers sutured.

Skull base fracture

This needs to be repaired surgically if it is large or associated with prolonged rhinorrhoea (Sakas *et al.* 1998). These fractures are best visualized on coronal thin-section CT (Figure 15.7). The repair requires a uni- or bicoronal craniotomy, division of the dura, elevation of the frontal lobe

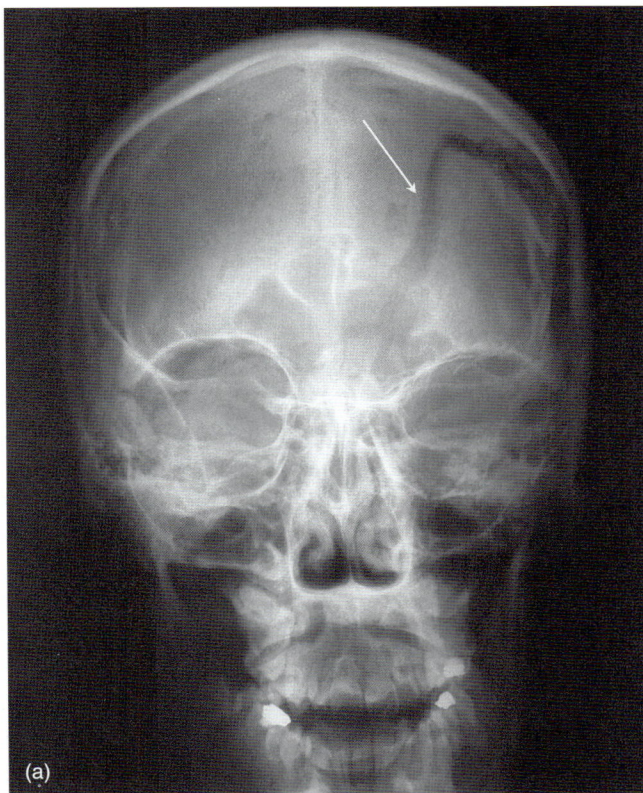

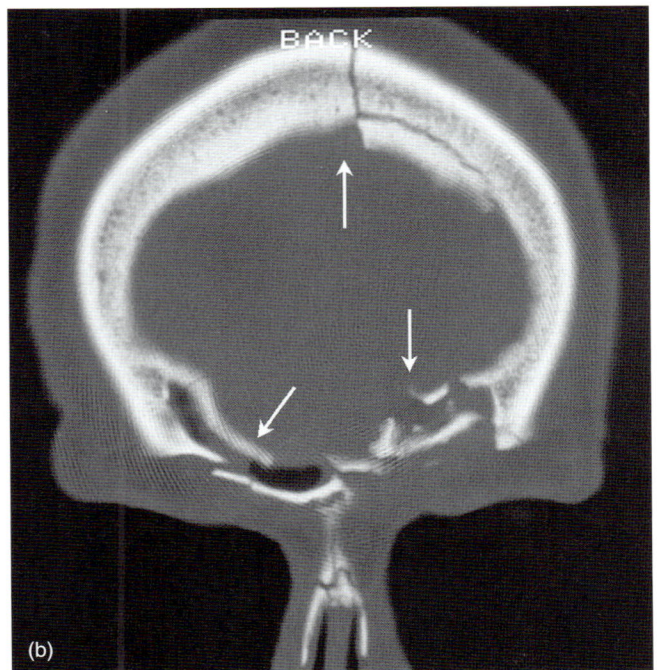

Figure 15.7 Skull fractures through the frontal sinuses. (a) Skull radiograph; (b) coronal thin-section computed tomography (arrows).

and visualization of the fracture on the cranial base. Then, the cranial base opening is plugged with methyl acrylate and a layer of pericranium placed to overlie this and secured in place with biological glue. The craniotomy is closed as described previously. In all craniotomies for evacuation of traumatic lesions, it is essential that a drainage tube is placed either subcutaneously or extradurally.

OUTCOME OF HEAD INJURY: SHORT- AND LONG-TERM COMPLICATIONS INCLUDING EPILEPSY

Complications of head injuries

The head injuries can be associated with short- and long-term complications.

SHORT-TERM COMPLICATIONS
The short-term complications include: early post-traumatic epilepsy, CSF rhinorrhoea or otorrhoea, meningitis and cranial nerve damage, post-traumatic coagulopathy and pulmonary complications.

Early post-traumatic epilepsy
This term describes epileptic seizures that occur in the first week after the injury. Of all cases of head injury admitted to hospital, 5 per cent will develop early epilepsy. The risk increases if the injuries include depressed skull fractures or cerebral parenchymal lesions (Annegers *et al.* 1980).

CSF rhinorrhoea–otorrhoea
This develops when the head injury creates a cranial defect and, hence, a communication between the intracranial compartment and the paranasal sinuses or less frequently the middle ear. If the CSF leak is not obvious and persistent, one may fail to detect this potential complication. In injuries associated with brain swelling or high ICP, a part of the brain may plug the defect for several days. When the swelling subsides, the defect may get unplugged, leaving the patient exposed to the risk of meningitis (Sakas *et al.* 1998). The following clinical signs may help to suspect or detect a skull base fracture: periorbital subcutaneous haematoma (the characteristic appearance of 'racoon eyes'), perimastoid subcutaneous haematoma (Battle's sign) or subconjunctival haematoma. Radiological signs include: air–fluid level in the frontal sinuses on plain skull radiograph, intracranial air on radiograph or CT, or opacity of air sinuses on radiograph or CT.

Cranial nerve injury
Head trauma may be complicated by injury to any of the following cranial nerves:

- Olfactory: this presents with anosmia and is frequently associated with anterior cranial fossa fracture and CSF rhinorrhoea.
- Optic: this gets injured at the optic foramen and presents with visual loss.
- Oculomotor: this gets damaged usually because of tentorial herniation and presents with papillary dilatation, ptosis and loss of reactivity to light.
- Trochlear: this gets damaged by fractures involving the superior orbital fissure or cavernous sinus and presents with disturbance of ocular movements.
- Trigeminal: this gets damaged with or without fractures of the cranial base, particularly petrous or sphenoid fractures, and presents with jaw drop, difficulty in mastication and loss of facial sensation.
- Abducens: this gets damaged by fractures involving the petrous or sphenoid bone. As it is the longest intracranial nerve, it can get damaged by a rise of ICP without a fracture. It presents with medial deviation of the eye.
- Facial: this gets damaged by fractures of the petrous bone and presents with peripheral facial palsy and otorrhoea.
- Acoustic: this gets damaged by petrous fractures which can affect not only the nerve but also the cochlea or ossicles. It may be associated with otorrhoea or haemotympanum and presents with vertigo, hearing loss, dizziness or tinnitus.

With respect to cranial nerve injuries, evaluation of the pupillary size and reactivity to light is very important at the first clinical evaluation, and for further monitoring of the head-injured patient. As the neuroanatomical region responsible for governing eye movements lies adjacent to the area within the brain stem that controls consciousness, eye movement evaluation is especially important when assessing the extent of the injury. Expanding intracranial haematomas or raised ICP may cause temporal lobe herniation; this will compress the third nerve and result in pupillary dilatation, almost always ipsilateral to the side of the haematoma. The pupil at first dilates and remains reactive to light, but subsequently becomes sluggish and finally does not respond to light at all. As this process progresses the contralateral pupil undergoes the same change. One should always check for a possible orbital fracture. This, as well as direct trauma to the eye, may cause a traumatic mydriasis; in these cases, the dilated pupil should be distinguished from that which has resulted from a third cranial nerve palsy.

Post-traumatic coagulopathy

After a severe head injury, post-traumatic coagulopathy can follow and may increase the risk for the patient of delayed haematoma.

Pulmonary complications

Patients with decreased consciousness following head injury frequently aspirate. After head trauma, pulmonary oedema can develop without direct pulmonary injury (neurogenic pulmonary oedema). Another serious potential complication is pulmonary thromboembolism.

LONG-TERM COMPLICATIONS

The most important long-term complications of head injuries are: late epilepsy, hydrocephalus, chronic encephalopathy, somatic neurological disability, postconcussive syndrome and post-traumatic encephalopathy.

Late post-traumatic epilepsy

Late post-traumatic seizures are described as those that develop after the first week from the trauma. They usually develop in the first year, although infrequently the first seizure may present several years later. In patients with severe head injury, it has been estimated that 10 per cent will suffer from late post-traumatic epilepsy within the first 2 years with generalized seizures being the most common type. The incidence is higher in open than in closed head injury (McQueen *et al.* 1983, Temkin *et al.* 1991).

Hydrocephalus

This is the result of malfunction of the arachnoid villi and, hence, it is a communicating hydrocephalus. In a proportion of patients, it can be quite severe and associated with the risk of intellectual deterioration or even visual loss and, therefore, it may require shunting.

Post-traumatic chronic encephalopathy

This term describes a constellation of motor, cognitive and psychiatric symptoms that can occur after repeated head injuries in athletes such as boxers. When mild, the main symptoms include lack of dexterity, decreased concentration, memory loss or mood changes. The very severe form is known as the 'punch-drunk syndrome'. The pathology is covered in Chapter 11. CT or MRI reveals cerebral atrophy.

Somatic neurological disability

This includes primarily hemiparesis and spasticity. The hemiparesis will require physiotherapy and the spasticity close management including possibly surgery to prevent contractions and improve patient care or mobility.

Postconcussive syndrome

This includes a variety of symptoms after minor head trauma, such as headache, dizziness, difficulty in concentration and memory, fatigue, emotional difficulties (irritability, depression) and other personality changes. It is unclear to what extent psychological factors such as the pre-trauma personality and the prospect of secondary gain play a more important part than the real physical problems.

Post-traumatic endocrinopathy

This is uncommon and may follow closed head injury or skull base fractures. The hormone deficiencies include: cortisol, growth hormone, thyroid-stimulating hormone, corticotrophin, raised prolactin and diabetes insipidus (Dimopoulou *et al.* 2004).

VARIABLES AFFECTING OUTCOME

There are numerous variables that can affect the outcome of head injury (Marshall *et al.* 1979). The following are the most important:

- Force of the impact.
- The mechanism of injury.
- Patient's age: of all the host-related factors that affect prognosis after head injury, the most important is the patient's age. The fifth decade appears to be the crucial age, after which mortality rapidly rises. Patients older than 65 years are expected to have a poor outcome in severe brain injury, especially when the patient became comatose immediately after the injury rather than some time later.
- Distance from the medical centre: a favourable outcome strongly correlates with the time interval from injury to definitive care – the shorter the interval, the better the outcome.
- Appropriate management in the first 24 h: this includes primarily the prevention and management of hypoxia and hypotension. Shock and hypoxia seem to be the most important conditions associated with severe head injury that can be treated in the pre-hospital phase. Hypoxia and hypotension can reduce responsiveness and aggravate an existing deficit. Patients with hypotension and hypoxaemia after head injury have higher rates of mortality and morbidity than patients with equivalent GCS scores without such a systemic insult. The main cause of hypotension in head-injured patients is hypovolaemia resulting from blood loss from extracranial sources, but it also may be secondary to brain-stem injury. A single episode of hypotension (i.e. systolic blood pressure <90 mmHg) virtually doubles the mortality of the injury.

- Pupillary response to light: absence of response to light, particularly bilaterally, is a prognosticator of poor outcome, especially if it lasts for more than 6 h after the injury (Sakas *et al*. 1995).
- Intracranial pressure: values of ICP above 20 mmHg for prolonged periods are a prognosticator of poor outcome. The level of ICP is an important indicator, not only in prognosis but also in tailoring the therapy of each individual patient. The level of ICP on admission tends to correlate with hospital survival. It is not clear, however, whether high ICP causes or merely reflects the severity of brain damage.
- CT findings: patients with obliterated perimesencephalic cisterns have been found to have a fourfold risk of poor outcome, compared with those with preserved basal cisterns (controlling for similar GCS scores) (Toutant *et al*. 1984). Midline shift of >15 mm is a reliable prognosticator of poor outcome.

Of the possibly salvageable patients who die with a head injury, the usual cause of death is airway obstruction or aspiration causing acute hypoxia. Of the head-injured patients who survive the primary injury and remain comatose for at least 6 h, 40 per cent die within 6 months. In the survivors, the quality of recovery depends on the timely management of complications and an appropriate rehabilitation programme (Kaye 1991).

ASSESSMENT OF OUTCOME

The outcome of head injury is assessed with the GCS (Jennet and Bond 1975) and is classified into five levels: (1) dead, (2) vegetative, (3) severely dependent, (4) partially dependent and (5) independent.

Dependency is the main variable to evaluate outcome. Patients are described as partially dependent if they have a disability that allows them to look after themselves, get dressed, cook and return to work totally or partially. Patients are described as severely dependent when they have severe disability, cannot look after themselves and need the support of family members or other carers.

Long-term disabilities

The post-traumatic disabilities can have a wide spectrum and be: (1) somatic such as hemiparesis, quadriparesis, ataxia, poor coordination or dysphasia, (2) cognitive such as memory and mental processing difficulties, dysphasia, dysarthria, (3) senses related, i.e. visual or hearing loss, anosmia, and (4) behavioural such as emotional and personality problems. It is particularly important to prevent speech difficulties, muscle contractures, ossifying myositis

and spasticity. To achieve this one needs to subject the patient to physiotherapy, occupational therapy and speech and language therapy.

It is essential to start rehabilitation as soon as possible with a multidisciplinary group of specialists (psychiatrists, speech and language therapists, occupational therapists, social workers, psychologists). Counselling is essential to the patient and family members.

REFERENCES

Adams, J.H., Graham, D.I. and Jennett, B. 2000. The neuropathology in the vegetative after acute brain insults. *Brain*, **123**, 1327–38.

Annegers, J.F., Grabow, J.D., Groover, R.V. *et al*. 1980. Seizures after head trauma: a population study. *Neurology*, **30**, 683–9.

Borovich, B., Braun, J., Guilburd, J.N. *et al*. 1985. Delayed onset of traumatic extradural hematoma. *J Neurosurg*, **63**, 30–4.

Dimopoulou, I., Korfias, S., Dafni, U. *et al*. 2003. Protein S100-b serum levels in trauma-induced brain-death. *Neurology*, **60**, 947–51.

Dimopoulou, I., Tsagarakis, S., Kouyialis, A. *et al*. 2004. Hypothalamic–pituitary–adrenal axis dysfunction in critically ill patients with traumatic brain injury: Incidence, pathophysiology and relationship to vasopressor dependence and peripheral interleukin-6 levels. *Crit Care Med*, **32**, 404–8.

Gean, A.D. 1995. *Imaging of Head Trauma*. New York: Raven Press.

Gennarelli, T.A., Thibault, L.E., Adam, J.H. *et al*. 1982. Diffuse axonal injury and traumatic coma in the primate. *Ann Neurol*, **12**, 564–74.

Goldstein, M. 1990. Traumatic brain injury: a silent epidemic. *Ann Neurol*, **27**, 327.

Greenberg, M.S. 2001. *Handbook of Neurosurgery*, 5th edn. New York: Thième.

Jennett, B. and Bond, M. 1975. Assessment of outcome after severe brain damage: A practical scale. *Lancet*, i, 480–4.

Jennett, B., Adams, J.E., Murray, L.S. *et al*. 2001. Neuropathology in the vegetative and severely disabled patients after head injury. *Neurology*, **56**, 486–90.

Kaye, A. 1991. *Essential Neurosurgery*. Edinburgh: Churchill Livingstone.

Lidsay, K. and Bone, I. 1997. *Neurology and Neurosurgery Illustrated*, 3rd edn. Edinburgh: Churchill Livingstone.

Lobato, R., Rivas, J., Gomez, P.A. *et al*. 1991. Head injured patients who talk and deteriorate into coma. Analysis of 211 cases studied with CT. *J Neurosurg*, **75**, 256–61.

McCrory, P. and Berkovic, S.F. 1998. Second impact syndrome. *Neurology*, **50**, 677–83.

McCrory, P. 2001. Does second impact syndrome exist? *Clin Sport Med* **11**, 144–9.

McQueen, J.K., Blackwood, D.H.R., Harris, P. *et al.* 1983. Low risk of late posttraumatic seizures following severe head injury. *J Neurol Neurosurg Psychiatry*, **46**, 899–904.

Marshall, L.F., Smith, R.W. and Shapiro, H.M. 1979. The outcome with aggressive treatment in severe head injuries. *J Neursurg*, **50**, 26–30.

Sakas, D.E., Bullock, M.R. and Teasdale, G. M. 1995. One-year outcome following craniotomy for traumatic hematoma in patients with fixed dilated pupils. *J Neurosurg*, **82**, 961–5.

Sakas, D.E., Beale, D.J., Ameen, A.A. *et al.* 1998. Compound anterior cranial base fractures: Classification using computerized tomography scanning as a basis for selection of patients for dural repair. *J Neurosurg*, **88**, 471–7.

Teasdale, G. and Jennett, B. 1974. Assessment of coma and impaired consciousness. A practical scale. *Lancet*, **ii**, 81–4.

Temkin, N.R., Dikmen, S.S. and Winn, H.R. 1991. Posttraumatic seizures. *Neurosurg Clin N Am*, **2**, 425–35.

Toutant, S.M., Klauber, M.R., Marshall, L.F. *et al.* 1984. Absent or compressed basal cisterns on first CT scan: ominous predictors of outcome in severe head injury. *J Neurosurg*, **61**, 691–4.

16

The role of the expert witness

Paul Watson and Christopher M Milroy

THE ADMISSIBILITY OF OPINION EVIDENCE

The role that the expert plays in a tribunal will vary between jurisdictions. Common law jurisdictions have developed complex rules of evidence. This chapter primarily discusses the law in England, with its adversarial trial system. However, the principles that apply to experts are likely to apply across jurisdictions and other common law jurisdictions have laws relating to expert evidence.

Although, in general, evidence that does not relate to matters within the immediate experience of the witness (namely, what he has seen, heard, etc.) is inadmissible in English law, there are of necessity exceptions to that principle. Among the exceptions is the rule that, if a witness can be said to have a sufficient level of expertise in a particular field of knowledge that can be said to be outside the experience or knowledge of the tribunal of fact, that witness may furnish the court with such information as is relevant to the issues of fact before the court, and may assist the court by the giving of evidence of opinion as to matters that fall within the witness's expertise. This is a rule of some antiquity. In the case of *Buckley v Rice-Thomas* 1554, Saunders J said:

If matters arise in our law which concern other sciences or faculties we commonly apply for the aid of that science or faculty which it concerns. This is a commendable thing in our law. For thereby it appears that we do not dismiss all other sciences but our own, but we approve of them and encourage them as things worthy of commendation... In an appeal of mayhem the Judges of our law have used to be informed by surgeons whether it be a mayhem or not, because their knowledge and skill can best discern it.

Although expert evidence is often discussed under the general heading 'opinion evidence', expert evidence defines the role many practitioners play, as their evidence often involves a combination of both fact and opinion, with such witnesses having a major role in criminal and civil proceedings.

The topics on which expert evidence has been received in accordance with this principle are manifold and it is clear that the list is not closed: as Steyn LJ said in *R v Clarke* 1995, at p. 430:

It would be entirely wrong to deny the law of evidence the advantages to be gained from new techniques and ... advances in science

and in recent times expert evidence has been received as to ear prints, facial mapping, voice identification and video reconstruction, and in relation to all manner of technological advances, in particular concerning the use (and misuse) of information technology.

The approach to the admissibility of new scientific evidence has varied from jurisdiction to jurisdiction. In the famous American case *Frye v United States* 1923, a conservative approach was adapted to the acceptance of new scientific techniques, where it was stated:

... while the courts will go a long way in admitting expert testimony deduced from a well-recognized scientific principle or discovery, the thing from which the deduction is made nust be sufficiently established to have gained general acceptance in the particular field to which it belongs.

This conservative approach has now been overruled in many jurisdictions in the USA following the case of *Daubert v Merrell Dow Pharmaceutical Inc* 1993, but the principles in Frye have been acknowledged in a number of judgements in Commonwealth countries. In Australia, Frye has been followed in a number of cases, for example in *R v Karger* 2001, *R v J no. 2* 1994 (Roberts and Zuckerman 2004). In England, Frye has attracted approving comments in *R v Gilfoyle* 2001 *and R v Dallagher* 2002, although it has been questioned whether a general acceptance rule has ever been applied in England. This conservative approach has now been rejected in Canada (see *R v Beland* 1987, *R v Diffenbaugh* 1993). In *R v Robb* 1991, Bingham LJ indicated boundaries of expertise when he stated:

The old academically established sciences such as medicine, geology or metallurgy and established professions... present no problem. The field will be regarded as one in which expertise may exist and any qualified member will be accepted without question as an expert. Expert opinions may be given of the quality of commodities, or the literary, artistic, scientific or other merit of works alleged to be obscene. Yet while receiving this evidence the courts would not accept the evidence of an astrologer, soothsayer, a witch-doctor or an amateur psychologist and might hesitate to receive evidence of attributed authorship on stylometric analysis.

The general rule now is that the key to the admissibility of such evidence is, first, whether the subject matter of the proposed evidence is such that a person without instruction or experience in the area of knowledge concerned would be able to form a sound judgment on the matter without the assistance of a witness possessing special knowledge in the area and, second, whether the subject matter of the opinion forms part of a body of knowledge or experience that is sufficiently recognized to be accepted as a reliable body of experience on which the opinion of the expert would be of assistance to the court.

Of course, once a court has found that expert evidence *is* admissible, there follows the further, and quite separate, question as to whether the proposed expert is in fact possessed of the relevant degree of knowledge to render his or her evidence and opinion of sufficient value in resolving the issues before the court. Whether or not an expert is possessed of the relevant degree of expertise to be competent to give such evidence is a matter for the judge to determine. Once decided that such a witness is competent, the issue as to the *weight* to be given to the witness's evidence and/or opinion is a matter for the tribunal of fact. This two stage process is exactly mirrored by the US Federal code, Rule 702 of which states:

If scientific, technical, or other specialized knowledge will assist the trier of fact to understand the evidence or to determine a fact in issue, a witness qualified as an expert by knowledge, skill, experience, training, or education, may testify thereto in the form of an opinion or otherwise, if (1) the testimony is based upon sufficient facts or data, (2) the testimony is the product of reliable principles and methods, and (3) the witness has applied the principles and methods reliably to the facts of the case.

It is not proposed, in this chapter to go into further detail about questions of the admissibility of either evidence of expertise or the qualification of an expert witness. These topics are dealt with comprehensively in standard works on the law of evidence (Tapper 2004) and the assumption is made that, in cases involving evidence of forensic neuropathology, it will be rare for the court to question the need for evidence of expertise or the qualifications of the relevant witness.

The statement of the general rule is, however, helpful in focusing, as it does, on the role that the expert witness has to play in providing the Court with information and informed opinion that it otherwise would be denied.

THE ROLE OF THE EXPERT

Although it has to be acknowledged that the justice system in England is essentially based on an adversarial procedure, it is submitted that the role of the expert witness enjoys something of a unique status within that system. It is the function of the expert to assist the fact-finding tribunal as to the issues that it has to decide. The notion, if it ever existed, that the role of the expert witness was to assist the party whose cause he or she had espoused is surely all but extinct. This view is now enshrined, in the context of civil litigation, in Part 35.3 of the Civil Procedure Rules, which provides as follows:

(1) it is the duty of an expert to help the court on the matter within his expertise
(2) this duty overrides any obligation to the person from whom he has received instructions

and in the context of family litigation (in cases involving children) it has been said that an expert witness bears a heavy responsibility to be as objective as possible, and to make clear the extent to which his or her opinion is based on hypothesis, especially where this view departs from the consensus opinion (see *Re AB (A Minor)* 1995).

In criminal litigation, the adversarial nature of the procedure is at its most pronounced. In the past there has been a tendency for expert witnesses to 'adopt' the cause of either the prosecution or the defence, and the language of the expert opinion (if not always the substance) was of polarization. In more recent times, however, the culture has moved towards neutrality and objectivity, with the result that expert reports in criminal cases now routinely contain declarations about the duties enshrined in Rule 35.3 above, even though the provision has no application to criminal cases. In practice, now, the duty of the expert 'to furnish the judge or jury with the necessary scientific criteria for testing the accuracy of their conclusions, so as to enable the judge or jury to form their own independent judgement as to the accuracy of their conclusions' (*Davie v Magistrates of Edinburgh* 1953) apply with equal force in whichever jurisdiction the evidence is given. This Scottish case also reconfirmed the principle that an expert may adopt statements made in scientific works as part of his testimony and have such works put to him in cross-examination.

In any event, it is submitted that the weight that will be attached to the evidence of an expert witness will invariably be greater in those cases in which the expert has (before being confronted with an alternative view in cross-examination) demonstrated such an awareness. It will surely not betray any trade secret to reveal that the cross-examiner's most difficult foe is the expert witness, who will make the appropriate concessions while maintaining his or her primary conclusions! On the other hand, the expert who has demonstrated partisanship, a personal agenda or an unblinking adherence to a particular hypothesis may too easily be discredited in the eyes of the fact-finding tribunal, so that, even if there is some significant merit in the view to which he or she holds, it may be significantly undermined or even rejected outright.

The need for objectivity applies with perhaps greater force to evidence of medical expertise than to any other area of knowledge. Medical expert opinion is often readily accepted by the tribunal of fact, in particular juries, precisely because of the respectable and respected body of opinion from which it derives. There tends to be less suspicion, on the part of either judges or jurors, as to the quality and credibility of medical expertise than there is in relation to other sources of opinion evidence, and it is for this very reason, it is submitted, that the expert needs to be aware of the need for self-critical objectivity and an awareness of the weaknesses, as well as the strengths, of the view that he or she feels able to express. Moreover, if that point applies to expert evidence arising from general medical practice, it applies with greater force in the more specific and technical field of neuropathology, in which failure to present all relevant information may lead to conclusions by the judge or jury that are not, in the context of all the other evidence in the case, justified. As an example, if a view is expressed about the *timing* in the context of an subdural haematoma, without reference to the difficulty of formulating an opinion based solely on pathological appearances, the tribunal of fact may be encouraged to reject other (perhaps more reliable) evidence of timing on the ground that it conflicts with the pathological evidence.

It is submitted that what is required, in such cases, is not just caution on the part of the expert expressing a view, but the importance of drawing to the attention of the court the need for caution and the reasons for it.

OPINION ABOUT THE ULTIMATE ISSUE

Although it will be clear from the above that the function of the expert is to help the court to draw conclusions from all of the available evidence, only part of which comprises evidence of expertise, it is now well settled that it is permissible, where appropriate, for the expert to give evidence about the 'ultimate issue', i.e. the very issue that the court has to decide. Of course it does not follow that the court is bound to accept the expert view about the issue, only that his or her view is admissible, e.g. see the case of *R v Mason* 1911 in which expert opinion was admitted as to the likelihood that the victim's wounds were self-inflicted or, more recently, *R v Hookway* 1999, in which the only evidence against the defendant was of expert opinion as to the results of a 'facial mapping' technique.

A recent example of an expert being permitted to give evidence as to the ultimate issue is the case of *R v Dallagher* 2002, in which the opinion of an expert to the effect that he was sure that the marks at the scene (ear prints) were made by the defendant, was held, on appeal, to have been admissible. The important feature, in each such case, is that the jury are reminded that it is their assessment of the totality of the evidence that matters and that they are not bound to accept the expert's opinion (see *R v Stockwell* 1993).

In the case of forensic neuropathology, the question as to evidence of the ultimate issue will most commonly be encountered in cases where it is contended that head injuries are the result of an accident. In such cases the expert is permitted and, indeed, should be prepared to give evidence of his or her opinion (for it is never more than that) as to the likelihood of, for example, an accidental fall. Provided that the expert has satisfied him- or herself that this view is based on solid scientific findings and a careful evaluation of the evidence relevant to the investigation, there can be no objection to evidence to the effect that such an explanation is wholly improbable or even, in an appropriate case, inconceivable.

The same is true with regard to a positive assertion ('I believe this to be so') as it is to the negative ('I do not believe this to be so'). Frequently, the expert witness is presented with an allegation or a series of allegations and is requested to review the evidence scientifically to determine whether there is support for a particular proposition. It is submitted that, in such cases, it is the duty of the expert to express a view, perhaps even in clear or strong terms, as to the extent of support that can be derived from the scientific evidence. Testimony to the effect that a particular injury is 'consistent' with a particular proposition is, it is suggested, of limited assistance to the court. If the expert believes that a proposition is a remotely possible explanation for a particular scientific finding, it may be said to be nevertheless 'consistent' with it, but such a description does not assist the tribunal of fact in its task, and may even be said to mislead.

Similarly, if the expert is of the view, perhaps because there are so many indications from the scientific evidence towards a certain conclusion, that such a conclusion is either highly probable or, even, virtually certain, it may be said to be 'consistent' with the evidence. Once again, however, such a description is so broad as to be of little value to the tribunal. No criticism can be made of the expert who expresses the view that a certain mechanism is the likely (or highly probable or even virtually certain) cause of a particular injury *provided* that he or she demonstrates that he or she has considered all the available evidence and examined alternative explanations with the required degree of objectivity, and that the conclusions are the result of a scientific reasoning process and not 'conclusion pulled' from the proposition that was given in the first instance.

PROOF OF FACTS ON WHICH THE EXPERT OPINION IS BASED

Before a court can assess the value of an opinion it must know the facts upon which it is based. If the expert has been misinformed about the facts or has taken irrelevant facts into consideration or has omitted to consider relevant ones, the opinion is likely to be valueless. In our judgement, counsel calling an expert should in examination-in-chief ask his witness to state the facts upon which his opinion is based. It is wrong to leave the other side to elicit the facts by cross-examination.

per Lawton LJ, *R v Turner* 1975 at p. 840

Today, this statement of general principle can be expanded to refer to the desirability of experts including in their reports the facts on which their opinion is based and, as importantly, the *source* of that information.

Of course, particularly in neuropathology cases, much of the relevant material that will form the basis of expert opinions will derive from observations and findings made by the experts themselves. To a perhaps lesser extent, the expert may rely on what he or she has been told by others. Such evidence is admissible as being relevant to the process leading to the formulation of his opinion but, if no direct evidence as to the truth of those facts is called, or if the quality of the evidence of those facts is poor, the value of the opinion may suffer correspondingly. For this reason, if for this reason only, it is submitted that prudent experts will make some critical enquiry as to the information that they are given before placing undue reliance on it in formulating their view.

It should go without saying that, where experts rely not on evidence of observation or information supplied by a third party, but on *assumptions* made for the purpose of the preparation of their report, they should say so in the body of that report and not await questioning as to the source of the information at a later stage.

It is now clear that expert witnesses may make references to their own research, the work of others, articles in respected publications, research papers and other such materials in forming their views. In the case of *R v Abadom* 1983, evidence was permitted to be adduced of unpublished statistics of the Home Office Central Research Establishment, it being noted by the Court of Appeal that part of the value of expert witnesses lies in their knowledge of such *unpublished* material; there can be no reason for not permitting experts to rely on such material, provided that they refer to it in their evidence so that the value of their conclusions can be tested by reference to it. Such reference does not infringe the rule against hearsay because it does no more than provide the basis on which the conclusions of the expert are drawn. Wherever reference is made to such material, experts should specify the source material in the body of their report – preferably by means of a short bibliography.

PRE-TRIAL DISCLOSURE IN CRIMINAL CASES

It is in relation to the topic of pre-trial disclosure that the difference between experts on opposing sides emerges most clearly. The expert giving evidence (or proposing to do so) for the prosecution is under an obligation, quite independently of the obligation on the police or the prosecution authorities, to bring to the attention of the solicitor instructing him or her, or the expert advising the other party, the records of all experiments and tests carried out by him or her, whether or not the results of those experiments or tests are of assistance to the prosecution (see *R v Ward* 1993). The obligation exists regardless of whether the expert witness is subsequently relied on by the prosecution to give expert evidence and is separate from his or her duty to assist the court. In this context the words of Kay LJ in the case of *R v Clark* 2003 provide a salutary lesson to the pathologist, where at paragraph 168 he stated:

… doctors reviewing the matter at a later stage are dependent upon the pathologist who conducted the original postmortem to draw to their attention not only any material which justifies the original pathologist's conclusion but also any which reveals any abnormality that might need to be considered before being discounted. Where tests have been carried out and reported upon to the pathologist, his responsibility to make that material available for consideration by others is clear one and his failure to do so may well mislead them into thinking that there has been negative findings when that is not the case.

For the defence in criminal cases expert witnesses are under no such obligation. In most cases involving neuropathology, the role of defence experts will be to provide an audit of the methodology and conclusions of the prosecution pathologist. In the event that they may criticize either, their obligation is only to draw their views to the attention of the party instructing them. It is then a matter of that party as to whether or not to rely on the expert.

In the event that defence pathologists do take issue with either the methodology or the conclusions of prosecution experts, it is incumbent on them to define the issues between the two with sufficient clarity to enable the tribunal of fact to be able to resolve such issues as easily as possible. In this context defence experts should always bear in mind that, if their evidence is to be relied on at trial, it will be subject to the Crown Court (Advance Notice of Expert Evidence) Rules 1987 (requiring pre-trial disclosure of defence expert evidence to the prosecution) and will thus form part of the judge's pre-trial reading. His or her understanding of the relevant issues may well influence the way in which the issues develop before the jury, and failure by the defence pathologist to identify those issues may result in the point becoming diluted or, worse still, lost.

It is also now relevant to note that by virtue of section 6D of the Criminal Procedure and Investigations Act 1996 (as inserted by section 35 of the Criminal Justice Act 2003) if an accused instructs a person with a view to his providing an expert opinion for possible use as evidence at the trial he must give to the court and the prosecutor a notice specifying that person's name and address. Thus it follows that the prosecution will know of the instruction of an expert (though they will not have any report he may have provided – that remains privileged) and may well wish to interview the expert, particularly if the defence are to rely on his report and if it is not served upon them as required. In that event, the expert must remain conscious of the fact that he may have been sent material which is privileged from disclosure and avoid inadvertent disclosure to a party who is not entitled to such material.

THE ROLE OF THE PATHOLOGIST

The common law has developed a different system for medicolegal investigation from civil law jurisdictions. The central figure in the common law systems is the coroner,

an office established in England in the twelfth century in the reign of Richard the Lionheart, among whose duties was the investigation of death. Some jurisdictions, notably in North America, have replaced the coroner with a medical examiner system. In civil jurisdictions the examining magistrate performs the role of the coroner. In Scotland, the procurator fiscal carries out this examining magistrate role. The first contact that the expert is likely to have is with one of these officials. Whichever system the practitioner is involved in, it is important that appropriate authority to perform a postmortem examination is obtained. This should be given in writing, although in England this has not been usual practice, particularly in cases of suspicious death. Where whole organs or tissue are retained, this should be for medicolegal reasons, unless appropriate consent has been obtained, and the need to keep tissue should be recorded and the next of kin informed. Whoever authorizes the postmortem examination, the pathologist should be aware of his or her duties and should conduct the postmortem examination so as to provide the best available evidence for presentation before any subsequent tribunal.

In the English legal system the postmortem findings may be used for any subsequent prosecution or civil proceedings, and the pathologist should retain all appropriate material on which his or her report is based. Although in the UK a second examination of the body on behalf of a defendant is common, this is not the case in many other countries. This material may need to be available to other experts and proof of its continuity will be required. Any notes, photographs, radiological material, histology or results of other tests should be appropriately stored for subsequent reexamination.

EVIDENCE-BASED MEDICINE

Medicine has moved away from basing practice on anecdote to trying to base it on a solid evidence base, i.e. evidence-based medicine (EBM). In recent years past dogma has been criticized because the evaluation of previous work and the practical application of proposed formulae have been found wanting. Notable examples of areas in forensic medicine that have been challenged include timing of death and dating of bruises. The need for further study and evaluation of the evidence base of forensic medicine has been raised (Milroy 2003). In a paper in 2003 examining the evidence base in the literature of the 'shaken baby syndrome', the author, Donohue, pointed out that, of the 54 articles he reviewed on 'shaken baby syndrome', only five had control groups and only one was prospective. Using a scale of I–IV to analyse the evidence (where I is consistent evidence obtained from either more than two independent, randomized and controlled studies or two independent, population-based epidemiological studies, II is evidence from two randomized controlled studies from independent centres, a single multicentre randomized study or a population-based epidemiological study, III involves either studies from one research group, limited studies or single case reports, or where studies conflict, and where IV is consensus opinions according to clinical experience or descriptive reports), Donohoe stated that the EBM of 'shaken baby syndrome' did not reach level II, let alone level I. Although it is not the place here to argue for the existence of the 'shaken baby syndrome', Donohue's paper does illustrate the lack of a systematic approach to controversial areas in forensic medicine. Opinions should not be based on personal hunches or suppositions and then presented as solid science.

In the recent case of *R -v- Angela Cannings* (2004), for example, the Court of Appeal identified the risks inherent in the use of terminology such as 'syndrome' without identifying fairly and accurately what is meant by the expression. As Judge LJ put it

Treating the problem as a syndrome [in the context of SIDS] tends to obscure the fact that sudden unexplained infant deaths occur in different circumstances, and some may be multifactorial, the result of a co-incidence of processes which, taken in isolation would not necessarily cause death.

In that case the Court was concerned that failure to acknowledge the potential for multiple potential causes of SIDS could lead to what the Court called the 'lightning does not strike three times in the same place' process of reasoning. That reasoning had led, as one witness had said, to the 'current dogma' that an 'unnatural cause has been established unless it is possible to demonstrate an alternative natural explanation'. The Court unhesitatingly felt that line of reasoning to be fallacious, leaving the 'route to a finding of guilt wide open'. Far more preferable, in the Court's view was the alternative approach

to start with the same fact, three unexplained deaths in the same family are indeed rare, but thereafter to proceed on the basis that if there is nothing to explain them, in our current state of knowledge at any rate, they remain unexplained, and still, despite the known fact that some parents do smother their infant children, possible natural deaths.

Of what another witness described as the 'fashion nowadays that if you have more than one sudden infant death the next one must have been killed deliberately' the Court said quite starkly:

If that is the fashion it must now cease.

It is submitted that an evidence based approach ('what is the objective evidence that *this* death was caused unlawfully') would inexorably lead toward the second, correct, approach to the evaluation of such evidence and would tend to avoid what the Court called the 'hidden trap' of taking the wrong starting point.

PREPARING FOR COURT

Whether 'for' the prosecution or the defence, or in a civil trial, it should be a matter of invariable routine that, in cases involving expert evidence, the expert should have had a conference in advance of the trial with the advocate who will actually be conducting the case. Frequently, in the course of such meetings (even where the evidence seems on its face to be relatively non-controversial), problems arise relating, for example, to terminology or a difference in approach by the pathologist and the lawyer respectively, which would otherwise have to be resolved – somewhat unedifyingly – in the witness box.

Best evidence requires careful thought and planning. Pathology material can be distressing for juries and photographs may not be the most appropriate method to demonstrate findings. In some jurisdictions photographs from postmortem examination are considered prejudicial. Consideration should be given to use of other methods of display, such as diagrams or computer graphics. Meetings conducted for the prosecution in England are disclosable to the defence. Meetings of experts for different parties in civil cases are common and a normal practice in child care proceedings, where any expert opinion obtained for child care proceedings must be disclosed, no matter which party requested the report.

Before attending court, the expert should familiarize him- or herself with where the court is and when they are required to give evidence. They should attend court with all appropriate documents. In some jurisdictions, such as in England, witnesses typically stand to give evidence, but in others they sit. In giving oral evidence, the expert should present a professional appearance and give clear evidence that can be heard by judge and jury. Technical terms should be explained and the expert should not stray beyond his or her field. Any temptation towards dogma, particularly if the state of scientific knowledge is limited or incomplete is to be eschewed (see the observations of the Court of Appeal in the *Cannings* case – see above).

Cross-examination techniques may appear annoying to the witness, but losing one's temper, becoming sarcastic or arguing with counsel is ultimately detrimental to the evidence that needs to be given. The judge has the role to see that questioning is appropriate and fair, not the witness. The stages of giving evidence have been summarized as dress up, turn up, stand up, speak up and shut up.

CONCLUSION

In conclusion effective expert witnesses need to prepare well, maintain objectivity in the evidence, not favouring the side that instructed them and present their evidence in a professional manner. When challenged as to a part of their evidence the words of Oliver Cromwell in his letter to the General Assembly of the Church of Scotland in 1650 should be remembered, when he wrote to the Kirk:

I beseech you, in the bowels of Christ, think it possible you may be mistaken. There may be a carnal confidence upon mistaken and misapplied concepts.

REFERENCES

Donohue, M. 2003. Evidence-based medicine and shaken baby syndrome. Part 1: Literature review 1966–1998. *Am J Forensic Med Pathol*, **24**, 239–42.

Milroy, C.M. 2003. Medical experts and the criminal courts. *BMJ*, **326**, 294–5.

Roberts, P. and Zuckerman, A. 2004. *Criminal Evidence*. Oxford: OUP.

Tapper, C. 2004. *Cross and Tapper on Evidence*, 10th edn. London: Butterworths.

TABLE OF CASES

AB (A Minor) (1995) 1FLR 181

Buckley v Rice-Thomas (1554) 1 Plowden 118

Daubert v Merriell Dow (1993) 125 L Ed 2d 469; 113 S Ct 2786

Davie v Magistrates of Edinburgh (1953) SC 34

Frye v US (1923) 54 App DC 46, 47; 293 F 1013, 1014

R v Abadom (1983) 1 WLR 126

R v Beland (1987) 2 SCR 398

R v Clarke (1995) 2 Cr App R 425

R v Clark (2003) EWCA Crim 1020

R v Cannings (2004) EWCA Crim 01

R v Dallagher (2002) Crim L R 821

R v Dallagher (2002) EWCA Crim 1903

R v Diffenbaugh (1993) 80 CCC (3d) 97

R v Gilfoyle (2001) 2 Cr App R 57

R v Hookway (1999) Crim L R 750

R v J no.2 (1994) 75 A Crim R 522

R v Karger (2001) 83 SASR 1; [2001] SASC 64

R v Mason (1911) 7 Cr App R 67

R v Robb (1991) 93 Cr App R 161

R v Stockwell (1993) 97 Cr App R 260

R v Turner (1975) QB 834

R v Ward (1993) WLR 619

Index